T0396757

Surface Engineering and Performance of Biomaterials

Surface Engineering and Performance of Biomaterials

Editor

Chander Prakash
Chandigarh University, India

World Scientific

NEW JERSEY · LONDON · SINGAPORE · BEIJING · SHANGHAI · TAIPEI · CHENNAI

Published by

World Scientific Publishing Co. Pte. Ltd.
5 Toh Tuck Link, Singapore 596224
USA office: 27 Warren Street, Suite 401-402, Hackensack, NJ 07601
UK office: 57 Shelton Street, Covent Garden, London WC2H 9HE

British Library Cataloguing-in-Publication Data
A catalogue record for this book is available from the British Library.

SURFACE ENGINEERING AND PERFORMANCE OF BIOMATERIALS

Copyright © 2025 by World Scientific Publishing Co. Pte. Ltd.

All rights reserved. This book, or parts thereof, may not be reproduced in any form or by any means, electronic or mechanical, including photocopying, recording or any information storage and retrieval system now known or to be invented, without written permission from the publisher.

For photocopying of material in this volume, please pay a copying fee through the Copyright Clearance Center, Inc., 222 Rosewood Drive, Danvers, MA 01923, USA. In this case permission to photocopy is not required from the publisher.

ISBN 978-981-98-1256-1 (hardcover)
ISBN 978-981-98-1257-8 (ebook for institutions)
ISBN 978-981-98-1258-5 (ebook for individuals)

For any available supplementary material, please visit
https://www.worldscientific.com/worldscibooks/10.1142/14295#t=suppl

Typeset by Stallion Press
Email: enquiries@stallionpress.com

Preface

Advancements in biomaterials and surface engineering have revolutionized biomedical applications, particularly in orthopedic and neurosurgical procedures. The present volume, *Surface Engineering and Performance of Biomaterials*, is a compilation of high-quality research articles that contribute to this evolving field. It provides insights into experimental investigations, parametric optimizations, tribological studies, surface modifications, and electrochemical evaluations that enhance the performance and longevity of biomaterials.

This volume presents a collection of research contributions covering various aspects of surface engineering and biomaterial performance. The first chapter delves into the experimental investigation and parametric optimization of neurosurgical bone grinding, focusing on enhancing precision under bio-mimic conditions. Subsequent studies explore fabrication and surface modification techniques for orthopedic implants, addressing biocompatibility and wear resistance. Research on tribological performance investigates the effects of loads and bio-lubricants on zirconia and zirconia-toughened alumina for hip prostheses. Investigations into wire electrical discharge machining highlight the influence of superficial layer formation on corrosion resistance and bioactivity.

Further contributions examine innovative coating technologies. Studies on Sr-doped hydroxyapatite biomedical coatings and thermally sprayed hydroxyapatite coatings offer insights into improving biointegration and mechanical stability. Deep cryogenic treatment of high-carbon steel blades is analyzed for its impact on tribological and economic aspects. Research on nanocomposite coatings explores their biological evaluation and application in stainless steel implants. Titanium alloy performance comparisons provide valuable insights into their suitability for biomedical applications, while electrophoretic deposition techniques are studied for bioglass coatings. Additionally, investigations into silver coatings on Ti15Mo evaluate corrosion, tribocorrosion, and antibacterial properties. The final chapter discusses magnesium hybrid nanocomposites, emphasizing their mechanical properties and corrosion resistance as biodegradable implants.

This book is an essential resource for researchers, academicians, and professionals working in the field of biomaterials and biomedical engineering. The comprehensive studies compiled here will serve as a foundation for further advancements in surface engineering and biomedical applications. We extend our sincere gratitude to all contributing authors and reviewers for their efforts in bringing this volume to fruition.

We hope that this book will inspire new research directions and contribute significantly to the growing field of biomaterials and surface engineering.

Editor
Surface Engineering and Performance of Biomaterials

Contents

Preface	v
Experimental investigation and parametric optimization of neurosurgical bone grinding under bio-mimic environment	1
Atul Babbar, Vivek Jain, Dheeraj Gupta and Chander Prakash	
Fabrication and surface modification of biomaterials for orthopedic implant: A review	29
MD Manzar Iqbal, Amaresh Kumar, Rajashekhara Shabadi and Subhash Singh	
Effect of loads and bio-lubricants on tribological study of zirconia and zirconia toughened alumina against Ti6al4v for hip prosthesis	77
S. Shankar, R. Nithyaprakash, R. Naveen Kumar, R. Aravinthan, Alokesh Pramanik and Animesh Kumar Basak	
Morphological, micro-mechanical, corrosion and *in vitro* bioactivity investigation of superficial layer formation during wire electrical discharge machining of Ti-6Al-4V alloy for biomedical application	93
Sandeep Malik and Vineet Kumar	
Preparation and characterization of Sr-doped HAp biomedical coatings on polydopamine-treated Ti6Al4V substrates	113
Gurmohan Singh, Abhineet Saini and B. S. Pabla	

Performance of thermally sprayed hydroxyapatite coatings for biomedical implants: A comprehensive review 129
Gaurav Prashar and Hitesh Vasudev

Deep cryogenic treated high carbon steel blades: Tribological, morphological, and economic analysis 187
Chander Jakhar, Anil Saroha, Parvesh Antil, Vishal Ahlawat, Asha Rani, Dharam Buddhi and Vinay Kumar

Preparation and biological evaluation of PLD-based forsterite-hydroxyapatite nanocomposite coating on stainless steel 316L 211
P. Shakti Prakash, Suryappa Jayappa Pawar and Ravi Prakash Tewari

Comparison and performance of α, $\alpha + \beta$ and β titanium alloys for biomedical applications 233
Pralhad Pesode and Shivprakash Barve

Electrochemical evaluation of Ti45Nb coated with 63s bioglass by electrophoretic deposition 287
Yakup Uzun

Comparison of corrosion, tribocorrosion and antibacterial properties of silver coatings on Ti15Mo by magnetron sputtering 307
E. Meletlioglu and R. Sadeler

Influence of mechanical and the corrosion characteristics on the surface of magnesium hybrid nanocomposites reinforced with HAp and rGO as biodegradable implants 333
Venkata Satya Prasad Somayajula, Shashi Bhushan Prasad and Subhash Singh

Experimental investigation and parametric optimization of neurosurgical bone grinding under bio-mimic environment*

Atul Babbar[†,‡], Vivek Jain[‡], Dheeraj Gupta[‡] and Chander Prakash[§,¶]

[†]Mechanical Engineering Department,
Shree Guru Gobind Singh Tricentenary University,
Gurugram 122505, Haryana, India
[‡]Mechanical Engineering Department,
Thapar Institute of Engineering and Technology,
Patiala 147003, Punjab, India
[§]School of Mechanical Engineering,
Lovely Professional University, Phagwara,
Punjab 144411, India
[¶]chander.mechengg@gmail.com

Bone grinding is a craniotomy procedure which is used to remove a bone flap from the skull to expose and create an access for the dissection of tumors. In this study, a computer-controlled neurosurgical bone grinding has been used to explore the effect of various neurosurgical bone grinding parameters, such as cutting forces, torque, grinding force ratio, and temperature generated during bone grinding have been investigated. Bone samples after grinding have been assessed for morphological analysis. Based on the outcomes, a multi-attribute decision-making methodology based on grey relational analysis has been adopted. Regression models have been developed and then validated to ensure the adequacy of the developed models. Subsequently, a comparative

---

*To cite this article, please refer to its earlier version published in Surface Review and Letters, Vol. 30, No. 1 (2023) 2141005 (13 pages) DOI: 10.1142/S0218625X21410055
*Corresponding author.

analysis of experimental and predicted results have been presented. It is revealed that grinding forces and torque decreased with the escalation of rotational speed from 35,000 revolutions per minute (rpm) to 55,000 rpm. The optimum combination of process parameters found as rotational speed of 55,000 rpm, feed rate of 20mm/min, and depth of cut of 0.50 mm.

Keywords: Bone grinding; forces; temperature; infrared thermography; optimization.

1. Introduction

Bone grinding is a common procedure used in neurosurgical operations for removing a part of the bone to reach the benign and cancerous tumor lesions present beneath the bone. Other applications of bone grinding include ossification of posterior longitudinal ligament (OPLL) surgery, trigeminal neuralgia surgery, bone decompression surgery, craniomaxillofacial surgery, and expanded endonasal approach (EEA). The EEA approach uses nostril as a passage to use miniature grinding burr to operate on the base of the skull and craniometrical junction to dissect the lesions affecting the normal functioning of the human body.[1] Endonasal assists in seeing an internal target region by providing light and lens which helps in transmitting the images from the target region to the display unit. The innovative EEA technology has been developed which allows the surgeons to reach the regions which were earlier inoperable without removing the major part of the skull or face. During EEA, when a miniature grinding burr is rotated at high revolutions per minute (rpm) and comes in contact with the hard tissue, i.e. bottom part of the skull bone then it causes abrasion. Owing to the abrasion during bone grinding, cutting forces and torque comes into the picture and plays a significant role in whole grinding operation in terms of temperature rise, microcracks, and quality of bone around the grinding site. So, the quantification of the cutting forces and torque becomes necessary to perform bone grinding efficiently with minimal disturbance to the surrounding tissues and bone.

The rise in cutting forces and temperature during bone grinding is a major concern for the neurosurgeons worldwide.[2] Different clinical studies were carried out to evaluate the effect of the rise in heat during osteotomy on hard and soft tissues. Vrind et al.[3] investigated the thermal effect on rat sciatic nerve during bone osteotomy. The results revealed that an increase in temperature beyond threshold caused complete sensory and functional loss with a decrease in the amplitude of motor and reflex responses. On the same context, Xu and Pollock[4] studied thermal injury to the sciatic nerve of a cat at 47°C and 58°C. Higher temperature-induced axonal degeneration, loss of myelinated fibers, heat-induced angiopathy, and reduction in nerve blood flow were found. Similarly, Tamai et al.[5] investigated nerve root injury in a rabbit model during bone drilling. McDannold et al.[6] predicted the brain tissue damage due to temperature variations in 26 rabbits at 63 different locations and tissue damage was seen at every location. The threshold temperature for tissue damage is observed to be 47°C. *In vitro* investigations were also executed by past researchers to determine the effect of different process parameters on temperature generated during bone osteotomy. Call[7] studied temperature elevations and bone surface osteocytes necrosis during bone drilling. It was noticed that temperature generated may injure facial nerves but the risk of thermal damage to bone and nerves can be minimized by using irrigation. In another research work,[8] a diamond surgical burr (5 mm) was used at a high speed of 60,000 rpm to predict the thermal injury to nerve roots during bone grinding. The results showed that beyond 60 s of drilling, the temperature around the nerve root reached up to 52°C. The heat generated during bone grinding could cause blood coagulation, loss of vision due to optic nerve damage, and thermal necrosis.[9-11] So, measuring and mitigating temperature rise during grinding is important to avoid thermal damage to the bone.

To the best knowledge of the authors, no study has been carried out which quantitively measures the cutting forces along with the temperature generated during bone grinding. Furthermore, an optimized set of parametric combination is required to mitigate thermal and mechanical damage during bone grinding. So, in this study, the effect of different bone grinding parameters has been investigated in terms of thrust force, tangential force, grinding force ratio, torque, and temperature generated during skull bone grinding. The bone samples after grinding have been characterized using scanning electron microscopy (SEM) to study the surface morphology of bone. Furthermore, the process parameters are further optimized using grey relational analysis. Regression models have been developed and are validated to ensure the adequacy of the developed model.

2. Materials and Methods

2.1. *Experimentation procedure*

The bone grinding experiments are conducted on the pig skull bone using a miniature diamond abrasive burr having a diameter of 4 mm. The reason for choosing the animal skull bone is owing to the reason that human skull bone is not easily available. Regarding ethical concerns, it should be noted that no animal was harmed for experimentation and bone is acquired from the local slaughterhouse. The experiments were immediately performed after slaughter. Bone samples are prepared and cut in the dimensions of 25 × 15 × 10 mm.[3] The experimental attributes and their levels have been selected after consultation with neurosurgeons available at All India Institute of Medical Science (AIIMS), New Delhi, India and literature studies.[2,8,12–22]

An in-house experimental setup is developed to perform bone grinding experiments on skull bone. The sequential manner followed in performing bone grinding experiments has been shown in Fig. 1. The whole bone grinding setup, which consists of air pencil grinder, fixture, bone, burr, and coolant was mounted on the CNC machine (Chandra+, Bharat Fritz Werner Ltd., India). Further, the feed rate to the burr and bone is provided using a computer-controlled mechanism. A sterile solution (0.9% sodium chloride (NaCl) in 1L of H_2O) is used to irrigate the grinding site; thereby,

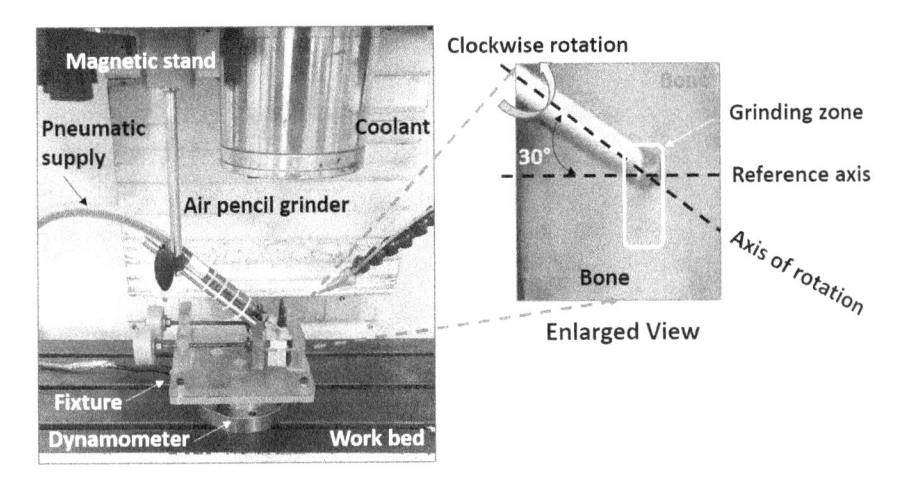

Fig. 1. Sequential manner followed in performing bone grinding experiments (Color online).

Table 1. Bone grinding parameters and their levels with response characteristics.

Parameters	Level 1	Level 2	Level 3	Response characteristics
Rotational speed (rpm)	35,000	45,000	55,000	Tangential force Thrust force
Feed rate (mm/min)	20	40	60	Grinding force ratio
Depth of cut (mm)	0.50	0.75	1.00	Torque Temperature

removal of bone chips and heat from the grinding zone. A continuous pneumatic supply is supplied to the air pencil grinder to achieve high-speed of the grinding burr. The high-speed air pencil grinder is inclined at an angle of 30° as used in osteotomy during removal of a bone flap from a skull bone.[8] Authors have tried to mimic the actual surgical condition based upon neurosurgeon's guidance. A full-factorial design consists of 27 unique and total of 81 trials has been used adopted to study the allpossible set of parametric combinations. Process parameter has been selected after consultation with neurosurgeons. Table 1 shows the process parameters and their levels with output characteristics selected for bone grinding experiments. The cutting forces developed during grinding has been quantitatively measured using a dynamometer (2852A-02, Kistler group, Switzerland), while the temperature generated on the bone's surface is measured using infrared thermography (U5855A, Keysight technologies, USA). Subsequently, bone samples are characterized using SEM to observe the microcracks over the surface of a bone. Subsequently, optimized set of parametric combination have been given to mitigate thermal damage during bone grinding.

2.2. *Statistical analysis and optimization procedure*

2.2.1. *Quadratic regression model*

The regression modeling is done utilizing polynomial functions if the real functional relationship between input and output parameters is unknown.[23,24] A least square method is used to perform the regression modeling and best fit is applied to the available regression data with continuous efforts to reduce the residuals (SS: a sum of a square) between the measured and response value. The regression equation is combined with the measured data to establish an empirical relationship between the input variables and target variables.

A general quadratic regression model is written as

$$y = \beta_0 + \beta_1 x_1 + \beta_2 x_2 + \beta_3 x_3 + \beta_4 x_1^2 + \beta_5 x_2^2 + \beta_6 x_3^2 \\ + \beta_7 x_1 x_2 + \beta_8 x_2 x_3 + \beta_9 x_3 x_1 + \varepsilon, \quad (1)$$

where y is the response characteristic/output variable, $\beta_0, \beta_1,...,\beta_9$ are the regression coefficients, and $x_1, x_2,...,x_3$ are the input variables. The key aspect of such a regression model is to predict the response value for a given set of input values. Equation (1) has been used to develop the quadratic regression models for thrust force, tangential force, grinding force ratio, torque, and temperature.

2.2.2. *Grey relational analysis*

Grey relational analysis is a normalization-based method established by professor Deng in 1982.[25] The uniqueness of this technique is that it converts the problem with multiple objectives into a response characteristic function, i.e. grey relational grade (GRG).[26,27] GRA is performed to reach an optimum set of parameters for real-world problems. It is an effective technique to solve the responses having a complex relationship among them.[28] To evaluate their dynamic and relative influence, it has proved its competences. The comprehensive argument on the calculation of GRG from the experimental data is discussed in the following manner.

The experimental data is normalized in range 0–1 with the help of Eq. (1).

$$x_i(j) = \frac{\max y_i(j) - y_i(j)}{\max y_i(j) - \min y_i(j)}, \qquad (2)$$

where $x_i(j)$ represents values after normalization, and max $y_i(j)$ and min $y_i(j)$ represent maximum and minimum value, respectively, for $y_i(j)$ corresponding to the jth term.

For example, in tangential force calculation of pre-processed data of Experiment 1, max $y_i(k) = 22.11$, min $y_i(j) = 14.68$ and $y_i(j) = 17.55$. So, $x_i(j) = (22.11 - 17.55)/(22.11 - 14.68) = 0.614$.

Similarly, data has been normalized for other combinations. Once we normalize the experimental data, the next move is to calculate the grey relational coefficient (GRC) $\xi i(j)$ with the help of Eq. (2), which is shown below

$$i(j) = \frac{\Delta_{\min} + \lambda \Delta_{\max}}{\Delta_{0i}(j) + \lambda \Delta_{\max}}, \qquad (3)$$

where $\Delta_{\min} = \forall j^{\min} \in i \forall j^{\min} \|x_0(j) - x_j(j)\|$ = lowest value of Δ_{0i}; $\Delta_{\max} = \forall j^{\max} \in i \forall j^{\max} \|x_0(j) - x_j(j)\|$ = largest value of Δ_{0i}; $\Delta_{0i}(j) = \|x_0(j) - x_j(j)\|$ = positive difference between $x_0(j)$ and $x_j(j)$; λ = characteristic coefficient (ranges between zero and one). $\xi i(j)$ for Experiment 1 is calculated as: $\Delta_{\min} = 0$, $\Delta_{\max} = 1$, $\lambda = 0.5$, $\Delta_{0i}(j) = 0.386$ and finally $\xi_i(j) = ((0 + 0.5 \times 1)/(0.386 + 0.5 \times 1)) = 0.564$. Afterward, GRG γ_i is determined using Eq. (4).

$$\gamma_i = \frac{1}{n} \sum_{j=1}^{m} \xi_i(j), \qquad (4)$$

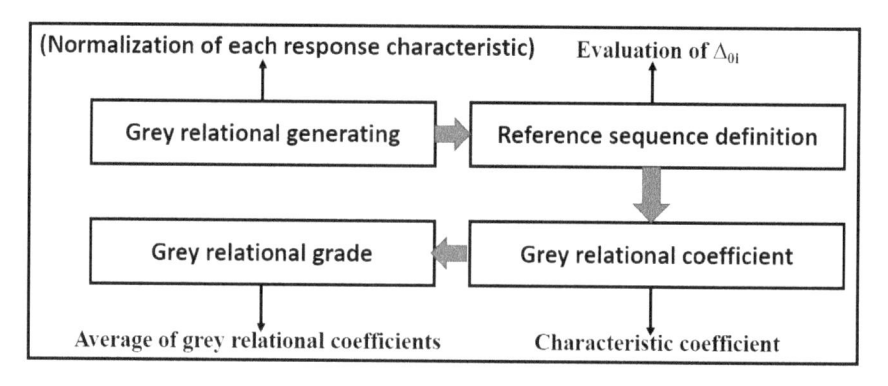

Fig. 2. A working sequence of grey relational analysis (Color online).

where "m" represents the number of inputs and $\gamma_i = (0.564 + 0.526 + 0.424 + 0.579 0.498)/5 = 0.518$ (Experiment 1). Then ranking is carried out from high to low value.

3. Results and Discussion

3.1. *Parametric effect on response characteristics*

Figure 3 illustrates the main effect plot for the tangential force which acts on the bone during grinding. It can be seen that as the rotational speed is increased from 35,000 rpm to 55,000 rpm, the tangential force significantly decreases from 19.26 N to 16.48 N. This may be because the reduction in the size of bone chips and mean friction coefficient leads to a reduction in the tangential force at a higher speed.[29,30] The results found are in-line with the trend results reported by Gupta and Pandey.[31] However, the trend shown by feed rate is just opposite of the rotational speed. Increase in the feed rate from 20 mm/min to 60 mm/min leads to the escalation of tangential force from 16.53 N to 19.25 N. The results are in consensus with trend results reported by Soriano et al.[32] The increase is owing to the reason deeper cuts are made at higher feed rate which requires more shearing energy and increased cutting forces.[33,34] It is revealed that higher feed rate leads to the higher cutting forces which in turn leads to development of microcracks over the surface

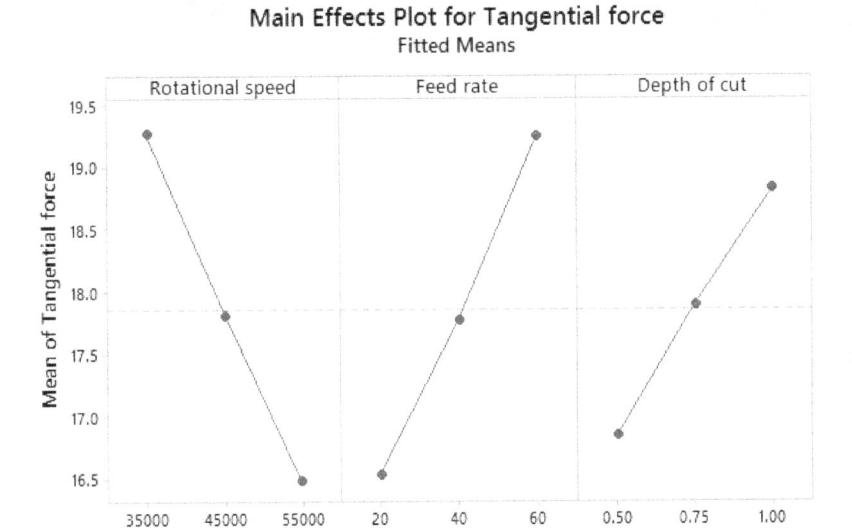

Fig. 3. Main effect plot for the tangential force generated during bone grinding (Color online).

of bone, as evidently in Fig. 4. The cutting streaks are also visible in grinding zone as can be seen in Fig. 4(a). Microcracks produced on the surface of bone after grinding has been demonstrated in Fig. 4(b). Subsequently, delamination caused on the edges of groove's surface after grinding has been depicted in Figs. 4(c) and 4(d).

The mechanism of the bone removal during grinding involves three possible stages of chips removal from the bone: (i) elastic–plastic deformation, i.e. rubbing, (ii) pile upstage in which bone chips start accumulating near the edges of abrasive grits owing to ploughing action and (iii) cutting which leads to the removal of chips from the surface of a bone. Three types of grain and bone interactions simultaneously exist in the grinding namely rubbing, ploughing, and cutting.

The increase in rotational speed from level 1 (35,000 rpm) to level 2 (45,000 rpm) decreases the thrust force from 8.36 N to 7.36 N and on further increasing the speed from 45,000 rpm to 55,000 rpm, decreases the thrust force from 7.36 N to 6.25 N as evidently in

Fig. 4. Surface morphology of bone specimens after grinding. (a) Cutting streaks at 1000 × magnification. (b) Microcracks over the surface of the bone at 2000 × magnification. (c) and (d) Delamination at end edges of groove after bone grinding at 30 × and 50 × magnification (Color online).

Fig. 5. On contrary, the increase in feed rate from level 1 to level 2, increases the thrust force from 6.52 N to 7.25 N and then 8.18 N at a feed rate of 60 mm//min. The increase in the depth of cut leads to the increase in the thrust force from 7.02 N to 7.03 N and then 7.64 N at depth of cut of 0.50, 0.75, 1.00 mm, respectively.

Figure 6 shows the main effect plot of torque generated during grinding of the skull bone. A similar trend has been shown by the torque as that of tangential and thrust force. It was observed that the increase in the rotational speed from level 1 to level 2, decreases the torque from 3.39 N · mm to 2.53 N · mm which further decreases to 2.14 N · mm when rotational speed reaches to level 3. On another side, the torque increases from 1.76 N · mm to 3.70 N · mm on increasing the feed rate from level 1 to level 3. The increase in the depth of cut from 0.50 mm to 1.00 mm also increases the torque from 2.37 N · mm to 3.00 N · mm.

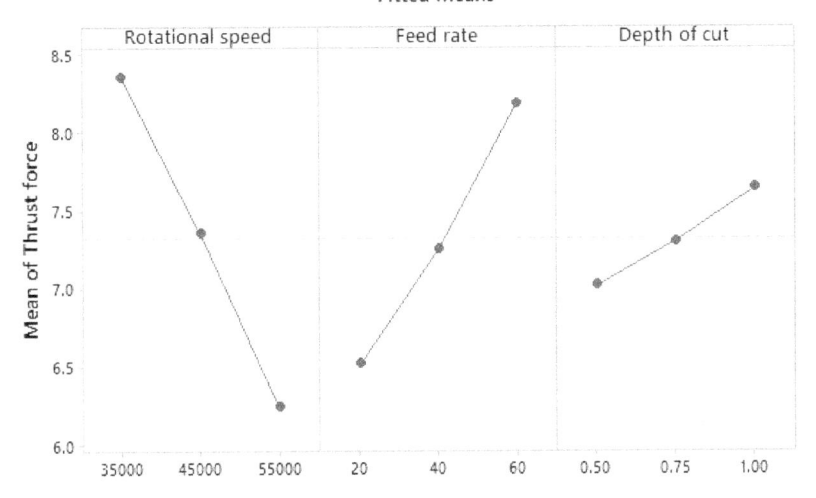

Fig. 5. Main effect plot for thrust force generated during bone grinding (Color online).

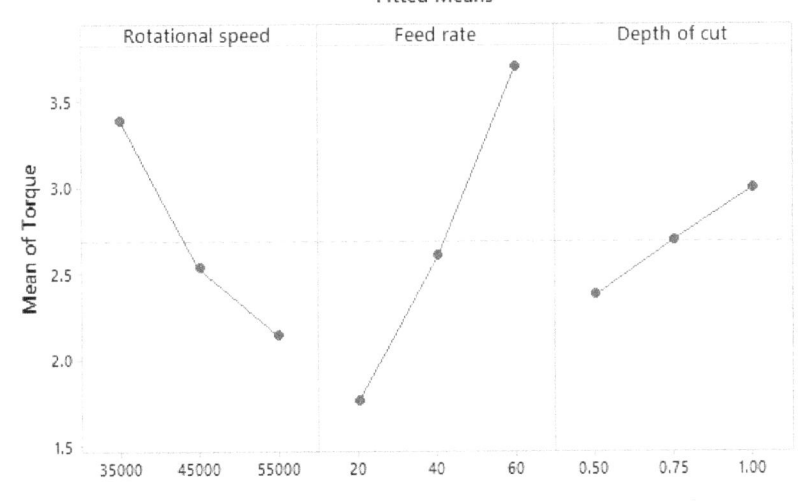

Fig. 6. Main effect plot for torque generated during bone grinding (Color online).

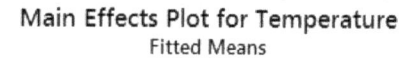

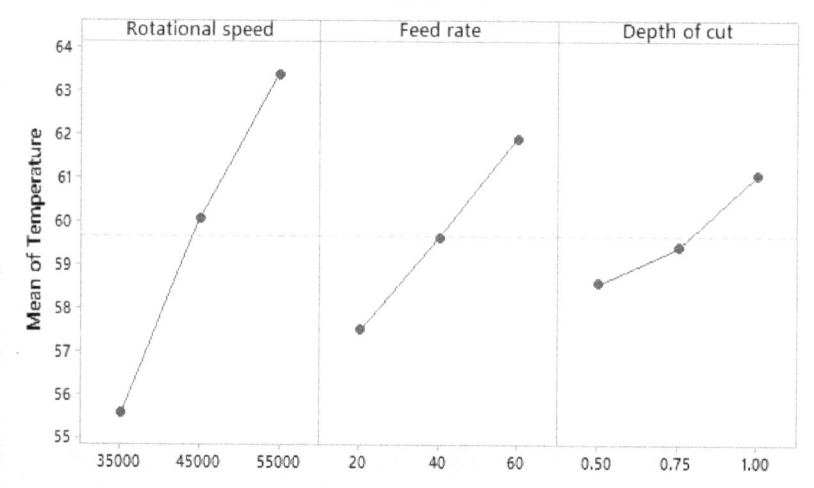

Fig. 7. Main effect plot for temperature generated during bone grinding (Color online).

The temperature increase during bone grinding is owing to the generation of heat leads to the thermogenesis. It was observed that as the rotational speed is increased from 35,000 rpm to 55,000 rpm, the temperature keeps on increasing and generates temperature from 55.57°C to 63.38°C as shown in Fig. 7.

It could be because at higher level of rotational speed, more energy is transferred into the bone at higher speed which ultimately raises the temperature of bone at the grinding site.[35] The increase in feed rate from 20 mm/min to 40 mm/min increases the temperature from 57.48°C to 59.61°C and then to 61.89°C at a feed rate of 60 mm/min. The increase in the depth of cut also leads to the rise in temperature as it is noticed that temperature changes from 58.56°C to 59.39°C and then 61.04°C at 0.50, 0.75, and 1.00 mm, respectively. The increase in temperature could cause blood coagulation, bone cell's carbonization, damage to osteons and bone matrix.[17,36] Furthermore, comparison of the temperature results as reported by past researchers and present study of bone grinding has been accentuated in Table 2.

Table 2. Comparison of the temperature results as reported by past researchers and our study of bone grinding.

Researcher	Workpiece material	Temperature/change in temperature (ΔT)[a]
Zhang et al.[17]	Bovine femur bone	$\geq 200°C$
Shakouri and Mirfallah.[19]	Bovine femur bone	$53.2°C\ (\Delta T)$
Mizutani et al.[20]	Bovine femoral compact bone	$\leq 100°C$
Yang et al.[21]	Bovine femur bone	$41.6 \pm 1.95°C$
		$37.4 \pm 1.40°C$
		$29.4 \pm 1.83°C$
		$26.2 \pm 1.65°C$
Enomoto et al.[22]	Bovine femoral cortical bone	$\geq 100°C$
This study	Pig skull bone	$\leq 65.8°C$

[a]Room/initial temperature is different for the above-mentioned studies.

3.2. Multi-attribute decision making

On the completion of experiments, the results found for the different bone grinding experiments have been shown in Table 3. The measured values of the tangential force, thrust force, grinding force ratio, torque and temperature have been recorded and plotted. Table 4 shows the pre-processed, i.e. normalized data of the response characteristics using Eq. (1). Table 5 shows the deviation sequence of response characteristics. Table 6 shows the GRC, GRG, and ranking is performed to given rank to different GRG. The highest value of the GRG has been assigned as rank 1 whereas lowest GRG value has been assigned as rank 27. It is clear that optimized set of conditions for performing bone grinding is the rotational speed of 55,000 rpm, a feed rate of 20 mm/min and depth of cut of 0.50 mm as it leads to the tangential force of 14.68 N, thrust force of 5.26 N, grinding force ratio of 0.36, the torque of 0.70 N · mm, and temperature of 56.3°C. The minimum cutting forces and temperature (below threshold temperature of 47°C) leads to reduced

Table 3. Bone grinding process parameters and recorded values of response characteristics.

Exp. No.	Bone grinding input parameters			Response characteristics				
	Rotational speed (rpm)	Feed rate (mm/min)	Depth of cut (mm)	Tangential force (N)	Thrust force (N)	Grinding force ratio	Torque (N·mm)	Temperature (°C)
1	45,000	40	0.75	17.55	7.34	0.42	2.1	58.6
2	35,000	60	0.75	21.32	9.25	0.43	4.12	57.3
3	35,000	40	0.5	17.78	8.05	0.45	3.1	54.1
4	55,000	40	0.5	15.87	5.85	0.37	1.84	59.5
5	55,000	20	0.5	14.68	5.26	0.36	0.7	56.3
6	45,000	40	0.5	16.89	7.1	0.42	1.98	58.1
7	45,000	20	0.5	15.14	6.23	0.41	1.33	54.7
8	45,000	60	1	20.23	8.34	0.41	4.1	62.4
9	35,000	40	1	20.54	8.44	0.41	3.67	55.2
10	55,000	20	1	16.22	5.65	0.35	1.54	58.8
11	35,000	20	1	18.44	7.98	0.43	2.88	54.2
12	55,000	20	0.75	15.77	5.44	0.34	0.99	57.4
13	45,000	60	0.5	18.44	7.96	0.43	3.47	60.5
14	55,000	40	0.75	16.14	6.13	0.38	2.33	60.6
15	55,000	40	1	17.54	6.45	0.37	2.5	61.4
16	55,000	60	1	18.24	7.44	0.41	3.44	65.8

17	35,000	60	1	22.11	9.88	0.45	4.55	58.4
18	55,000	60	0.5	16.36	6.88	0.42	2.75	60.3
19	45,000	20	0.75	16.33	6.54	0.40	1.65	55.3
20	35,000	40	0.75	19.17	8.23	0.43	3.45	54.7
21	35,000	20	0.75	17.66	7.55	0.43	2.54	53.7
22	35,000	60	0.5	19.48	8.68	0.45	3.88	56.7
23	45,000	60	0.75	19.57	8.11	0.41	3.85	61.3
24	35,000	20	0.5	16.88	7.14	0.42	2.32	52.5
25	45,000	40	1	18.46	7.69	0.42	2.44	59.7
26	45,000	20	1	17.64	6.9	0.39	1.87	56.4
27	55,000	60	0.75	17.49	7.12	0.41	3.2	62.7

Table 4. Normalization of response characteristics values.

Exp. No.	Tangential force (N)	Thrust force (N)	Grinding force ratio	Torque (N · mm)	Temperature (°C)
1	0.614	0.550	0.320	0.636	0.541
2	0.106	0.136	0.175	0.112	0.639
3	0.583	0.396	0.000	0.377	0.880
4	0.840	0.872	0.781	0.704	0.474
5	1.000	1.000	0.876	1.000	0.714
6	0.703	0.602	0.300	0.668	0.579
7	0.938	0.790	0.383	0.836	0.835
8	0.253	0.333	0.376	0.117	0.256
9	0.211	0.312	0.388	0.229	0.797
10	0.793	0.916	0.969	0.782	0.526
11	0.494	0.411	0.186	0.434	0.872
12	0.853	0.961	1.000	0.925	0.632
13	0.494	0.416	0.196	0.281	0.398
14	0.803	0.812	0.677	0.577	0.391
15	0.615	0.742	0.789	0.532	0.331
16	0.521	0.528	0.416	0.288	0.000
17	0.000	0.000	0.055	0.000	0.556
18	0.774	0.649	0.299	0.468	0.414
19	0.778	0.723	0.485	0.753	0.789
20	0.396	0.357	0.217	0.286	0.835
21	0.599	0.504	0.234	0.522	0.910
22	0.354	0.260	0.067	0.174	0.684
23	0.342	0.383	0.356	0.182	0.338
24	0.704	0.593	0.276	0.579	1.000
25	0.491	0.474	0.336	0.548	0.459
26	0.602	0.645	0.571	0.696	0.707
27	0.622	0.597	0.424	0.351	0.233

Table 5. Evaluation of Δ_{0i} (deviation sequence) of response characteristics values.

Exp. No.	Tangential force (N)	Thrust force (N)	Grinding force ratio	Torque (N·mm)	Temperature (°C)
1	0.386	0.450	0.680	0.364	0.459
2	0.894	0.864	0.825	0.888	0.361
3	0.417	0.604	1.000	0.623	0.120
4	0.160	0.128	0.219	0.296	0.526
5	0.000	0.000	0.124	0.000	0.286
6	0.297	0.398	0.700	0.332	0.421
7	0.062	0.210	0.617	0.164	0.165
8	0.747	0.667	0.624	0.883	0.744
9	0.789	0.688	0.612	0.771	0.203
10	0.207	0.084	0.031	0.218	0.474
11	0.506	0.589	0.814	0.566	0.128
12	0.147	0.039	0.000	0.075	0.368
13	0.506	0.584	0.804	0.719	0.602
14	0.197	0.188	0.323	0.423	0.609
15	0.385	0.258	0.211	0.468	0.669
16	0.479	0.472	0.584	0.712	1.000
17	1.000	1.000	0.945	1.000	0.444
18	0.226	0.351	0.701	0.532	0.586
19	0.222	0.277	0.515	0.247	0.211
20	0.604	0.643	0.783	0.714	0.165
21	0.401	0.496	0.766	0.478	0.090
22	0.646	0.740	0.933	0.826	0.316
23	0.658	0.617	0.644	0.818	0.662
24	0.296	0.407	0.724	0.421	0.000
25	0.509	0.526	0.664	0.452	0.541
26	0.398	0.355	0.429	0.304	0.293
27	0.378	0.403	0.576	0.649	0.767

Table 6. GRC, GRG and ranking of response variables.

Exp. No.	Tangential force (N)	Thrust force (N)	Grinding force ratio	Torque (N·mm)	Temperature (°C)	GRG	Rank
1	0.564	0.526	0.424	0.579	0.522	0.523	14
2	0.359	0.367	0.377	0.360	0.581	0.409	25
3	0.545	0.453	0.333	0.445	0.806	0.517	16
4	0.757	0.797	0.695	0.628	0.487	0.673	5
5	1.000	1.000	0.801	1.000	0.636	0.888	1
6	0.627	0.557	0.417	0.601	0.543	0.549	12
7	0.890	0.704	0.448	0.753	0.751	0.709	4
8	0.401	0.429	0.445	0.362	0.402	0.408	26
9	0.388	0.421	0.450	0.393	0.711	0.473	20
10	0.707	0.856	0.941	0.696	0.514	0.743	3
11	0.497	0.459	0.380	0.469	0.796	0.520	15
12	0.773	0.928	1.000	0.869	0.576	0.829	2
13	0.497	0.461	0.383	0.410	0.454	0.441	22
14	0.718	0.726	0.607	0.541	0.451	0.609	8
15	0.565	0.660	0.703	0.517	0.428	0.574	10
16	0.511	0.514	0.461	0.413	0.333	0.446	21
17	0.333	0.333	0.346	0.333	0.530	0.375	27
18	0.689	0.588	0.416	0.484	0.460	0.527	13
19	0.692	0.643	0.493	0.670	0.704	0.640	6
20	0.453	0.438	0.390	0.412	0.751	0.489	17
21	0.555	0.502	0.395	0.511	0.847	0.562	11
22	0.436	0.403	0.349	0.377	0.613	0.436	23
23	0.432	0.448	0.437	0.379	0.430	0.425	24
24	0.628	0.551	0.409	0.543	1.000	0.626	7
25	0.496	0.487	0.429	0.525	0.480	0.484	18
26	0.557	0.585	0.538	0.622	0.630	0.586	9
27	0.569	0.554	0.465	0.435	0.395	0.484	19

damage to the skull bone which in turn reduces the post-operative healing time.

3.3. *Quadratic regression analysis*

The quadratic regression models have been developed for thrust force, tangential force, grinding force ratio, torque, and temperature using Eq. (1). All the influential parameters have been considered during modeling. The developed models have been presented as Regression model for GRG

$$\begin{aligned}
\text{GRG} = {} & 0.936 - 0.000013 \text{ rotational speed} \\
& - 0.00183 \text{ feed rate } 0.177 \text{ depth of cut} \\
& \times 0.000000 \text{ rotational speed} \times \text{rotational speed} \\
& + 0.000038 \text{ feed rate} \times \text{feed rate} + 0.0311 \text{ depth of cut} \\
& \times \text{depth of cut} + 0.000000 \text{ rotational speed} \\
& \times \text{feed rate} + 0.000004 \text{ rotational speed} \\
& \times \text{depth of cut} + 0.003317 \text{ feed rate} \times \text{depth of cut}.
\end{aligned} \quad (5)$$

Figure 8 shows the residual plots for GRG which are obtained during the development of empirical models. The normal probability plot of the GRG indicates that all residual points lie in a straight-line pattern and maximum residuals lie on the upper side of the fitted line and only a few residuals are away from the fitted straight line. Residuals versus and fitted values verify the hypothesis of residuals with equal variance. It has been observed that residuals for all 27 observations are randomly distributed on both sides of the zero line. Uniform distribution of the residuals about the zero lines indicates the goodness of the developed model. Histogram plot shows that no outlier exists in the observed data and hence developed model has better accuracy. Residual versus the observation order determines the input set of grinding parameters which influences the results of response characteristics.

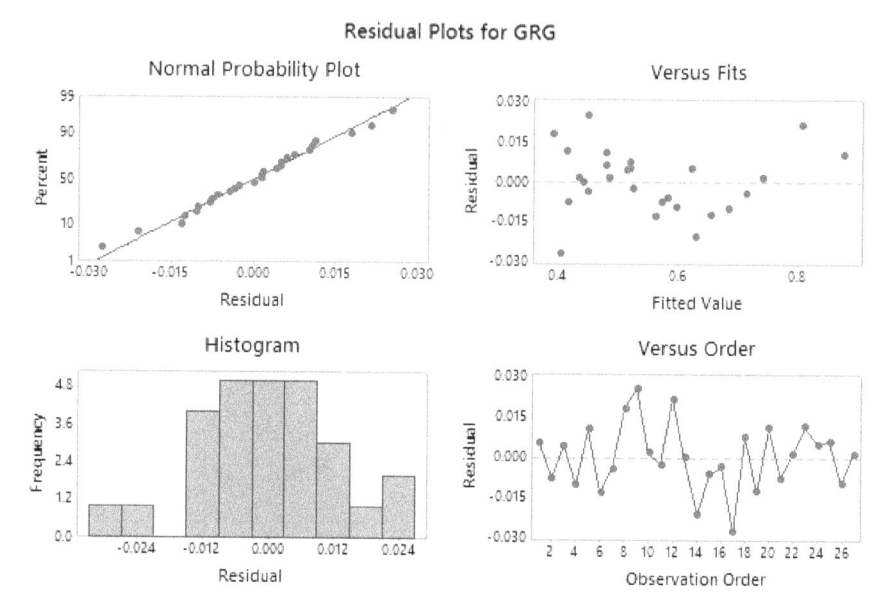

Fig. 8. Residual plots for GRG (Color online).

Figure 9 shows the three-dimensional (3D) surface plots and box plots of GRG generated in the MATLAB considering the developed quadratic regression model using Eq. (5). The interaction of the rotational speed and feed rate has been highlighted and it is revealed that GRG starts increasing the increase in feed rate from 20 mm/min to 60 mm/min. However, the escalation of the rotational speed shows the opposite trend and GRG decreases with the increase in the rotational speed from 35,000 rpm to 55,000 rpm. Figure 10 shows the 3D surface plots of GRG against feed rate and depth of cut (refer Fig. 10(a)) and box plot of GRG versus feed rate (refer Fig. 10(b)). It has been observed that GRG increases with the increase in depth of cut and decreases with the increment in feed rate from 20 mm/min to 60 mm/min. Figure 11 indicates the trend of GRG against rotational speed and depth of cut. It is noticed that as the rotational speed and depth of cut is increased, the GRG shows a decreasing trend. Further, the box plot presented

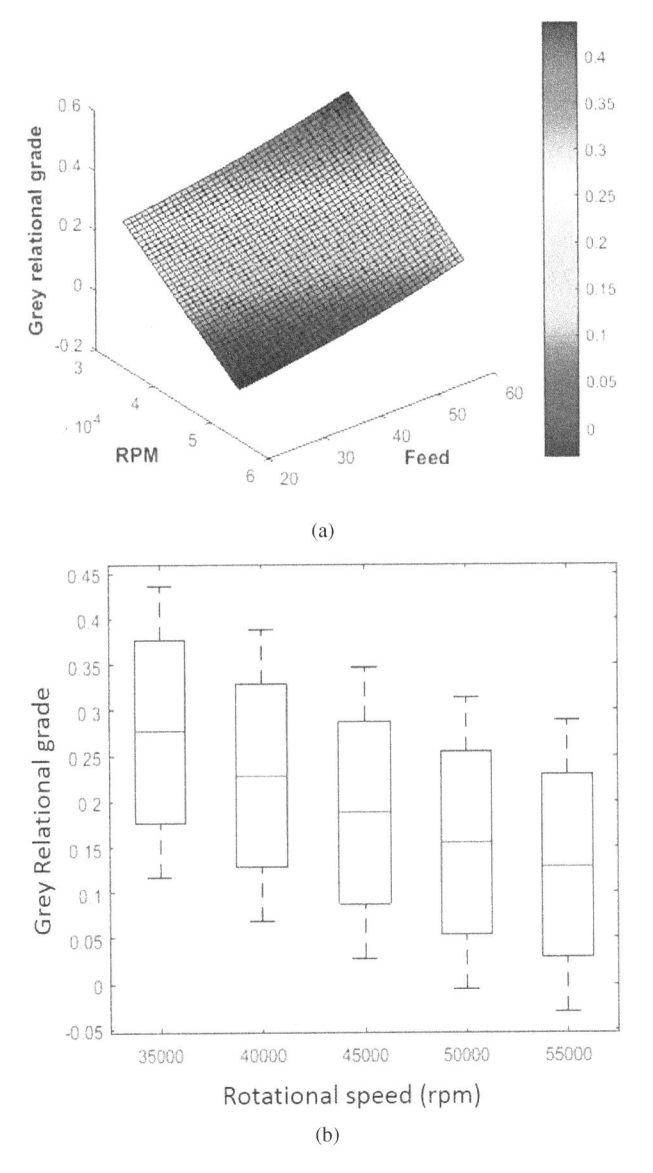

Fig. 9. (a) 3D surface plots of GRG (*Z*-axis) versus rotational speed (*Y*-axis) and feed rate (*X*-axis) generated using developed quadratic regression models and (b) Box plots for GRG versus rotational speed (Color online).

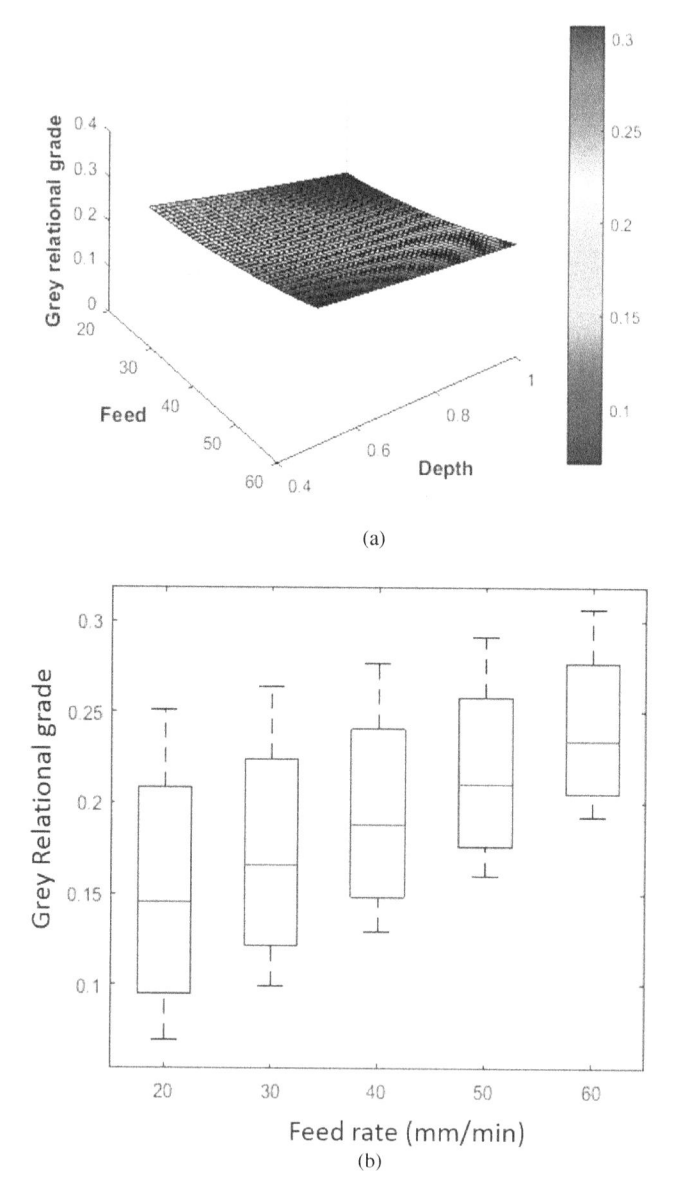

Fig. 10. 3D surface plots of GRG (Z-axis) versus feed rate (Y-axis) and depth of cut (X-axis) generated using developed quadratic regression models and (b) Box plots for GRG versus feed rate (Color online).

Investigation and parametric optimization 23

(a)

Depth of cut (mm)

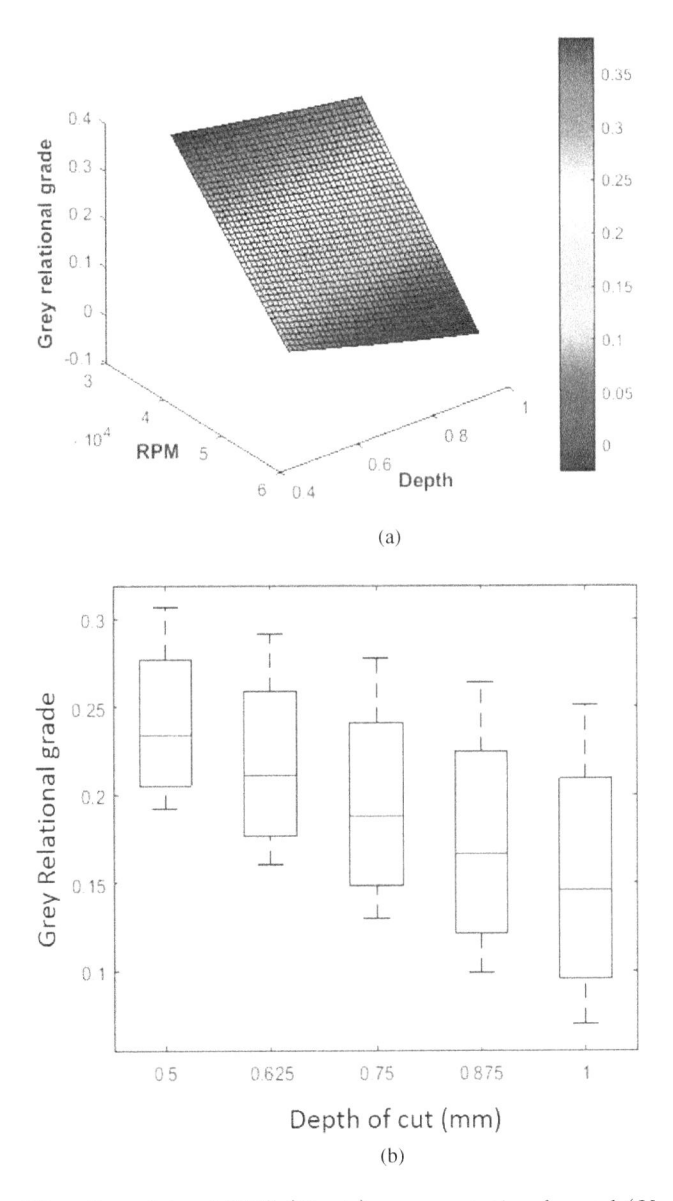

(b)

Fig. 11. 3D surface plots of GRG (Z-axis) versus rotational speed (Y-axis) and depth of cut (X-axis) generated using developed quadratic regression models and (b) Box plots for GRG versus depth of cut (Color online).

in Fig. 11(b) also shows that escalation of depth of cut from 0.50 to 1.00 mm caused a significant decrease in GRG.

3.4. Validation of developed quadratic regression model

The model developed for the GRG has been validated with the experimental results. All 27 sets of parametric combinations have been used as the input for the developed model and value of GRG is calculated, as evidently in Table 7. The predicted results using the empirical model have been compared with the experimental results and are plotted, as shown in Fig. 12. It is clear that the predicted results lie very close to the experimental results. Table 8 shows the optimized set of bone grinding parameters. The predicted and

Table 7. Experimental and predicted results of grey relational grade.

Exp. No.	Experimental	Predicted
1	0.523	0.518
2	0.409	0.417
3	0.517	0.513
4	0.673	0.684
5	0.888	0.878
6	0.549	0.562
7	0.709	0.713
8	0.408	0.390
9	0.473	0.448
10	0.743	0.741
11	0.520	0.523
12	0.829	0.808
13	0.441	0.441
14	0.609	0.630
15	0.574	0.580

Table 7. (*Continued*)

Exp. No.	Experimental	Predicted
16	0.446	0.450
17	0.375	0.403
18	0.527	0.520
19	0.640	0.653
20	0.489	0.478
21	0.562	0.570
22	0.436	0.435
23	0.425	0.414
24	0.626	0.621
25	0.484	0.478
26	0.586	0.596
27	0.484	0.483

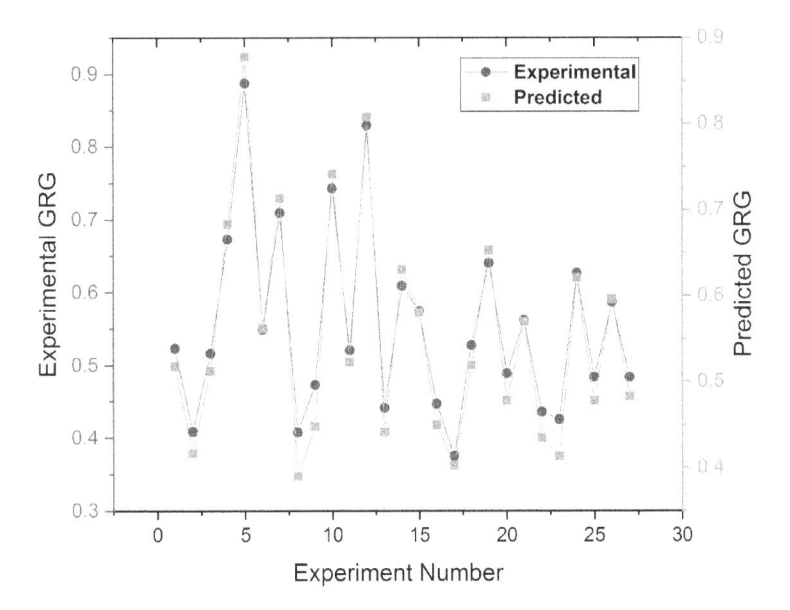

Fig. 12. Comparative plot for experimental and predicted results (Color online).

Table 8. Validation of the developed regression model at optimized conditions.

Optimized conditions	Parametric value	Predicted	Experimental	Error
Rotational speed (rpm)	55,000	—	—	—
Feed rate (mm/min)	20			
Depth of cut (mm)	0.5	—	—	—
GRG	A3B1C1	0.878	0.888	0.010

experimental results of GRG at optimized conditions is presented which validates the accuracy of the models. Improvement in the GRG has been noticed which indicates that the developed empirical model can be used for the prediction of response characteristics.

4. Conclusion

The study investigates the effect of various bone grinding parameters on the tangential force, thrust force, grinding force ratio, torque, and temperature generated during skull bone grinding. It is revealed that grinding forces and torque decreased with the escalation of rotational speed from 35,000 to 55,000 rpm. However, cutting forces and torque significantly increased by increasing the feed rate from 20 mm/min to 60 mm/min. On contrary, the temperature increased on increasing the rotational speed, feed rate, and depth of cut. 3D surface and box plots indicate a significant influence of interaction terms on response characteristics. Furthermore, it is revealed that higher cutting forces and torque adversely affect the bone's surface quality and leads to the generation of microcracks over the surface of bone. Regression models have been successfully developed for response characteristics and are also validated with the experimental findings. The bone grinding parameters are optimized using GRA analysis and optimal combination of process parameters found as rotational speed of 55,000 rpm, feed rate of 20 mm/min, and depth of cut of 0.50 mm, considering all response characteristics simultaneously.

Acknowledgment

The authors would like to thank Dr. Deepak Agrawal and his team at the All-India Institute of Medical Science (AIIMS) in New Delhi for the up-to-date guidance and assistance in providing information on neurosurgical bone grinding operations to thoroughly understand the entire bone grinding process to investigate.

References

1. P. Cappabianca, L. M. Cavallo and E. de Divitiis, *Neurosurgery* **55** (2004) 933.
2. L. Zhang, B. L. Tai, A. C. Wang and A. J. Shih, *CIRP Ann.* **62** (2013) 367.
3. H. H. De Vrind, J. Wondergem and J. Haveman, *J. Neurosci. Meth.* **45** (1992) 165.
4. D. Xu and M. Pollock, *Brain* **117** (1994) 375.
5. K. Tamai *et al.*, *Bone Joint J.* **99B** (2017) 554.
6. N. McDannold, N. Vykhodtseva, F. A. Jolesz and K. Hynynen, *Magn. Reson. Med.* **51** (2004) 913.
7. W. H. Call, *Ann. Otol. Rhinol. Laryngol.* **87** (1978) 43.
8. B. L. Tai *et al.*, *Med. Eng. Phys.* **35** (2013) 1545.
9. K. I. A. Al-Abdullah, H. Abdi, C. Peng and W. A. Yassin, *Measurement* **116** (2018) 25.
10. R. K. Pandey and S. S. Panda, *Measurement* **59** (2015) 9.
11. R. K. Pandey and S. S. Panda, *Measurement* **47** (2014) 386.
12. A. Babbar, V. Jain and D. Gupta, *Proc. Inst. Mech. Eng. H, J. Eng. Med.* **234** (2020) 626.
13. A. Babbar, V. Jain and D. Gupta, *Mater. Today Proc.* **33** (2020) 1458.
14. A. Babbar, V. Jain and D. Gupta, *Biomanufacturing* (Springer, Cham, 2019), pp. 137–155.
15. B. Atul, J. Vivek and G. Dheeraj, *Grey Syst. Theory Appl.* **10** (2020) 479.
16. A. Babbar, V. Jain and D. Gupta, *J. Braz. Soc. Mech. Sci. Eng.* **41** (2019) 401.
17. L. Zhang *et al.*, *Med. Eng. Phys.* **35** (2013) 1391.

18. B. L. Tai, D. A. Stephenson and A. J. Shih, *J. Manuf. Sci. Eng.* **134** (2012) 21006.
19. E. Shakouri and P. Mirfallah, *Proc. Inst. Mech. Eng. H, J. Eng. Med.* **233** (2019) 648.
20. T. Mizutani, U. Satake and T. Enomoto, *Precis. Eng.* **56** (2019) 80.
21. M. Yang *et al.*, *Mater. Manuf. Process.* **32** (2017) 589.
22. T. Enomoto, H. Shigeta, T. Sugihara and U. Satake, *CIRP Ann.* **63** (2014) 305.
23. P.-Y. Zhao, M. Zhou, Y.-J. Zhang and G.-C. Qiao, *J. Cent. South Univ.* **25** (2018) 277.
24. M. K. Gupta, P. K. Sood and V. S. Sharma, *Mater. Manuf. Process.* **31** (2016) 1671.
25. N. Tosun, *Int. J. Adv. Manuf. Technol.* **28** (2006) 450.
26. P. Kao and H. Hocheng, *J. Mater. Process. Technol.* **140** (2003) 255.
27. C. L. Lin, *Mater. Manuf. Process.* **19** (2004) 209.
28. J. Lin and C. Lin, *Int. J. Mach. Tools Manuf.* **42** (2002) 237.
29. W. Wang, Y. Shi, N. Yang and X. Yuan, *Med. Eng. Phys.* **36** (2014) 261.
30. K. Alam, A. V. Mitrofanov and V. V. Silberschmidt, *Med. Eng. Phys.* **33** (2011) 234.
31. V. Gupta and P. M. Pandey, *Proc. Inst. Mech. Eng. B, J. Eng. Manuf.* **232** (2018) 1549.
32. J. Soriano *et al.*, *Mach. Sci. Technol.* **17** (2013) 611.
33. J. E. Lee, C. L. Chavez and J. Park, *J. Biomech.* **71** (2018) 4.
34. R. Kumar and S. S. Panda, *J. Clin. Orthop. Trauma* **4** (2013) 15.
35. J. Lee, O. R. Ozdoganlar and Y. Rabin, *Med. Eng. Phys.* **34** (2012) 1510.
36. B. J. O'Daly *et al.*, *J. Mater. Process. Technol.* **200** (2008) 38.

Fabrication and surface modification of biomaterials for orthopedic implant: A review*

MD Manzar Iqbal[‡], Amaresh Kumar[‡], Rajashekhara Shabadi[‡] and Subhash Singh[†,§]

[‡]Department of Production and Industrial Engineering,
National Institute of Technology,
Jamshedpur, Jharkhand 831014, India
[‡]Faculty of Science and Technology,
University of Lille, Villeneuve d'Ascq 59650, Lille, France
[§]subh802004@gmail.com

An upsurge in demand and extensive effort in orthopedic implants directed toward innovative biomaterials for orthopedic applications. Orthopedic implants are significantly used in mature alternatives to retain, restore or modify the defective bone or tissue. However, exhaustive research in the past reveals various health-associated problems that can be effectively overcome by inventing newer kinds of biomaterials. The selection of optimal materials and the fabrication process are crucial challenges enforced by numerous novel materials that could be made for orthopedic applications. This paper intends to systematically assess the processing method employed in manufacturing the biomaterials for orthopedic applications. However, the success of biomedical implants in orthopedic are commonly restricted owing to insufficient bone-implant integration, wear debris induced osteolysis, and implant-associated infections. Nevertheless, the endeavor has also been intended

---

*To cite this article, please refer to its earlier version published in Surface Review and Letters, Vol. 30, No. 1 (2023) 2141008 (24 pages) DOI: 10.1142/S0218625X21410080
[§]Corresponding author.

to enhance the biological properties of the biomaterials by surface modification process while retaining their strength and hardness. Furthermore, various surface modifications have been comprehended. This review conferred contemporary advancements in surface coating approaches in orthopedic to enhance their osteointegration, improve corrosion resistance and accomplish antibacterial performance, clinical success and long-term service. The insight review has revealed the current outcomes in the field of engineering biomaterials concerning surface modifications of metallic implants or composite for enhancing their biological properties.

Keywords: Biomaterials; osteointegration; composite materials; biodegradable materials.

1. Introduction

During the global industrial innovation back in the 19th century, medical implants were directed toward the metal industry's early expansion. The progress of metal implants was directly proportional to the demand for metal implants, generally used in long bone fracture internal fixation. Professor Williams first defined biomaterials as "materials intended to interface with biological systems to evaluate, treat, augment, or replace any tissue, organs, or function of the body".[1-9] The first surgical suture application has been observed in 3000 BC, whereas the first dental implant was obtained in Europe around 900 AD. The study of biomaterials is generally known as biomaterial science or biomaterial engineering, which comprises biology, chemistry, medicine, material science and tissue engineering. Biomaterials are manufactured either from nature or produced directly from the laboratory or through several chemical approaches using metals, polymers, ceramics, composite materials or combinations. These materials are utilized for medical functions that involve the whole or a portion of the living systems, thus performing, enhancing or replacing a natural process of human physiology. These materials are incredibly sophisticated and sensitive during their applications. Therefore, medical implants must regulate the following prospects: biocompatibility, bio adhesion, bifunctionality and corrosion resistance from the aspect of bone tissues.[10,11] Moreover, the mechanical properties of bioimplant must be equivalent to the autogenous tissue without releasing harmful effects.[12–14] Enormous efforts have been addressed on biomaterials and orthopedic applications presented in Fig. 1, extracted from the Scopus database, which adds a new dimensional thought to the scientific research. The electronic literature database presented with a systematic and objective means to evaluate the emerging scientific literature in any area critically. Scopus data provided all data related to the biomaterial in the orthopedic

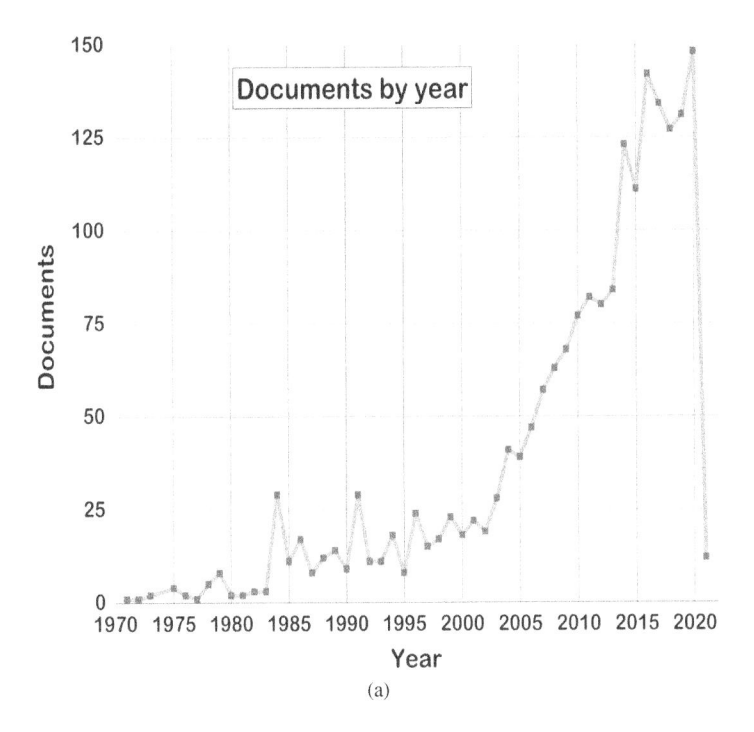

Fig. 1. (a) Illustration of the number of scientific publications from 1970 to 2021 using the search terms "magnesium" and "orthopedic applications". Data analysis was conducted on 19 January 2021 using Scopus search system. (b) The pie chart displays material science (27.0%) area is the most dominant in this research, followed by engineering (24.6%), biochemistry, genetics and molecular biology (12.7%), chemical engineering (9.8%) and medicine (8.1%) (Color online).

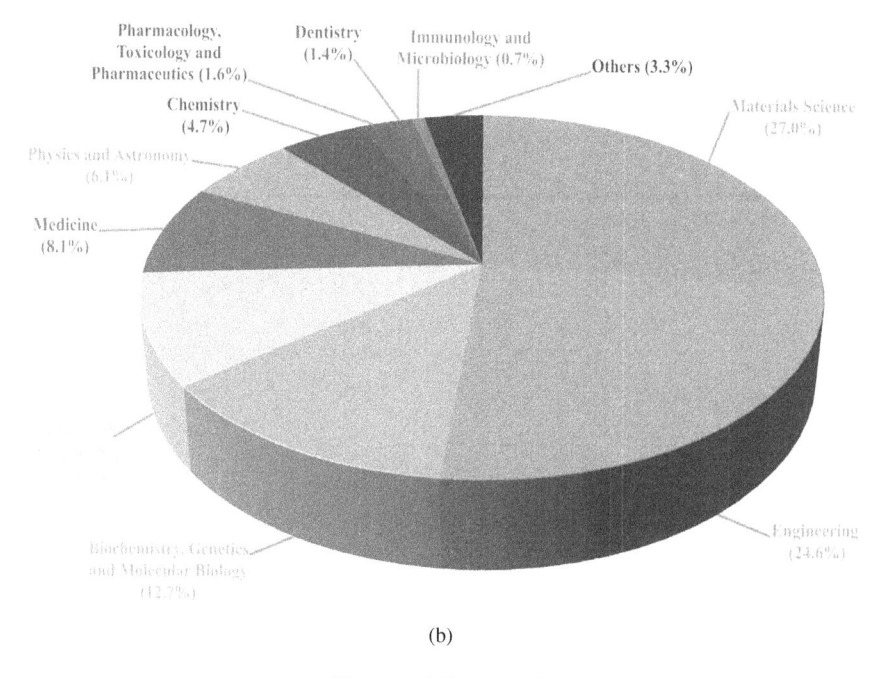

Pharmacology, Toxicology and Pharmaceutics (1.6%)
Dentistry (1.4%)
Immunology and Microbiology (0.7%)
Others (3.3%)
Chemistry (4.7%)
Materials Science (27.0%)
Physics and Astronomy (6.1%)
Medicine (8.1%)
Biochemistry, Genetics and Molecular Biology (12.7%)
Engineering (24.6%)

(b)

Fig. 1. (*Continued*)

application. The advancement of biomaterials has been growing year by year. Moreover, a noteworthy increase in publications reveals the growing importance of biomaterials, as exhibited in Fig. 1(a) and dominated fields confined to orthopedic applications presented in Fig. 1(b). In the last couple of decades, a significant shift is noticed at the molecular level from bioinert response to stimulating specific cellular response. This is attributed to the 3rd generation of biomaterial and facilitates to render its direct response through monitoring the surface characteristics. Moreover, an in-depth understanding of host tissue-implant interactions was coined to develop opportunities for material scientists to attain specific biological responses by tuning their surface properties.

Concurrently, researchers are also trying to develop a novel class of bone biomaterial, which is biocompatible, has mechanical properties significantly closer to a natural bone, antibacterial activities, strong bond between implant–host tissues. However, the success of long-term biomaterial implants still necessitates being focused because releasing of ions, poor osseointegration, uncontrolled degradation, stress shielding and inflammation. Faster recovery of the bone-implant interface is attributed to enhancing implant stability, longevity and shortening hospitalization. To overcome these issues, there is a requirement for surface modification of the fabricated implants. Therefore, surface modification is a feasible solution for enhancing the integration of bones with implants and further accelerating bone recovery. Various surface modification techniques are employed for improving the surface of implants.

Although many review papers are available, only few have extensively discussed the fabrication of various biomaterials by various processes and surface modification of the fabricated implants. Moreover, numerous biocompatible materials, such as stainless steel (SS), cobalt–chromium (Co–Cr) alloy, titanium base alloy, polymers, ceramics, biodegradable materials, porous materials and composite materials are extensively discussed for potential applications. Since most of the as-fabricated materials are not potentially favorable due to releasing ions, insufficient osseointegration, antibacterial properties, uncontrolled degradation, stress shielding and inflammation, respectively, thus could be revived by surface modification. Thus, this paper addresses various surface modification techniques to conquer the problems mentioned above with visual contrast. Moreover, *in vitro* and *in vivo* studies have been carried out for their clinical success.

2. Basic of Bone Science

Bone is a connective tissue that protects the organs and endows mechanical support, hematopoiesis and mineral.[15–18] Human bone may be categorized based on composition and structure. Based on composition, human bone is made up of composite materials of polymer and ceramic materials.[15] The different bone compositions have been illustrated in Table 1, further classified into organic and inorganic components based on composition. Inorganic components,

Table 1. Numerous composition of bone.[19,20]

Inorganic phase	wt.%	Organic phase	wt.%
Hydroxyapatite	60	Collagen	20
Carbonate	4	Water	9
Citrate	0.9	Noncollagenous proteins	3
Sodium	0.7		
Magnesium	0.5		
Other traces: Cl^-, F^-, K^+, Sr^{2+}, Pb^{2+}, Zn^{2+}, Cu^{2+}, Fe^{2+}		Other traces: Polysaccharides, lipids, cytokines Primary bone cells: osteoblasts, osteocytes, osteoclasts	

such as Type I collagen, fibrillin, glycoproteins and proteoglycans strengthen the bone. Inorganic components, such as hydroxyapatite, which is placed in a periodic array in the bone, provides the bone's toughness.

Based on the structure, bones are categorized as compact bone (cortical bone) and trabecular bone (cancellous bone), as shown in Fig. 2. Further, it is subdivided into macroscale, microscale, sub-microscale, nanoscale and sub-nanoscale. These divisions play a crucial role in improving mechanical properties, such as stiffness, strength, creep, fatigue and elastic modulus. The different biomechanical properties of bone are listed in Table 2. The compact bone is generally more robust and stiffer with less porous (5% to 10%), while the properties of trabecular bone are anisotropic and more absorbent (50% to 90%) in nature. For supporting bone and bone replacement, biomechanical properties must contain a critical parameter that is mainly considered. It is reported in the literature that every year the fracture of the bone is escalating globally. Therefore, there is still a requirement deeply scrutinized in the design and fabrication of biomaterials for orthopedic applications.

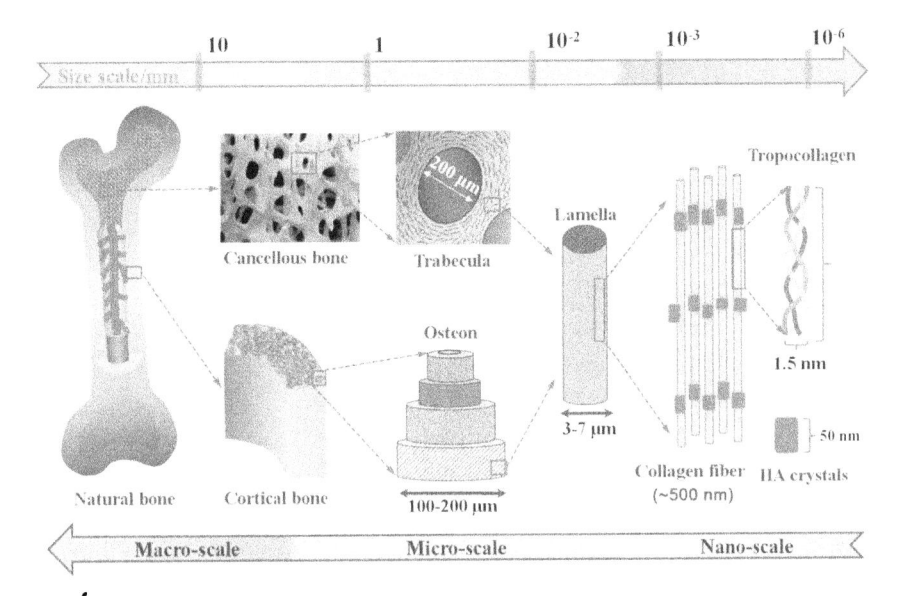

Fabrication and surface modification of biomaterials 35

Fig. 2. Chemical composition of bone, from micro to nanoscale. Reproduced with permission from Ref. 15 (Color online).

Table 2. Various biomechanical properties of bone.[19,20]

		Measurements	
Properties	Unit	Cortical bone	Cancellous bone
Compressive strength	MPa	170–193	7–10
Tensile strength	MPa	50–150	10–20
Density	g/cm^3	18–22	0.1–1.0
Young's modulus	GPa	14–20	0.05–0.5
Fracture toughness	$MPa \cdot m^{1/2}$	2–12	0.1
Surface/bone volume	mm^2/mm^3	2.5	20
Elongation	%	1–3	—
Strain to failure	—	1–3	5–7
Total bone volume	mm^3	1.4×10^6	0.35×10^6
Total internal surface	—	3.5×10^6	7.0×10^6

3. Biomaterials for Orthopedic Applications

Orthopaedical problems, including bone fractures, osteoarthritis and osteoporosis, are the main confronts in health care organizations, which significantly stimulate implants' growth. The specific application of biomaterials in orthopedic must fulfill numerous major requirements, like biocompatible materials, which means that solid affinity shows the host's cells with implant materials. It exhibits excellent mechanical and physical properties with ease of fabrication and is essential to achieve the desired function. Nowadays, newer materials have been discovered, which show excellent mechanical and physical properties but cannot be used as a bioimplant because of their poor biocompatibility and the toxicity with bone tissue is unacceptable by the body. The material does not have any adverse effect by releasing various ions or other harmful products, which may cause allergies, inflammation, cancer, necrosis and calcification. Moreover, corrosion resistance is the essential property of the materials, very much associated with biocompatibility and toxicity, determining the implant's life.[21] For permanent implants, corrosion resistance should be very high. Moreover, the material should have high strength and ductility as well as high fatigue strength. Even though the Young's modulus of elasticity should be matched with the natural bone, higher elastic modulus causes stress shielding on the bone-implant interface. The Young's modulus may be defined as the stiffness during elastic deformation. However, the elastic modulus of bone is 15–25 GPa.

4. Classification of Biomaterials for Orthopedic Materials

In general, materials that are employed in orthopedic are categorized as metals, polymers, ceramics and composites.[22] Furthermore, it can be subdivided with external appearance and internal chemical composition.[23] From the design point of view, the material used in some particular applications must satisfy a specific application requirement. The biomaterials most commonly employed in orthopedic implant purposes and their essential properties are listed in Table 3. However, in applying biomaterial, the biological requirements must be incorporated into the common properties (physical, chemical and mechanical). While designing implantable devices, wear resistance, stress shielding effect, biocompatibility, bioactivity and osteoinduction characteristics are added.[24] The biomechanical properties of the selected materials for orthopedic applications are listed in Table 4.

4.1. Metallic materials

Metallic biomaterials have been extensively employed as an orthopedic implant material for a long time due to their outstanding mechanical properties and significant biocompatibility. Despite several biomaterials, metallic biomaterials possess excellent mechanical properties, like superior tensile strength, toughness, excellent wear resistance and high corrosion resistance.[25] Metals have also exhibited excellent biocompatibility to human bone. Hence, metallic biomaterials are mainly used in the fabrication of bone plates or rods.[26]

4.1.1. Stainless steel

In metals, among other metallic implant materials, SS is most regularly employed for producing bone plates or rods because of its superior biocompatibility, mechanical strength, implant deformation during surgery and less cost. After the implant, the biomaterial is immediately covered with adsorbed protein. Protein and its concentration play a vital role in the corrosion of biomaterial with the time domain. Austenitic steel (AISI 316 and AISI 316L classes) is the most commonly employed biomedical implant due to its superior machinability, high ductility and better fatigue strength. In austenitic steel, "L" denotes low carbon steel, which contains less percentage of carbon (C ≤ 0.030%). Besides, it consists of nickel, which promotes toxicity, allergy and sensitization. Hence, nowadays, nickel-free SS has been fabricated for orthopedic applications.

Table 3. Characteristics and application of biomaterials.[19,27–65]

Biomaterials	Advantages	Disadvantages	Applications	Examples
Metal	Very strong, high resistance to impact and wear, ductile	Variation in mechanical in biological tissues, corrosion in a physiological environment, dense, low biocompatibility	Plates and wires, joint prostheses, staples, load-bearing bone implants	SS, Co-based alloy, Ti alloy
Ceramics	Corrosion resistance, high biocompatibility, bioresorbable, bioinert, high resistance to wear, bioactive	Brittle, low toughness, weak tensile strength, difficulties in fabrication and processing, lack of resilience, properties challenging to reproduce	The coating on bioimplant, bone filler, bone tissue engineering	HA, alumina, bioglass, TCP, zirconia
Polymer	Easy to produce, flexible, low density, resilient	Low mechanical resistance, toxic of some degraded products, not strong	Bone screws, bone plates, pins, bone tissue scaffold	Polyethylene, collagen, polypropylene, PMMA, PCL
Composites	Tailorable mechanical properties, corrosion-resistant, high biocompatibility	Variation in the properties due to fabrication methodology	Bone tissue scaffold, a knee implant, artificial joints	Mg/HA, HA/gelatin, HA/PE

Table 4. Mechanical properties of the biomaterial.[19,66-72]

Biomaterials	Density (g/cm^3)	Elastic modulus (GPa)	Compressive strength (MPa)	Tensile strength (MPa)
Stainless steel	7.9	193–200	600	480–620
Co-based alloy	8.3–9.2	195–230	655	450–960
Ti alloy	4.2–4.5	110–120	550–860	550–985
Mg alloy	1.79–2.0	35–40	100–200	90–230
Polymer	1.12–1.31	1.8–3.3	45–90	38–80
Ceramic	3.15–3.99	80–110	400–900	500

Generally, SS is most appropriate for temporary fixation implant devices, for instance, fracture plates or rods and screws. However, several studies have been carried out to find the impact of protein and its concentration on steel corrosion behavior. Different bovine serum albumin (BSA) concentrations on austenitic SS (316L) in phosphate buffer saline (PBS) solution with time domain have been studied. It was observed that higher concentration and longer immersion time accelerated the corrosion rate owing to the corrosion product formation. Nevertheless, at low concentrations, steel containing nitrogen content enhances the corrosion resistance during the lime time. Even though, the protein accomplishes as a shielding layer on the metal's surface and prevents oxide formation. The presence of nitrogen in steel also improves corrosion resistance by forming ammonia or nitrate ions.[73]

4.1.2. Co–Cr alloy

Co–Cr alloys are extensively employed in load-bearing usages, including ankle joints, because of their tremendous wear resistance. Moreover, Co–Cr alloy shows very excellent corrosion resistance in the chloride medium.[74,75] The alloy also exhibits superior wear resistance as compared to SS and Ti alloys. Molybdenum (Mo) can be used as an alloying element for enhancing the mechanical properties. The elastic modulus of Co–Cr alloy is approximately 220–230 GPa, which is very close to SS (approximately 200 Gpa). Furthermore, the alloy exhibits elastic modulus better than cortical bone (20–30 GPa), which resulted in a stress shielding effect at the adjacent bone. The release of metal products (wear debris) may harm the human body. Although, high cost and poor fabricability are not suitable for broad application in bone plates. Co-based alloy's corrosion behavior depends on thermal treatment and the number of carbide inclusions.[76] However, the spark plasma sintering (SPS) process showed better tribological performances as compared to the other conventional methods.[77]

4.1.3. *Titanium and its alloys*

Titanium (Ti) and its alloy are being used as an orthopedic implant since the 1950s owing to its superior biocompatibility, low density and superior corrosion resistance.[78] Excellent biocompatibility and moderate elastic modulus make them to be preferred over SS and Co-based alloys.[27] Ti-based alloys exhibit long-term implanted devices, avoiding the risk of loosening due to tightly integrated human bone.[79] Commercially, Ti6Al4V alloy is most commonly utilized in orthopedic due to its superior mechanical properties and biocompatibility.[80] Despite Ti and its alloy having remarkable properties, however, their fabrication process, including machining, heat treatment or forging, is complicated. It was also found that the presence of vanadium (V) causes some toxic effects. Therefore, to overcome these demerits, V-free alloys are developing by adding Nb, Ta and Zr, which have elastic low modulus. Nevertheless, the progress of a new generation of Ti alloy is still under research. Ti and its alloy exhibit excellent mechanical properties as well as antibacterial properties. In another experiment, micro/nanostructured titanium oxide (MNT) has been manufactured via microarc oxidation (MAO). Many studies on antibacterial activity and macrophage performances were explored. However, silver nanoparticles (AgNPs)-decorated MNT (AgNPD-MNT) exhibited superior bacterial efficacy versus *Escherichiacoli* with no cytotoxicity of osteoblasts and mediated the inflammatory response. The inflammatory response was reduced after 7–10 days. The AgNPD-MNT sample revealed potential antibacterial activity and immunomodulatory properties, which can be used in bone implants for orthopedic surgery.[81] However, the Ti-6Al-4V alloy was fabricated through the electron beam melting (EBM) method. They had indicated that surface roughness and topography of the specimens affected the interaction between bioimplants and cells. The scaffold exhibited promising biocompatible implants for dental and orthopedic surgeries.[82] These metallic implants (SS, Co alloy and Ti Alloy) have excellent mechanical properties and significantly develop load-bearing orthopedic implants.[83,84] These biomaterials' service life is not up to mark and sometimes requires revision surgery, as shown in Fig. 3. The release of metallic ions causes inflammation, allergic reactions and pain, leading to implant failure.[85,86] To address these problems, researchers are trying to develop numerous functional coatings to reduce the release of metallic ions and wear debris while improving the durability and osseointegration of metallic implants.[87–90]

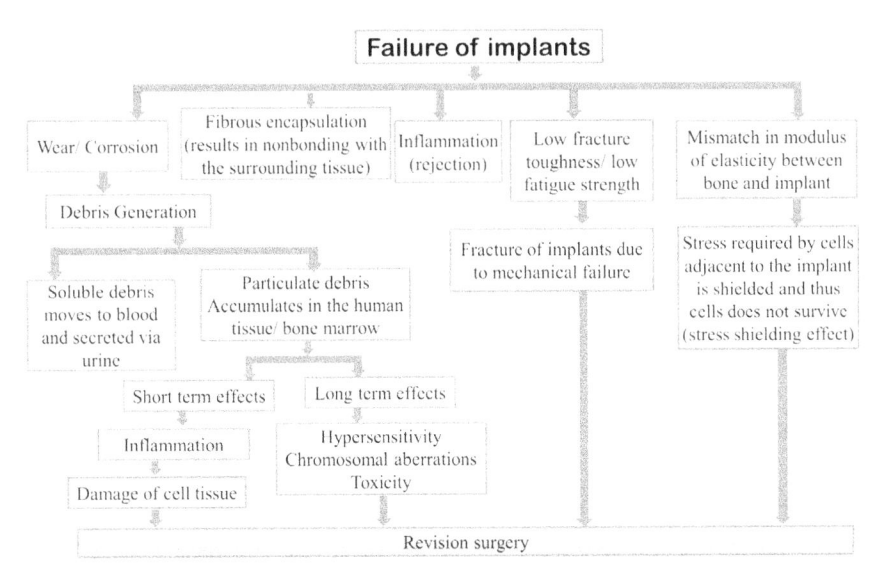

Fig. 3. Causes of revision surgery.[78,91] Reprinted with permission from Ref. 78 (Color online).

4.2. Polymers

Polymeric biomaterials are new materials that are broadly employed in bone tissue engineering applications because of their lightweight, biocompatibility, surface modifiability and design flexibility.[19] Furthermore, the elastic modulus is significantly lower, i.e. 0.5 GPa to 10 GPa. However, polymer materials have the drawback of their mechanical and chemical properties. It cannot be sustained at a higher load and gets fractured. Polymeric materials have been used in orthopedic applications since 1937, and polymethylmethacrylate (PMMA) was the first polymer. Various types of polymers are being used, such as silicone rubber, polyethylene (PE), acrylic resins, chitosan, polyurethanes, polypropylene (PP), polycaprolactone (PCL), polyglycolide (PGA), polylactide (PLA), polydioxanone (PDS), polyhydroxy butyrate (PHB) and hydrogels in the orthopedic implant. In contrast, an attempt is being made to refine the mechanical properties of dextran/chitosan polymer by incorporating nano-hydroxyapatite (HA) fabricated by the blending method. It was examined that the pores were interconnected. The addition of nano-HA reduced the size of the pores from 82% to 67%, and these materials may be used in tissue engineering.[92]

4.3. Ceramics

Ceramics, also known as bioceramics, are used for orthopedic applications since the 1960s. They are very hard, have higher compressive strength, excellent biocompatibility and are tissue responsive. Ceramics could be categorized as bioinert [alumina (Al_2O_3) and zirconia (ZrO_2)], bioactive [bioglass and HA] and bioresorbable (tricalcium phosphate; TCP).[93] Al_2O_3 and ZrO_2 were commercially used as bioceramic material for orthopedic in 1970 and 1985, respectively, because of their significant mechanical properties and outstanding biocompatibility.[19] Hence, HA is most commonly employed in biomedical implant applications because of its bioactivity, non-toxicity and its crystallographic structure is very similar to the human bone.[94-97] However, owing to low load-bearing capacity their application is restricted in orthopedic implants. Therefore, some materials have been incorporated for enhancing the mechanical properties.[98-104] Hence, HA–Al$_2$O$_3$–CNT hybrid composites were fabricated via SPS. It was found that the composite has promoted cell adhesion as well as proliferation on the surface and can be used as a suitable candidate for new biocompatible materials.[105] However, (Ti–13Nb–13Zr)–10HA composite was fabricated at several sintering temperatures (950–1150°C) via SPS. Scanning electron microscope (SEM) image confirmed the bone-like apatite layer comparable to the human bone was found on the composite. With increasing the temperature, corrosion resistance was increased due to higher corrosion potential and lowered corrosion current density. The composite materials had shown significant Vickers hardness and compressive strength. The elastic modulus rises remarkably at above 1050°C. The composite at 1000°C and 1050°C possess dynamic biomaterial properties for orthopedic implant applications.[106]

4.4. Composite materials

Composite materials have heterogeneous mixtures of two or more than two materials possessing distinct chemical and physical properties, and properties after the combination is superior to regular materials. There has been much study in composite materials for the bone implant, and further researches are currently being performed. The two distinct composites of iron (Fe) and Zinc (Zn) based Fe-TCP and Zn-TCP biomaterials have been fabricated through the codeposition method. It was studied that these materials have exhibited superior cytocompatibility and osteogenic properties and prospects a good candidate for medical application.[107] However, three-dimensional (3D) printing was employed for fabricating the (n-MS)/zein(ZN)/PCL composites. The influence of Zn contents in composite materials *in vitro* and *in vivo* was analyzed and concluded that pores were highly interconnected in a controllable way having a macroscopic size of around 500 μm. The results also revealed that the amount of Zn contents improved the degradation property, and therefore the composite material can be employed in tissue engineering.[108] However, the bottom–up assembly method developed the fabrication of high-performance biomaterials of graphene oxide/chitosan/calcium silicate (GO/CTS/CS). This composition's mechanical properties were comparable to the cortical bone due to the interface interaction layered microstructure and multilayered cylindrical structure. *In vitro* analysis showed significant enhancement in the properties of osteogenesis and angiogenesis. *In vivo* analysis exhibited good bone-forming ability.[109] The resistance to corrosion as well as bioactivity of Mg was enhanced by adding the different concentrations (10–40%) of CS particles synthesized by SPS. The optimum value of 20% (by wt.%) of CS had exhibited the highest compressive strength, structural compactness and enhanced the corrosion resistance in simulated body fluid (SBF). It was also revealed that this system could be used in an orthopedic implant application.[110] However, titanium/silicon carbide (Ti/SiC) metal matrix nanocomposite (MMNC) was fabricated via friction stir processing (FSP). SEM micrographs of the processed composite materials revealed that SiC had uniformly distributed over titanium, and intense interaction led to superior mechanical properties. The composite materials were suggested appropriate for stress-bearing applications. *In vitro* analysis also revealed that the proliferation, adhesion and osteogenic differentiation enhanced in rat BMSCs culture medium.[111]

4.5. *Biodegradable material*

The requirement for biodegradable surgical implants has been raised in recent years. The properties of Mg, such as being lightweight and high mechanical strength, are potentially amicable to natural bone, proposing it as a suitable material for orthopedic implants.[112] Adult human bodies have 1000 mmol or 24,000 mg of Mg, but it varies according to the age.[113] More than half of the Mg ions are stored in the bones.[114,115] Pure Mg gets easily degraded in the body fluids due to chlorides and could be absorbed in the human body without any toxicity.[116] Uncontrolled or rapid corrosion came into the picture when the engineered implant interact with chloride medium in the human body and consequent release of hydrogen ion and variation in pH.[117-124] Corrosion resistance could be boosted by modifying the composition and microstructure of base material through refining the grain size and texture.[121] Such approaches could be gratified by alloying, compositing and further establishing exceptional mechanical characteristics by optimizing fabrication and post-processing heat treatment methods.[124-126] Whereas, some alternative techniques include the fabrication of porous structures or composite[127] and surface coating.[128,129] Although, researchers are also trying to discover some biocompatible and nontoxic real earth elements (REEs) for developing the corrosion resistance of Magnesium. Furthermore, Mg-2.0Zn-xGd alloys are engineered by potentially varying Gd narrated as $x = 0.5$, 1.0, 1.5 and 2.0 wt.%, respectively, via resistance furnace, containing appropriate blending gases (99 vol.% of CO_2 with 1 vol.% of SF_6). Thereafter, the fabricated alloys were extruded under forward pressure at an extrusion ratio of 9:1. However, the as-extruded alloy indicated the most promising candidate for orthopedic implants because it possesses excellent biocompatibility, superior mechanical properties, and lower degradation rate *in vivo* (0.24 mm/year).[130] However, osteoinduction activity, antibacterial effects and controllable degradation of Mg alloy can be enhanced through the addition of Zn and Sr ions via a one-pot hydrothermal method. Furthermore, the material showed bioactivity and can be employed as a promising candidates for clinical application.[131] However, incorporating Ag in cast Mg–Zn–Y–Nd–Ag enhanced the grain structure and microhardness. It was also found that the as-extruded alloy encompassing 0.4 wt.% revealed enhanced antibacterial properties and better mechanical properties with high corrosion resistance.[132] Whereas, ternary magnesium (Mg–1.8Zn–0.2Gd) alloy was fabricated by resistance melting under the control environment by using appropriate blending of gases (99 vol.% of CO_2 with 1 vol.% of SF_6) showed nontoxic behavior with the different cells medium. In electrochemical testing, the degradation rate was found to be 0.12 mm/year. Furthermore, it revealed that the implant maintained its surface integrity for 2 months, and serious degradation was observed after six months. Therefore, the negative impact of the RE element was eliminated and suggested that the alloy is the most suitable for orthopedic implants.[133] However, Mg–Zn alloy was constructed by powder metallurgy method followed by secondary extrusion process revealed an establishment of superior mechanical properties with corrosion resistance encouraged by performing aging treatment.[134] These encouraging results may revive through grain refinement controlled by regular, systematic temperature rise and drop during manufacturing. However, Mg–Cu alloy was fabricated through the melting process in a graphite crucible with the appropriate blending of gases (99 vol.% of CO_2 with 1 vol.% of SF_6) and it was found that alloying of Cu increases mechanical properties and nontoxicity. The alloy exhibited prosperous stimulation to osteogenesis, and angio-genesis can be used in orthopedic applications.[135] However, the effect of REE holmium (Ho) on Mg–Zr–Sr was examined via the melting process. Cylindrical specimen with 10-mm diameter and 3-mm thickness was prepared by the EDM process and the results suggested that Ho increased the mechanical properties and decreased the degradation rate due to the development of the intermetallic phases, such as $MgHo_3$, Mg_2Ho and $Mg_{17}Sr_2$ They also investigated that Mg1Zr2Sr3Ho had significant biocompatibility and low corrosion rates compared to the base material and can be used in biomedical applications.[136] The effect of alloying elements (Sr and Zn) on Mg alloy was carried out by induction melting by using a coated steel crucible in a protective inert gas atmosphere. They found that alloying elements enhanced biocompatibility and proliferation of osteoblast. The composition was suitable for fracture treatment in orthopedic.[137] However, the antibacterial properties of Mg–Ca–Sr alloy were enhanced by the increment of Zn contents. They found that the Mg1Ca0.5Sr6Zn alloy had a strong antibacterial outcome among the different combinations, and also the alloy exhibited good biocompatibility and can be used as an orthopedic implant.[138] Moreover, Mg–Zr–Sr with different concentrations of zirconium and strontium by the casting process, they investigated the microstructure and degradation rate by different characterization techniques, such as SEM, X-ray diffraction (XRD) and energy dispersive X-ray spectroscopy (EDS). They found that over 5 wt.%

concentration leads to the formation of decreased biocorrosion resistance and biocompatibility of the alloy. The reason mentioned that excess zirconium led to the acceleration of anodic dissolution. Also, above 5 wt.% enhanced the galvanic effect formation between the Mg matrix and $Mg_{17}Sr_2$ phases. They also examined that Mg1Zr2Sr alloy significantly showed a degradation rate 2.74 times slower than that of Mg. The composition can be employed as a novel biodegradable implant.[139] The Mg–Nd–Zn–Zr alloy developed utilizing semi-continuous casting showed excellent biocompatibility, high corrosion resistance and antimicrobial properties.[140] However, Mg–Ca and Mg–Ca–Mn–Zn was fabricated at 760°C via a melting process with an inert argon atmosphere in an electrical resistance furnace. It was analyzed that Mg–2Ca–0.5Mn–2Zn alloy had significant mechanical properties, and the outstanding corrosion resistance makes it suitable for biodegradable implants.[141] The zone purifying solidification process was employed to develop Mg-1Zn-xSr alloys; then, the fabricated alloys were extruded under backward pressure. It was revealed that above 0.8 wt. % of Sr increased the degradation rate and mechanical properties. They also reported that the alloy exhibited good biocompatibility and was suitable for orthopedic implants.[142] However, incorporating Zn in Mg–Ca alloy enhanced the degradation rate by forming the phase ($Mg+Mg_2Ca$), which inhibited the development of the galvanic circuit. Further effective grain refinement occurred in the as-extruded alloy. Mg–5Ca–xZn (less than 3 wt.%) alloy showed promising properties for biodegradable implant application.[143] Furthermore, Mg–Zn and Mg–Zn–Gd alloys by incorporating varying wt.% of Zn and Gd were fabricated through induction melting in protective argon gas and it was found that corrosion resistance of the alloy was boosted by the addition of 3 wt.% Gd.[144] However, Mg–Y–Ca–Zr alloy was developed via conventional melting and casting process, and it was observed that the 4 wt.% of Y content was suitable for orthopedic implant biomaterials.[145] Furthermore, Mg–Sr alloy was prepared by the melting and casting process under the mixing of CO_2 with 0.5% and SF_6 with 99.5%. Whereas Mg–0.5Sr exhibited a lower degradation rate due to HA and Mg $(OH)_2$ modified layer formation. They concluded that Mg–0.5Sr alloys could be used in orthopedic applications.[146] Although, the synthesized Mg–Ca–Sr alloy at varying Ca (0.5–7.0 wt.%) and Sr (0.5 3.5 wt.%) by melting in high purity graphite crucibles sustained at a temperature of 725°C and 825°C, and argon gas was used for the protective atmosphere. It was concluded that Mg–1.0Ca–0.5Sr alloy is a furthermost favorable alloy suitable for orthopedic implants. Furthermore, the composition exhibited a low corrosion rate in Hank's solution, nontoxic and superior compressive strength.[147] However, Mg–Sr binary alloy system containing Sr (1.0–4.0 wt.%) were fabricated via melting process in a graphite crucible with the appropriate blending of gases (99 vol.% of CO_2 with 1 vol.% of SF_6), then rolled into the alloy. It was observed that Mg–2Sr alloy unveiled the highest mechanical properties and slower degradation rate, suggesting that the alloy can be applied as orthopedic implants.[148] Furthermore, Mg–2Zn–0.5Sr and Mg–4.0Zn–0.5Sr possess optimized both degradation rate and mechanical properties.[149] Besides, AZ31 and Mg–Nd–Zn–Zr alloys were developed by melting in steel with proper blending of gases (99 vol.% of CO_2 with 1 vol.% of SF_6). It was observed that the Mg–Nd–Zn–Zr alloy has better biodegradability and is a favorable candidate for degradable biomaterials.[150] Magnesium alloy (Mg–Zn) was fabricated by casting process and it was observed that the incorporation of Zn improved the corrosion resistance. Furthermore, the mechanical properties, such as tensile strength and elongation achieved 279.5 MPa and 18.8%, respectively.[151] Corrosion resistance of Mg–2Zn–0.2Mn alloy has been significantly improved due to the protective layer found on the $Mg(OH)_2$ surface and the alloys were identified as the most promising candidate for orthopedic applications.[152] The corrosion resistance and mechanical properties were enhanced by the addition of Gd up to 10 wt.%, but beyond this saturation limit the corrosion properties get decreased.[153]

4.6. *Porous metals*

Porous metals are usually utilized in orthopedic implants because they enhance osseointegration and overcome the stress shielding effect.[154] Therefore, Ti6Al4V alloy was fabricated by the sintering technique and their porosities were examined. It was found to possess more significant mechanical properties and porous, which accelerated bone ingrowth.[155] Whereas, Mg–Zn–Mn–HA composite was fabricated through mechanical alloying and SPS method. It was found that HA's addition could refine the grain and enhance the porosity, which led the osseointegration.[156] However, the salt-leaching process was used to fabricate the porous iron–manganese–HA (Fe–Mn–HA) biocomposites. For that, Fe–Mn–HA powders were mixed with salt (NaCl) particles and then pressed at a pressure of 350°C. The composites were compacted and sintered in an argon gas environment at 750°C for 3h, then cooled inside the furnace itself. It was found that incorporation of HA led to the secondary phase $Ca_2Mn_7O_{14}$ in the sintering process, which increased the degradation rate. Moreover, porous Fe–30Mn–10HA demonstrated more conducive to cell attachment, proliferation and bone-like apatite formation, which is a promising candidate for orthopedic applications.[157] Although, the novel porous Mg–Nd–Zn was fabricated by using Ti as a space holder. *In vitro* and *in vivo* studies showed that the biodegradation, biocompatibility and osteogenesis had been enhanced. They results suggested that the Mg–Nd–Zn scaffold is a promising candidate for bone repair materials.[158] Porous Ti6Al4V scaffold was successfully developed by powder metallurgy technique. Furthermore, recombinant human bone morphogenetic protein-2 (rhBMP-2) and thermosensitive hydrogels were incorporated in a porous scaffold, which promoted bone regeneration. It was observed that a higher cell proliferation rate, faster speed and more bone growth formation were achieved. Therefore, it was most suitable for orthopedic applications.[159] A 3D iron scaffold was developed through a template-assisted electrodeposition process and it was found that the pores are regularly interconnected. Furthermore, the scaffold showed a closer structure having mechanical strength comparable to the cancellous bone. Moreover, strontium-incorporated octa-calcium phosphate (Sr–OCP) was uniformly coated into the scaffold, enhancing the biocompatibility and promoting the cell adhesion property. The scaffold had more prospects in bone generation and repair.[160] However, Mg–8Er–1Zn was fabricated by direct-chill semi-continuous casting followed by post-processing process, heat-treatment as well as a hot-extrusion process. Mechanical properties, such as ultimate tensile strength, tensile yield strength and elongation were 318 MPa, 207 MPa and 21%, respectively, were obtained. Corrosion resistance was enhanced and uniform corrosion morphology appeared due to the formation of stacking faults (SFs) and the alloys exhibited promising characteristics for biomedical application.[161]

5. Surface Modification of Orthopedic Implant

Numerous types of orthopedic implant materials and their fabrication process have been discussed in the previous section. However, there are still specific issues of tissue inflammation, antibacterial properties and cell adhesion, which usually evolve at the host tissue and implant interface.[162] Moreover, it is challenging to control the uniform degradation of a biodegradable orthopedic implant. The reasons may be attributed to the orthopedic implant's surface is continuously exposed to the proactive bioenvironment. Although, different reactions occurred between the host tissue and implant surface due to implant surface contamination, surface topography and mechanical and chemical properties corresponding to the host tissue.[163] Therefore, surface coating techniques could play a vital role in overcoming these mentioned issues associated with the developed orthopedic implant. Moreover, corrosion and wear resistance could be significantly enhanced through surface modification. Although, the tailored implant can interact as well as adequately adhere to the surrounding bone tissue with significant osteointegration and biocompatibility.[164] However, research on surface treatment (surface modification) is still conducted by bioengineers, surgeons and material scientists.[165] Biocompatibility and surface properties are the only retort of biomaterials as reported earlier.[166] Therefore, it is a prime requirement for their surface to be engineered.[167] By amending the surface modification, the surface properties could be engineered without adjusting the bulk materials. Nevertheless, the help of physical, chemical or coating various materials could alter

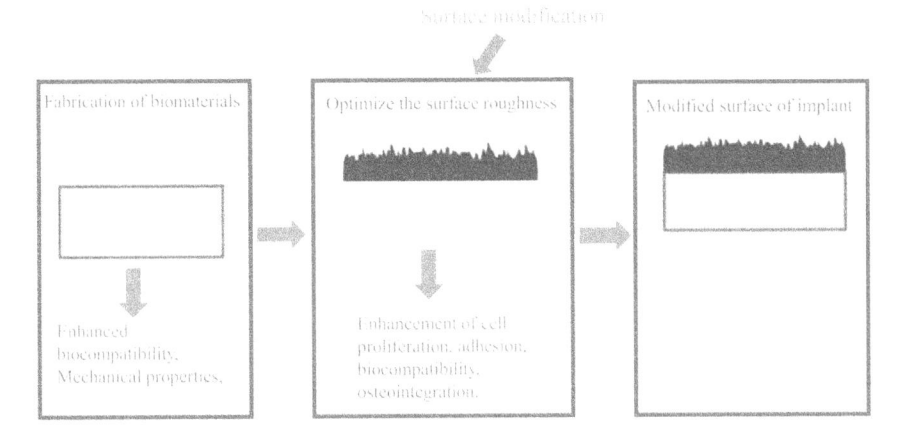

Fig. 4. Flow chart portraying the surface modification (Color online).

the molecules, atoms and compounds on the surface.[168] Moreover, the surface modification could enhance biocompatibility, corrosion resistance, tribological properties and cell adhesion. Figure 4 illustrates the flow chart of surface modification of the biomaterial implant. Figure 5 illustrates the various methods associated with the surface modification for the substrates. Figure 6 demonstrates a flow chart of surface treatment and biocompatible coating that can be incorporated on biomaterials.

5.1. *Surface treatment techniques*

Various modifications approaches have been employed to modify the surface of metallic materials that improves their cellular activities compared with the conventional fabricated materials.

5.1.1. *Plasma-immersed ion implantation*

Plasma immersion ion implantation (PIII) is an innovative surface modification technique which is initially known as plasma source ion implantation (PSII). The technique was discovered by Conrad and their coworkers at the University of Wisconsin. PIII is a

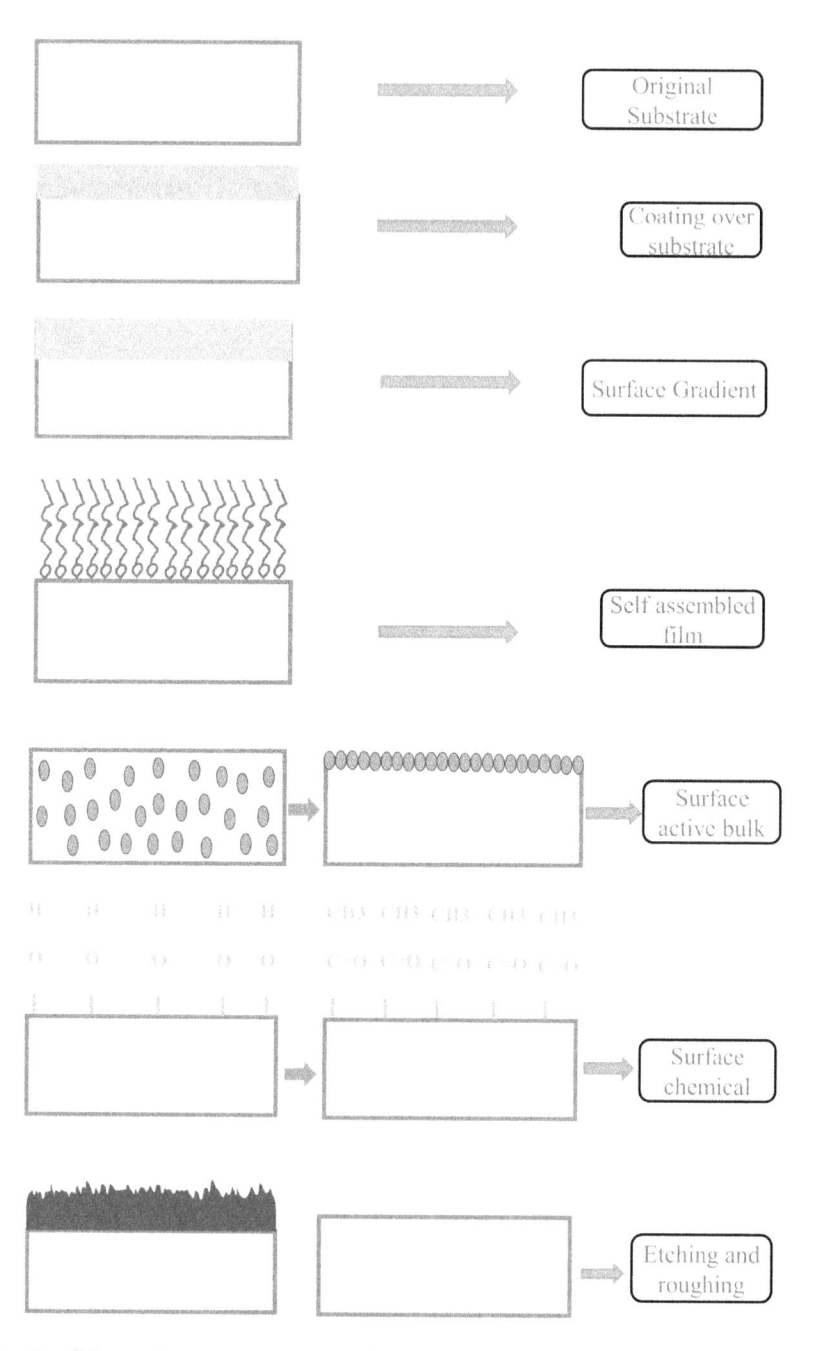

Fig. 5. Schematic representations of processes to modify surfaces of substrates (Color online).[166,169,170]

```
                        Surface
                      modification
                    ┌──────┴──────┐
              Surface          Biocompatible
             treatment            coating
          ┌─────┴─────┐        ┌─────┴─────┐
    Surface       Surface    Surface        Surface
 modification and modification modification and modification
tribological studies  only   tribological studies  only
```

Surface modification and tribological studies	Surface modification only	Surface modification and tribological studies	Surface modification only
▸ Thermal oxidation	▸ Micro-arc oxidation	▸ Hydroxyapatite	▸ Calcium Phosphate
▸ Plasma immersed	▸ Anodic oxidation	▸ Titanium Nitride	▸ Silicon dioxide
▸ Ion implantation	▸ Bioactive surface treatment	▸ Micronite	▸ Titanium dioxide
▸ Carburization		▸ Tantalum	▸ Poly (dimethylsiloxane)

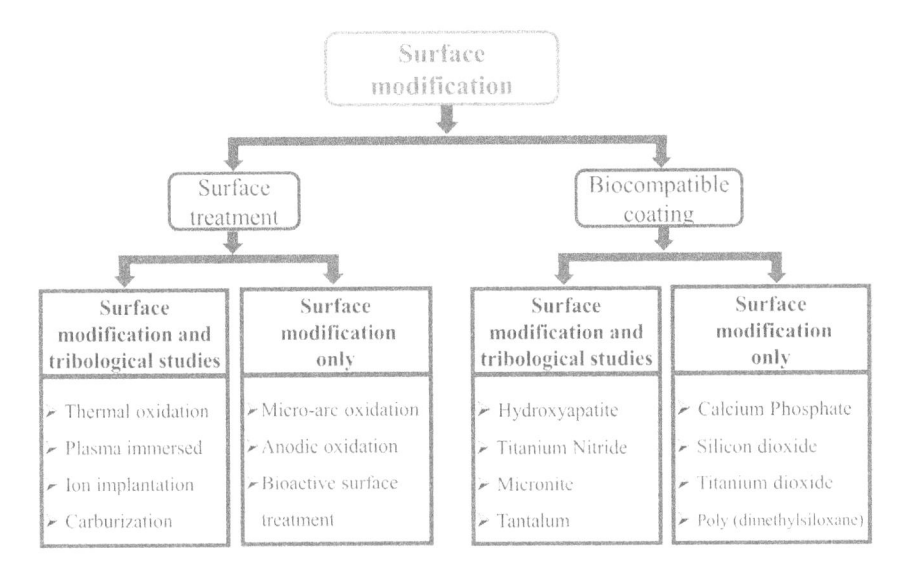

Fig. 6. Classification of biomaterials for orthopedic applications (Color online).[166,171–182]

technique encompassing the samples through a high-density plasma and pulsed-biased to a high negative potential compared to the chamber wall.[183] The ions generated in the plasma cover are accelerated across the sheath produced surrounding the samples and implanted usually into the targeted surface of the sample, as demonstrated in Fig. 7. It is a very prominent process because surface properties and selected biomedical properties, including cytocompatibility and corrosion resistance, could be improved while retaining the properties of the bulk material, such as strength and hardness.

Evidently, PIII is enormously appropriate for biomaterials because it is a non-line-of-sight technique conflicting to conventional beam-line ion implantation and can produce a sophisticated shape with better uniformity and conformality.[184] For instance, carbon and nitrogen PIII have enhanced the wear resistance and surface hardness of Ti-6Al-4V. Zhao et el.[185] modified the surface of the Ti-6Al-4V through nitrogen (N) and carbon plasma ion implantation (C-PIII and N-PIII) by developing the surface layers of TiC and TiN, respectively. It was found that both techniques enhanced

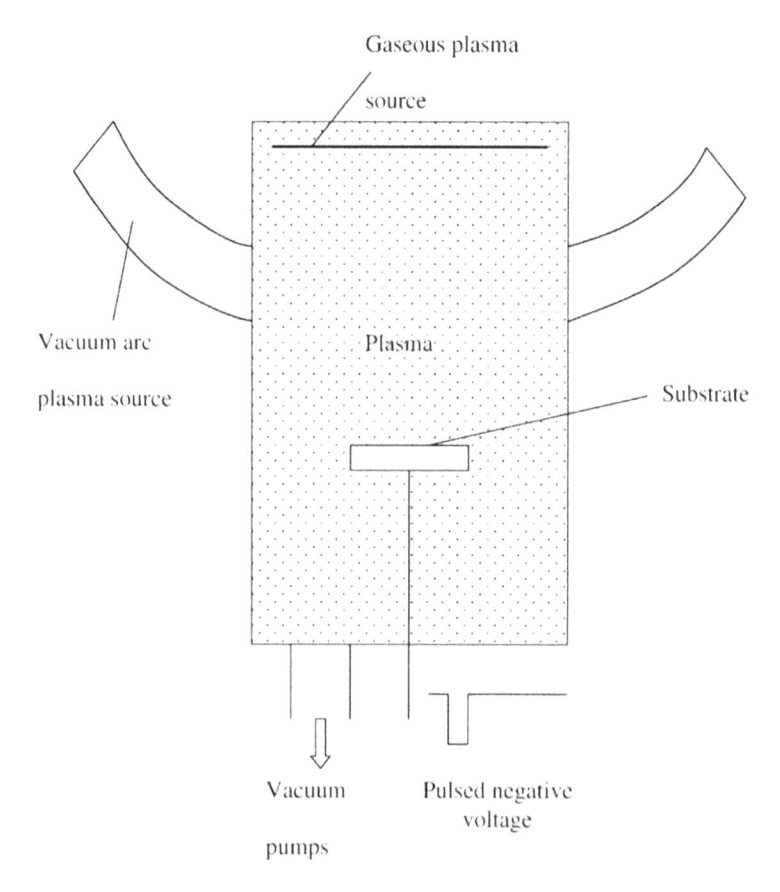

Fig. 7. Schematic diagram of PIII. Reprinted with permission from Ref. 184.

the corrosion resistance and surface roughness, while retaining the surface hydrophilicity. Furthermore, *in vitro* studies revealed the enhanced cell proliferation and adhesion of L929 fibroblast and MC3T3-E1 preosteoclasts after PIII. Moreover, N-PIII is most effective compared to the C-PIII and could be employed in medical-graded Ti-6Al-4V alloy for enhancing the surface biocompatibility. Furthermore, tribology characteristics could be enhanced by PIII techniques.[186] Wear test and tribo-corrosion were conducted in Hank's solution. It was found that wear resistance was vaguely enhanced due to the skinny layer of thickness formed during the PIII technique. Thus, the optimum thickness of coating needs to be addressed to enhance the wear resistance of the substrates. However, corrosion resistance was significantly enhanced.

5.1.2. *MAO coating*

MAO techniques are also called as plasma electrolyte oxidation processes. It is commonly employed to enhance the corrosion resistance of Mg alloys. In MAO, higher potentials are utilized to nurture the breakdown of dielectric potential for producing the oxide film due to discharging.[187] The discharges accountable of localized plasma reactions at intense, very high pressure and temperature are attributed to the tailoring of the growing oxide.[188] The MAO technique developed a comparatively thick, dense and hard oxide coating on the Mg alloys' surface. In general, MAO is a process of converting metal substrate into its oxide at outward and inward directions, which provides remarkable adhesion and enhances corrosion and wear. Li *et al.*[176] successfully modified the surface of the Ti implant through the MAO technique. Consequently, a porous layer was observed through SEM after the MAO process. The morphology and phase can be tailored by varying the applied voltage. The phase of TiO_2 from anatase to rutile can be transformed by increasing the voltage. Moreover, Ca and P ions were added to the oxide layer. However, the biocompatibility of the implant is directly associated with the chemical composition as well roughness of the Ti surface. The highest proliferation rate was found at lower voltage and then deceased with increasing the applied voltage. The effect of the various voltage parameters on the modification of Ti by MAO is displayed in the SEM image in Fig. 8. Furthermore, the influence of MG63 cell of unmodified Ti and modified Ti after cultured for 3 days is shown in the SEM image in Fig. 9. On the flip side, alkaline phosphatase (ALP) activity increased with increasing the applied voltage. Moreover, *in vivo* test was conducted in rabbits and it revealed that the capability of osseointegration enhanced

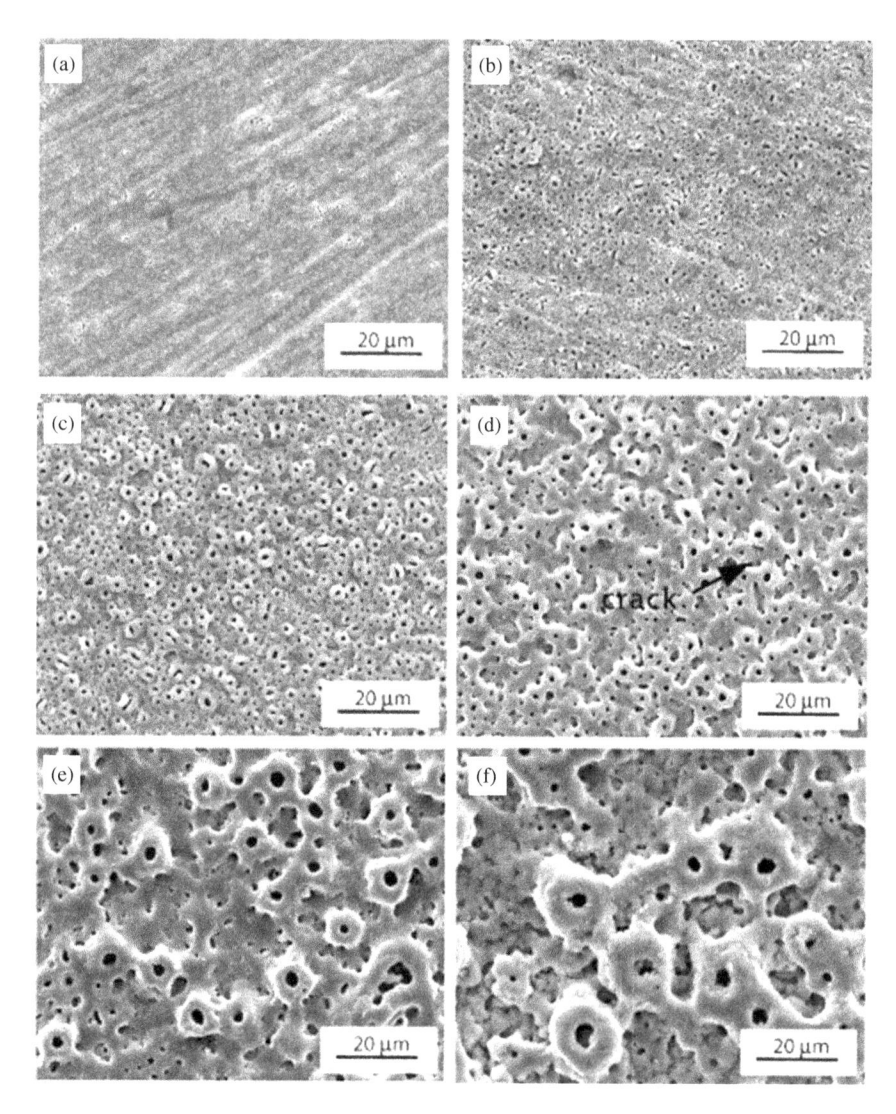

Fig. 8. Illustrated SEM image of Ti substrate modified by MAO at various voltages. (a) 190 V, (b) 230 V, (c) 270 V, (d) 350 V, (e) 450 V and (f) 600 V. Crack is clearly displayed in (d). Reprinted with permission from Ref. 176.

corresponding to the Ti implant. Many adhered chips were perceived on MAO-treated Ti which is shown in Fig. 10. Mechanical interlocking between bones and implant can be enhanced by increasing the surface roughness.[188]

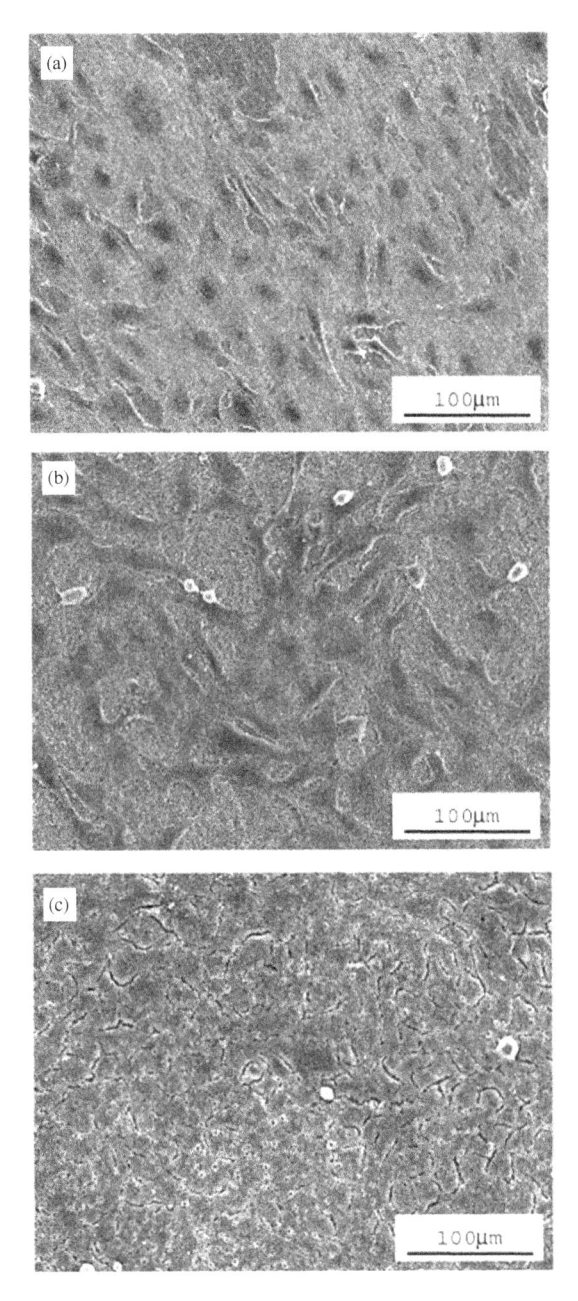

Fig. 9. SEM morphology is portraying the effect of MG63 cells after culturing for 3 days on (a) pure titanium and modified Ti@MAO at (b) 270 V and (c) 350 V. Reprinted with permission from Ref. 176.

58 *M. M. Iqbal et al.*

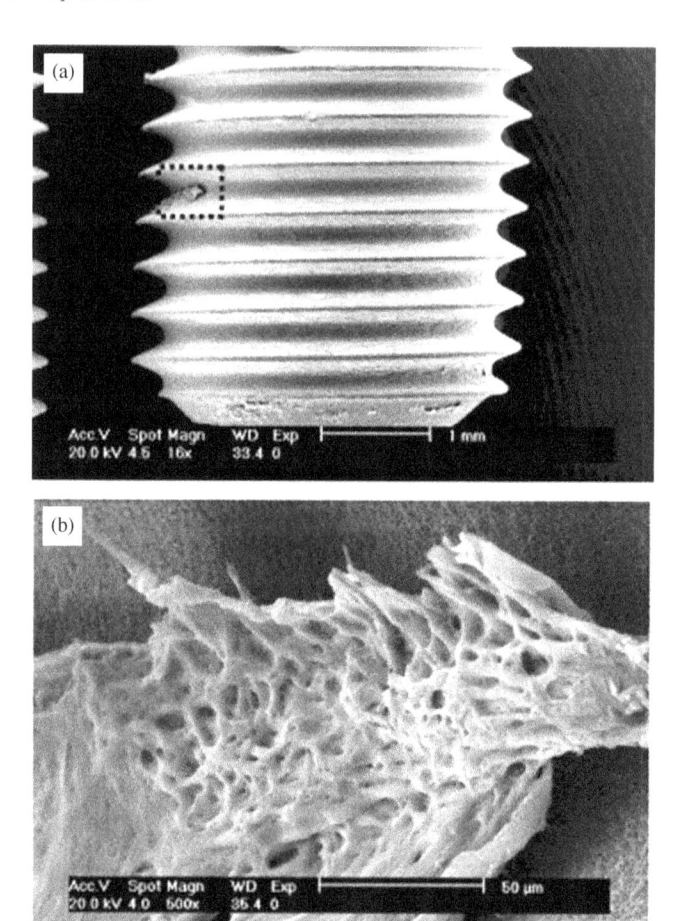

Fig. 10. (a) SEM image presenting the *in vivo* analysis of modified Ti implants at 270 V removed from the tibia of rabbits 4 weeks after their implantation and (b) SEM image at high magnification (×500) to analyze the effect of the adhered chip. Reprinted with permission from Ref. 176.

5.1.3. *Carburization*

The carburization technique is most commonly employed in face hardening of the steel. In this technique, low carbon steel is usually heated at very high-temperature ranges between 900°C and 1100°C in the existence of a carbon (C)-rich medium. Since, the presence of a C atom in the iron (face-centered cubic structure) diffused at temperature 910°C, layers of C were developed due to the occurrence of enriched C medium, and the thickness of the C layer could be monitored through the holding time. Finally, the carburized layer developed could be further hardened either at carburizing temperature or through reheating and quenching. Various mediums of carburizing could be employed, either liquid, solid or gas, which rely on the choice and nature of work. The prominent role of the carburizing medium is to release the carbon atoms at the carburizing temperature and will be absorbed interstitially into the substrates.[189]

5.1.4. Thermal oxidation

The surface characteristics of the Ti and its alloys could be enhanced through a thermal oxidation process. The crystalline oxide film is developed through oxidation at more than 200°C. High temperature tends to the formation of thick layer of oxide with oxygen beneath it. Therefore, the main advantage of thermal oxidation is the simultaneous formation of a thin layer as well as an oxygen diffusion zone.

5.2. Biocompatible coatings

The surface of the implant materials is also enhanced by a biocompatible coating of polymers, ceramics. The bifunctional titanium–polydopamine–zinc (Ti–PDA–Zn) coatings had superior antibacterial properties, excellent biocompatibility and enhanced osseointegration capability, making it ideal for application in orthopedic implants.[190] The corrosion resistance could be improved by providing a coating on the surface of Mg at the initial stage of the implant.[191,192] Additionally, biocompatibility could be enhanced by the coating process.[130,193] Wettability of matrix and reinforcement could be enhanced in the casting process through modification of reinforcement by coating process.[194] The thin layer of nanosized magnesium oxide (MgO) was successfully formed on the surface of TCP ceramic nanoparticles (m-β-TCP) through the mixing technology process. Then, the comparative study of Mg-based composites with modified (m-β-TCP) and unmodified (β-TCP) was investigated. It was confirmed through EDS that MgO coating was formed, and the size was found to be 25 nm with a spherical shape, as shown in Fig. 11. It was concluded that mechanical properties (hardness and tensile strength), corrosion resistance and cytocompatibility were enhanced in MgO coated on TCP composites (m-β-TCP) due to the superior wettability and uniform distribution of reinforcement of in the Mg.

However, the coating of PDA on HF-pre-treated magnesium alloy (AZ31) by a simple immersion method formed a protective

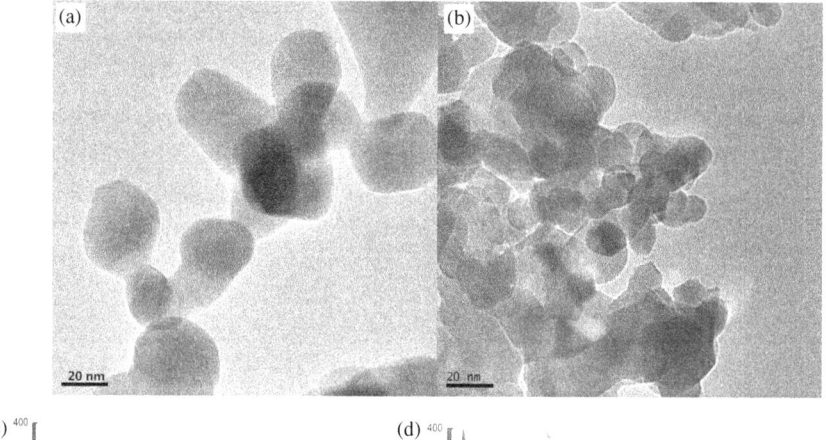

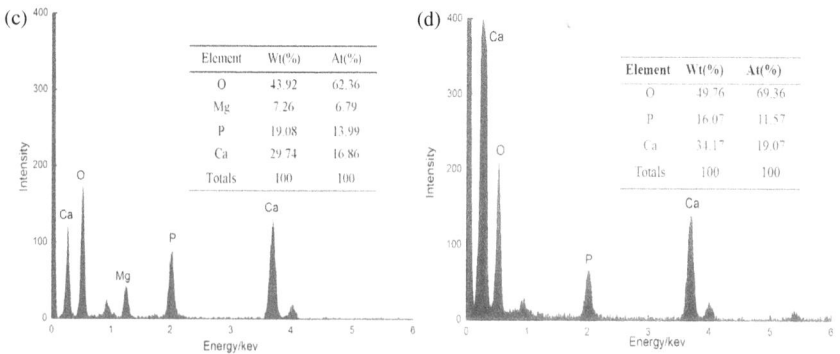

Fig. 11. Comparative analysis of TEM and EDS of modified and unmodified β-TCP nanoparticles. (a) and (c) m-β-TCP and (b) and (d) β-TCP nanoparticles, respectively. Reproduced with permission from Ref. 194.

Table 5. Surface modifications of the orthopedic implant biomaterial and their outcome.

Biomaterial implant	Surface modification technique	Coating materials	Experimental model	Influence of surface modification on biomaterials	Ref.
Titanium	MAO	Calcium (Ca) and Phosphorous (P) incorporated	In vitro: MG63 and human osteosarcoma (HOS) cell lines In vivo: Rabbits	Enhanced osseointegration capability Cell proliferation decreased, and ALP activities increased with increasing temperature	176
Ti-6A1-4V	PIII	Nitrogen and carbon plasma immersion ion implantation (N-PIII and C-PIII)	In vitro: L929 fibroblast and MC3T3-EI preosteoclasts In Vivo: Twelve-week-old male ICR rats	Superior cytocompatibility, resistance to corrosion and enhanced roughness	185
Magnesium	Chemical process	Calcium phosphate (Ca–P)	In vivo: Rabbit model	After 8 weeks, the implant was 100% fixed, with no inflammation, formation of new bones	209
316L SS	Dip coating–electrospinning process	Bioglass/gelatin/polycaprolactone (BG/GE/PCL)	In vitro: MG63 human osteoblast-like cells In vivo: Rabbits	Significantly enhanced the apatite formation and cell viability, no inflammatory response and toxicity	210

layer of magnesium fluoride (MgF_2). Silver/gold (Ag/Au) nanoparticles are uniformly distributed onto the surface by immobilized silver/gold ions *in situ*. This kind of multifunction composite coating exhibited corrosion resistance, enhanced antibacterial properties and accountable biocompatibility suitable for degradable implant applications.[195] Various combinations of minor alloying elements on Mg–5Ca and Mg-5Ca–1Zn were fabricated by chemical conversion and alternating dip-coating techniques. Moreover, Mg–PA coating led to bone-like apatite precipitation and unveiled osteoblast cell adhesion and proliferation.[196] Bioglass and HA are extensively used in coating agents on biometals or biocomposites and bone graft substitutes due to their superb bioactivity. Nonetheless, HA coating on the implant improved the osseointegration and bone tissue growth at a much faster rate.[157] Moreover, HA is a vital ingredient of human bone and is being broadly studied for bone biomaterials.[15] The degraded product and ions release help to enhance cell activity and accelerate bone defect healing.[197] Nevertheless, HA degraded very slowly *in vivo* because of its stable nature and high crystallinity.[198] On the flip side, the TCP's degradation rate is 10–20 times more than the HA. However, ceramic materials have some demerits, such as low strength, high brittleness and low toughness. To overcome these limitations, researchers have endeavored the application of ceramics as a surface coating agent,[199] secondary nanophases[200] and self-toughening process.[201] Whereas, the bilayer coating (ZA–CaP) on Mg–Sr-based alloy exhibited no harmful effect on osteoblast proliferation and induced tremendous osteogenic effects.[202] The bone remolding process is reasonably necessary for stable and favorable bone implants to the orthopedic implant interface. The galvanic couple can be easily formed in the ceramic/metal interface. However, it was reported that Mg/HA composites had not reflected good corrosion resistance. Additionally, the biocompatible dicalcium phosphate dihydrate (DCPD) coating has been carried out on Mg/HA composites. Electrochemical results revealed that DCPD and HA coating improved the corrosion resistance with a benchmark of 15 and 65 times higher relative to Mg/HA composites. Furthermore, immersion tests in SBF disclosed that the coatings enhanced the resistance to corrosion and biomineralization ability of the composites.[203] To increase corrosion resistance and biocompatibility, a series of Sr–HA coating was carried out in Mg alloy.[204] However, Mg porous scaffold was produced through titanium wire space holder process. These scaffolds exhibited enhanced corrosion resistance owing to the barrier of MgF_2, which acts as a coating formed during the fabrication process.

The scaffold indicated superior biocompatibility because the cell has grown and increased well. They suggested that the scaffold is suitable for a bone generation.[205] However, the different PCL and bioactive glass (BG) layers into the Mg scaffold were deposited using dip-coating method with a maintained low vacuum. It was found that the addition of PLC–BG into the Mg scaffold enhanced the porosity of the structure, improved corrosion resistance, increased bioactivity, as well as mechanical stability.[206] The HA-doped poly(lactic acid) coated on AZ31alloy enhanced the porous characteristics with improved cell attachment, cell growth and cell proliferation, and suggested that the materials can be used in implant materials in orthopedic implants.[207] Plasma electrolytic oxidation (PEO) was carried out to enhance the polymer and metals' adhesion properties. It was found that the properties of the magnesium screw was successfully improved by a coating of PCL, and it was investigated that 6 wt.% and four cycles group had an optimum condition.[208] Furthermore, to enhance the biocompatibility and cell adhesion of biomaterial, *in vivo* studies were performed and concluded that the biomaterials are suitable candidate for orthopedic applications.[209,210]

6. Summary and Future Perspectives

The current comprehensive work communicates a comparative, wide-ranging and vital review on biomaterials depicted its fabrication and performance evaluation of numerous kinds of biomaterials for orthopedic applications. Numerous findings in bone biomaterials, including biodegradable metals, biodegradable polymers and bioactive ceramics, are being extensively studied and discussed to comprehend their scope and application. Bioactive ceramics, which are significantly closer to human bone, possess excellent biocompatibility as well as bioactivity. Notwithstanding, its disadvantages, low toughness and inherent brittleness restrict their application in orthopedic applications. On the flip side, biodegradable polymers have significant toughness and can be modulated through fabrication and molecular design. However, the lack of mechanical properties and fewer toxic degraded products collectively constraint their extensive applications. Moreover, biodegradable metals, for instance magnesium and its alloys, are more rigid than bioactive ceramics and stronger than biodegradable polymer. Despite this, the loosening of mechanical integrity due to its uncontrolled degradation has significant challenges associated with them. The essential criteria for adjusting the equivalent young's modulus and yield strength of metallic implants via incorporating the material's average porosity are still needed. Therefore, metallic orthopedic implants' porous characteristics are very appropriate because of the tendency to restore bone function and promote bone tissue regeneration at defective sites. In addition to this, biomaterial should have the favorable properties of cell proliferation, attachment, as well as differentiation. Furthermore, the implant's porous structure is favorable for cell growth and sufficient nutrients and metabolic waste transportation. As defects in bones are prevented due to sports injuries, accidents and aging, there is an urgent requirement for biomaterial implants to support the bone or replacement of the bone. The foremost reasons for implant failure include stress shielding, inflammation owing to wear debris, corrosion fatigue, poor osseointegration and metal ion toxicity. Among these causes, most of them are surface phenomena, including osseointegration, wear and corrosion fatigue. Hence, surface modification plays a very vital role in developing the orthopedic implant. The success of metallic implants relies not only on their bulk properties but also on their surface properties that impact their interaction with the native host tissue. Numerous approaches to surface modification techniques have been reviewed. In PIII, the substrates are immersed in plasma and ions are implanted on the whole surface simultaneously, providing a faster and more cost-effective treatment accordingly. Since, the process is performed at low temperature, which minimizes the thermal deformation of materials. The surface coating tends to improve the cell adhesion, biocompatibility, corrosion resistance, wear-resistance and antimicrobial activities of the orthopedic implant. These approaches displayed promising consequences in orthopedics in terms of enhanced bone regeneration and repairing of bones. Regardless, an effort is being composed to the implant material to enhance the surfaces by surface modification process to render their compatibility between bone and implant without any adverse effects. Therefore, future endeavors are essential to evolve a more novel sustainable surface modification process in orthopedic implant perspectives. However, the option of materials explicitly monitoring their degradation behavior and surface properties is significant in the long-term reliability and demonstration of bioimplants.

Acknowledgments

The authors thankfully acknowledged the financial support from the National Institute of Technology Jamshedpur and Ministry of Human Resource & Development (MHRD), India (No. O.O.NO. NIT/ACAD/2018/443, dated 07/09/2018) to carry out the present research work.

References

1. H. Qu, H. Fu, Z. Han and Y. Sun, *RSC Adv.* **9** (2018) 26252.
2. M. Kaur and K. Singh, *Mater. Sci. Eng. C* **102** (2019) 844.
3. F. N. Alaribe, S. L. Manoto and S. C. K. M. Motaung, *Biologia (Bratislava)* **71** (2016) 353.
4. W. W. Thein-Han and R. D. K. Misra, *Acta Biomater.* **5** (2009) 1182.
5. D. F. Williams, *The Williams Dictionary of Biomaterials* (Liverpool University Press, 1999), p. 368.
6. N. S. Manam, W. S. W. Harun, D. N. A. Shri, S. A. C. Ghani, T. Kurniawan, M. H. Ismail and M. H. I. Ibrahim, *J. Alloys Compd.* **701** (2017) 698.

7. F. Barrère, T. A. Mahmood, K. de Groot and C. A. van Blitterswijk, *Mater. Sci. Eng. R, Rep.* **59** (2008) 38.
8. F. J. O'Brien, *Mater. Today* **14** (2011) 88.
9. D. F. Williams, *Biomaterials* **30** (2009) 5897.
10. B. D. Ratner, A. S. Hoffman, F. J. Schoen and J. E. Lemons, *Biomaterials Science,* 3rd edition (Elsevier, 2004).
11. V. Migonney, *Biomaterials* (ISTE Ltd. and John Wiley & Sons Inc, 2014).
12. J. Park and R. S. Lakes, *Biomaterials*, 3rd edition (Springer, 2007).
13. A. J. Smith, P. Dieppe, K. Vernon, M. Porter and A. W. Blom, *Lancet* **379** (2012) 1199.
14. S. Moniz, S. Hodgkinson and P. Yates, *Arthroplast. Today* **3** (2017) 151.
15. C. Gao, S. Peng, P. Feng and C. Shuai, *Bone Res.* **5** (2017) 17059.
16. L. Terranova, R. Mallet, R. Perrot and D. Chappard, *Acta Biomater.* **29** (2016) 380.
17. A. V. Maksimkin, F. S. Senatov, N. Y. Anisimova, M. V. Kiselevskiy, D. Y. Zalepugin, I. V. Chernyshova, N. A. Tilkunova and S. D. Kaloshkin, *Mater. Sci. Eng. C* **73** (2017) 366.
18. H. Lu, Y. Liu, J. Guo, H. Wu, J. Wang and G. Wu, *Int. J. Mol. Sci.* **17** (2016) 1.
19. R. Murugan and S. Ramakrishna, *Compos. Sci. Technol.* **65** (2005) 2385.
20. M. R. Shirdar, N. Farajpour, R. Shahbazian-Yassar and T. Shokuhfar, *Front. Chem. Sci. Eng.* **13** (2019) 1.
21. C. N. Elias, M. A. Meyers, R. Z. Valiev and S. N. Monteiro, *J. Mater. Res. Technol.* **2** (2013) 340.
22. N. Sezer, Z. Evis, S. M. Kayhan, A. Tahmasebifar and M. Koç, *J. Magnesium Alloy* **6** (2018) 23.
23. K. L. Ong, S. Lovald and J. Black, *Orthopaedic Biomaterials in Research and Practice,* 2nd edition (CRC Press, 2014).
24. D. Mareci, R. Chelariu, D. M. Gordin, G. Ungureanu and T. Gloriant, *Acta Biomater.* **5** (2009) 3625.
25. S. Wu, X. Liu, K. W. K. Yeung, H. Guo, P. Li, T. Hu, C. Y. Chung and P. K. Chu, *Surf. Coat. Technol.* **233** (2013) 13.
26. M. Niinomi, *J. Artif. Organs* **11** (2008) 105.
27. A. J. Festas, A. Ramos and J. P. Davim, *Proc. Inst. Mech. Eng. Part L, J. Mater. Des. Appl.* **234** (2020) 218.

28. A. Sáenz, E. Rivera, W. Brostow and V. O. Castaño, *J. Mater. Educ.* **21** (1999) 267.
29. A. M. Slaney, V. A. Wright, P. J. Meloncelli, K. D. Harris, L. J. West, T. L. Lowary and J. M. Buriak, *ACS Appl. Mater. Interfaces* **3** (2011) 1601.
30. N. Kurgan, *Mater. Des.* **55** (2014) 235.
31. I. Gurappa, *Surf. Coat. Technol.* **161** (2002) 70.
32. S. S. Kumar and B. Stucker, Development of a Co-Cr-Mo to tantalum transition using LENS for orthopedic applications, in *16th Int. Solid Freeform Fabrication Symp.* (2005), pp. 273-282.
33. V. K. Balla, P. D. DeVasConCellos, W. Xue, S. Bose and A. Bandyopadhyay, *Acta Biomater.* **5** (2009) 1831.
34. J. E. G. González and J. C. Mirza-Rosca, *J. Electroanal. Chem.* **471** (1999) 109.
35. X. M. Liu, E. P. Maziarz, E. Quinn and Y. C. Lai, *Int. J. Mass Spectrom.* **238** (2004) 227.
36. M. Koike, K. Martinez, L. Guo, G. Chahine, R. Kovacevic and T. Okabe, *J. Mater. Process. Technol.* **211** (2011) 1400.
37. M. M. Dewidar, K. A. Khalil and J. K. Lim, *Trans. Nonferrous Met. Soc. China* **17** (2007) 468.
38. T. J. Webster and J. U. Ejiofor, *Biomaterials* **25** (2004) 4731.
39. K. Beer-Lech and B. Surowska, *Eksploatacjai Nieza-wodnsc* **17** (2015) 90.
40. K. Hazlehurst, C. J. Wang and M. Stanford, *Mater. Des.* **51** (2013) 949.
41. K. Monroy, J. Delgado and J. Ciurana, *Proc. Eng.* **63** (2013) 361.
42. N. T. C. Oliveira and A. C. Guastaldi, *Acta Biomater.* **5** (2009) 399.
43. K. H. Chung, G. T. Liu, J. G. Duh and J. H. Wang, *Surf. Coat. Technol.* **188–189** (2004) 745.
44. L.-J. Chen, T. Li, Y. Li, Y. Min, H. He and Y.-H. Hu, *Trans. Nonferrous Met. Soc. China* **19** (2009) 1174.
45. H. C. Hsu, C. H. Pan, S. C. Wu and W. F. Ho, *J. Alloys Compd.* **474** (2009) 578.
46. S. A. Yavari, S. M. Ahmadi, J. van der Stok, R. Wauthle, A. C. Riemslag, M. Janssen, J. Schrooten, H. Weinans and A. A. Zadpoor, *J. Mech. Behav. Biomed. Mater.* **36** (2014) 109.
47. E. B. Taddei, V. A. R. Henriques, V. A. R. Silva and C. A. A. Cairo, *Mater. Sci. Eng. C* **24** (2004) 683.

48. J. Vaithilingam, S. Kilsby, R. D. Goodridge, S. D. R. Christie, S. Edmondson and R. J. M. Hague, *Mater. Sci. Eng. C* **46** (2015) 52.
49. W. F. Ho, T. Y. Chiang, S. C. Wu and H. C. Hsu, *J. Alloys Compd.* **468** (2009) 533.
50. R. Palanivelu, S. Kalainathan and A. Ruban Kumar, *Ceram. Int.* **40** (2014) 7745.
51. C. S. S. de Oliveira, S. Griza, M. V. de Oliveira, A. A. Ribeiro and Mô. B. Leite, *Powder Technol.* **281** (2015) 91.
52. T. Puskar, D. Jevremovic, R. J. Williams, D. Eggbeer, D. Vukelic and I. Budak, *Materials* **6** (2014) 6486.
53. S. A. S. Yavari, R. Wauthle, A. J. Böttger, J. Schrooten, H. Weinans and A. A. Zadpoor, *Appl. Surf. Sci.* **290** (2014) 287.
54. L. Reclaru, R. Lerf, P. Y. Eschler and J. M. Meyer, *Biomaterials* **22** (2001) 269.
55. C. C. Shih, C. M. Shih, Y. Y. Su, L. H. J. Su, M. S. Chang and S. J. Lin, *Corros. Sci.* **46** (2004) 427.
56. M. Dadfar, M. H. Fathi, F. Karimzadeh, M. R. Dadfar and A. Saatchi, *Mater. Lett.* **61** (2007) 2343.
57. F. A. España, V. K. Balla, S. Bose and A. Bandyo-padhyay, *Mater. Sci. Eng. C* **30** (2010) 50.
58. A. Ghiban and P. Moldovan, *UPB Sci. Bull. Ser. B Chem. Mater. Sci.* **74** (2012) 203.
59. J. Parthasarathy, B. Starly, S. Raman and A. Christensen, *J. Mech. Behav. Biomed. Mater.* **3** (2010) 249.
60. L. C. Zhang, D. Klemm, J. Eckert, Y. L. Hao and T. B. Sercombe, *Scr. Mater.* **65** (2011) 21.
61. G. Jiang, Q. Li, C. Wang, J. Dong and G. He, *Mater. Des.* **67** (2015) 354.
62. B. V. Krishna, S. Bose and A. Bandyopadhyay, *Acta Biomater.* **3** (2007) 997.
63. B. V. B. Krishna, W. Xue, S. Bose and A. Bandyo-padhyay, *Acta Biomater.* **4** (2008) 697.
64. K. B. Hazlehurst, C. J. Wang and M. Stanford, *Med. Hypotheses* **81** (2013) 1096.
65. W. F. Ho, C. H. Cheng, C. H. Pan, S. C. Wu and H. C. Hsu, *Mater. Sci. Eng. C* **29** (2009) 36.
66. G. E. J. Poinern, S. Brundavanam and D. Fawcett, *Am. J. Biomed. Eng.* **2** (2013) 218.

67. F. Witte, N. Hort, C. Vogt, S. Cohen, K. U. Kainer, R. Willumeit and F. Feyerabend, *Curr. Opin. Solid State Mater. Sci.* **12** (2008) 63.
68. Z. Li, X. Gu, S. Lou and Y. Zheng, *Biomaterials* **29** (2008) 1329.
69. X. N. Gu and Y. F. Zheng, *Front. Mater. Sci. China* **4** (2010) 111.
70. Y. Yang, C. He, E. Dianyu, W. Yang, F. Qi, D. Xie, L. Shen, S. Peng and C. Shuai, *Mater. Des.* **185** (2020) 108259.
71. S. Wu, X. Liu, K. W. K. Yeung, C. Liu and X. Yang, *Mater. Sci. Eng. R, Rep.* **80** (2014) 1.
72. X. Li, X. Liu, S. Wu, K. W. K. Yeung, Y. Zheng and P. K. Chu, *Acta Biomater.* **45** (2016) 2.
73. M. Talha, Y. Ma, Y. Lin, A. Singh, W. Liu and X. Kong, *New J. Chem.* **43** (2019) 1943.
74. M. Navarro, A. Michiardi, O. Castano and J. A. Planell, *J. R. Soc. Interface* **5** (2008) 1137.
75. J. J. Ramsden, D. M. Allen, D. J. Stephenson, J. R. Alcock, G. N. Peggs, G. Fuller and G. Goch, *CIRP Ann. Manuf. Technol.* **56** (2007) 687.
76. C. V. Vidal and A. I. Muñoz, *Electrochim. Acta* **54** (2009) 1798.
77. B. Patel, G. Favaro, F. Inam, M. J. Reece, A. Angadji, W. Bonfield, J. Huang and M. Edirisinghe, *Mater. Sci. Eng. C* **32** (2012) 1222.
78. M. Geetha, A. K. Singh, R. Asokamani and A. K. Gogia, *Prog. Mater. Sci.* **54** (2009) 397.
79. C. Prakash, S. Singh, S. Ramakrishna, G. Królczyk and C. H. Le, *J. Alloys Compd.* **824** (2020) 153774.
80. C. Prakash, S. Singh, C. I. Pruncu, V. Mishra, G. Królczyk, D. Y. Pimenov and A. Pramanik, *Materials* **12** (2019) 1006.
81. Y. Li, C. Yang, X. Yin, Y. Sun, J. Weng, J. Zhou and B. Feng, *J. Mater. Chem. B* **7** (2019) 3546.
82. J. Liu, F. Jin, M.-L. Zheng, S. Wang, S.-Q. Fan, P. Li and X.-M. Duan, *ACS Appl. Bio Mater.* **2** (2019) 697.
83. I. Nemcakova, I. Jirka, M. Doubkova and L. Bacakova, *Sci. Rep.* **10** (2020) 1.
84. K. S. Katti, *Colloids Surf. B, Biointerfaces* **39** (2004) 133.
85. C. Vermes, R. Chandrasekaran, J. J. Jacobs, J. O. Galante, K. A. Roebuck and T. T. Glant, *J. Bone Joint Surg. A* **83** (2001) 201.
86. V. Sansone, D. Pagani and M. Melato, *Clin. Cases Min. Bone Metab.* **10** (2013) 34.

87. L. Grausova, L. Bacakova, A. Kromka, S. Potocky, M. Vanecek, M. Nesladek and V. Lisa, *J. Nanosci. Nanotechnol.* **9** (2009) 3524.
88. I. Kopova, J. Kronek, L. Bacakova and J. Fencl, *Diam. Relat. Mater.* **97** (2019) 107456.
89. I. Kopova, V. Lavrentiev, J. Vacik and L. Bacakova, *PLoS One* **10** (2015) e0123680.
90. I. Kopova, L. Bacakova, V. Lavrentiev and J. Vacik, *Int. J. Mol. Sci.* **14** (2013) 9182.
91. Y. Li, C. Yang, H. Zhao, S. Qu, X. Li and Y. Li, *Materials* **7** (2014) 1709.
92. E. El-Meliegy, N. I. Abu-Elsaad, A. M. El-Kady and M. A. Ibrahim, *Sci. Rep.* **8** (2018) 12180.
93. C. R. Gautam, S. Kumar, S. Biradar, S. Jose and V. K. Mishra, *RSC Adv.* **6** (2016) 67565.
94. M. Topuz, B. Dikici and M. Gavgali, *J. Mech. Behav. Biomed. Mater.* **118** (2021) 104480.
95. B. Dikici, M. Niinomi, M. Topuz, Y. Say, B. Aksakal, H. Yilmazer and M. Nakai, *Prot. Met. Phys. Chem. Surf.* **54** (2018) 457.
96. V. T. Nguyen, T. C. Cheng, T. H. Fang and M. H. Li, *J. Mater. Res. Technol.* **9** (2020) 4817.
97. K. Castkova, H. Hadraba, A. Matousek, P. Roupcova, Z. Chlup, L. Novotna and J. Cihlar, *J. Eur. Ceram. Soc.* **36** (2016) 2903.
98. S. Singh, C. Prakash and H. Singh, *Surf. Coat. Technol.* **398** (2020) 126072.
99. K. Balani, R. Anderson, T. Laha, M. Andara, J. Tercero, E. Crumpler and A. Agarwal, *Biomaterials* **28** (2007) 618.
100. K. Balani, Y. Chen, S. P. Harimkar, N. B. Dahotre and A. Agarwal, *Acta Biomater.* **3** (2007) 944.
101. M. F. Morks, *J. Mech. Behav. Biomed. Mater.* **1** (2008) 105.
102. J. Ren, D. Zhao, F. Qi, W. Liu and Y. Chen, *J. Mech. Behav. Biomed. Mater.* **101** (2020) 103418.
103. S. Singh, K. K. Pandey, A. Islam and A. K. Keshri, *Ceram. Int.* **46** (2020) 13539.
104. L. Fu, K. A. Khor and J. P. Lim, *Surf. Coat. Technol.* **127** (2000) 66.
105. S. Kalmodia, S. Goenka, T. Laha, D. Lahiri, B. Basu and K. Balani, *Mater. Sci. Eng. C* **30** (2010) 1162.
106. Y. H. He, Y. Q. Zhang, Y. H. Jiang and R. Zhou, *RSC Adv.* **6** (2016) 100939.

107. L. Xie, Y. Yang, Z. Fu, Y. Li, J. Shi, D. Ma, S. Liu and D. Luo, *RSC Adv.* **9** (2019) 781.
108. J. Ru, Q. Wei, L. Yang, J. Qin, L. Tang, J. Wei, L. Guo and Y. Niu, *RSC Adv.* **8** (2018) 18745.
109. J. Xue, C. Feng, L. Xia, D. Zhai, B. Ma, X. Wang, B. Fang, J. Chang and C. Wu, *Chem. Mater.* **30** (2018) 4646.
110. Z. Huan, C. Xu, B. Ma, J. Zhou and J. Chang, *RSC Adv.* **6** (2016) 47897.
111. C. Zhu, Y. Lv, C. Qian, H. Qian, T. Jiao, L. Wang and F. Zhang, *Sci. Rep.* **6** (2016) 38875.
112. M. Khodaei, F. Nejatidanesh, M. J. Shirani, S. Iyengar, H. Sina and O. Savabi, *J. Bionic Eng.* **17** (2020) 92.
113. J. R. Weisinger and E. Bellorín-Font, *Lancet* **352** (1998) 391.
114. M. Nabiyouni, T. Brückner, H. Zhou, U. Gbureck and S. B. Bhaduri, *Acta Biomater.* **66** (2018) 23.
115. M. Diba, F. Tapia, A. R. Boccaccini and L. A. Strobel, *Int. J. Appl. Glass Sci.* **3** (2012) 221.
116. E. H. M. Nunes, V. F. C. Lins, P. H. R. Pereira, A. Isaac, T. G. Langdon and R. B. Figueiredo, *Materials* **12** (2019) 2609.
117. D. Zhao, F. Witte, F. Lu, J. Wang, J. Li and L. Qin, *Biomaterials* **112** (2017) 287.
118. Y. Sheng, L. Tian, C. Wu, L. Qin and T. Ngai, *Langmuir* **34** (2018) 10684.
119. D. Zhao, S. Huang, F. Lu, B. Wang, L. Yang, L. Qin, K. Yang, Y. Li, W. Li, W. Wang, S. Tian, X. Zhang, W. Gao, Z. Wang, Y. Zhang, X. Xie, J. Wang and J. Li, *Biomaterials* **81** (2016) 84.
120. R. Zeng, W. Dietzel, F. Witte, N. Hort and C. Blawert, *Adv. Eng. Mater.* **10** (2008) 3.
121. A. H. M. Sanchez, B. J. C. Luthringer, F. Feyerabend and R. Willumeit, *Acta Biomater.* **13** (2015) 16.
122. S. Virtanen, *Mater. Sci. Eng. B, Solid-State Mater. Adv. Technol.* **176** (2011) 1600.
123. C. Prakash, S. Singh, M. K. Gupta, M. Mia, G. Królczyk and N. Khanna, *Materials* **11** (2018) 1602.
124. R. Zeng, J. Zhang, W. Huang, W. Dietzel, K. U. Kainer, C. Blawert and W. Ke, *Trans. Nonferrous Met. Soc. China.* **16** (2006) s763.
125. C. Liu, Y. Xin, G. Tang and P. K. Chu, *Mater. Sci. Eng. A* **456** (2007) 350.
126. Y. Wang, G. Liu and Z. Fan, *Scr. Mater.* **54** (2006) 903.

127. N. Li and Y. Zheng, *J. Mater. Sci. Technol.* **29** (2013) 489.
128. J. Čapek and D. Vojtěch, *Mater. Sci. Eng. C* **33** (2013) 564.
129. H. Hornberger, S. Virtanen and A. R. Boccaccini, *Acta Biomater.* **8** (2012) 2442.
130. H. Miao, D. Zhang, C. Chen, L. Zhang, J. Pei, Y. Su, H. Huang, Z. Wang, B. Kang, W. Ding, H. Zeng and G. Yuan, *ACS Biomater. Sci. Eng.* **5** (2019) 1623.
131. G. Yang, H. Yang, L. Shi, T. Wang, W. Zhou, T. Zhou, W. Han, Z. Zhang, W. Lu and J. Hu, *ACS Biomater. Sci. Eng.* **4** (2018) 4289.
132. Y. Feng, S. Zhu, L. Wang, L. Chang, Y. Hou and S. Guan, *Bioact. Mater.* **3** (2018) 225.
133. D. Bian, J. Deng, N. Li, X. Chu, Y. Liu, W. Li, H. Cai, P. Xiu, Y. Zhang, Z. Guan, Y. Zheng, Y. Kou, B. Jiang and R. Chen, *ACS Appl. Mater. Interfaces* **10** (2018) 4394.
134. Y. Yan, H. Cao, Y. Kang, K. Yu, T. Xiao, J. Luo, Y. Deng, H. Fang, H. Xiong and Y. Dai, *J. Alloys Compd.* **693** (2017) 1277.
135. C. Liu, X. Fu, H. Pan, P. Wan, L. Wang, L. Tan, K. Wang, Y. Zhao, K. Yang and P. K. Chu, *Sci. Rep.* **6** (2016) 27374.
136. Y. Ding, J. Lin, C. Wen, D. Zhang and Y. Li, *Sci. Rep.* **6** (2016) 31990.
137. J.-L. Wang, S. Mukherjee, D. R. Nisbet, N. Birbilis and X.-B. Chen, *J. Mater. Chem. B* **3** (2015) 8874.
138. G. He, Y. Wu, Y. Zhang, Y. Zhu, Y. Liu, N. Li, M. Li, G. Zheng, B. He, Q. Yin, Y. Zheng and C. Mao, *J. Mater. Chem. B* **3** (2015) 6676.
139. Y. Ding, Y. Li, J. Lin and C. Wen, *J. Mater. Chem. B* **3** (2015) 3714.
140. H. Qin, Y. Zhao, Z. An, M. Cheng, Q. Wang, T. Cheng, Q. Wang, J. Wang, Y. Jiang, X. Zhang and G. Yuan, *Biomaterials* **53** (2015) 211.
141. H. R. Bakhsheshi-Rad, M. H. Idris, M. R. Abdul-Kadir, A. Ourdjini, M. Medraj, M. Daroonparvar and E. Hamzah, *Mater. Des.* **53** (2014) 283.
142. H. Li, Q. Peng, X. Li, K. Li, Z. Han and D. Fang, *Mater. Des.* **58** (2014) 43.
143. P.-R. Cha, H.-S. Han, G.-F. Yang, Y.-C. Kim, K.-H. Hong, S.-C. Lee, J.-Y. Jung, J.-P. Ahn, Y.-Y. Kim, S.-Y. Cho, J. Y. Byun, K.-S. Lee, S.-J. Yang and H.-K. Seok, *Sci. Rep.* **3** (2013) 2367.

144. J. Kubásek and D. Vojtěch, *J. Mater. Sci. Mater. Med.* **24** (2013) 1615.
145. D. T. Chou, D. Hong, P. Saha, J. Ferrero, B. Lee, Z. Tan, Z. Dong and P. N. Kumta, *Acta Biomater.* **9** (2013) 8518.
146. M. Bornapour, N. Muja, D. Shum-Tim, M. Cerruti and M. Pekguleryuz, *Acta Biomater.* **9** (2013) 5319.
147. I. S. Berglund, H. S. Brar, N. Dolgova, A. P. Acharya, B. G. Keselowsky, M. Sarntinoranont and M. V. Manuel, *J. Biomed. Mater. Res. B Appl. Biomater.* **100B** (2012) 1524.
148. X. N. Gu, X. H. Xie, N. Li, Y. F. Zheng and L. Qin, *Acta Biomater.* **8** (2012) 2360.
149. H. S. Brar, J. Wong and M. V. Manuel, *J. Mech. Behav. Biomed. Mater.* **7** (2012) 87.
150. Y. Zong, G. Yuan, X. Zhang, L. Mao, J. Niu and W. Ding, *Mater. Sci. Eng. B, Solid-State Mater. Adv. Technol.* **177** (2012) 395.
151. S. Zhang, X. Zhang, C. Zhao, J. Li, Y. Song, C. Xie, H. Tao, Y. Zhang, Y. He and Y. Jiang, *Acta Biomater.* **6** (2010) 626.
152. F. Rosalbino, S. De Negri, A. Saccone, E. Angelini and S. Delfino, *J. Mater. Sci. Mater. Med.* **21** (2010) 1091.
153. N. Hort, Y. Huang, D. Fechner, M. Störmer, C. Blawert, F. Witte, C. Vogt, H. Drücker, R. Willumeit and K. U. Kainer, *Acta Biomater.* **6** (2010) 1714.
154. C. Prakash, S. Singh, S. Sharma, J. Singh, G. Singh, M. Mehta, M. Mittal and H. Kumar, *Mater. Today Proc.* **21** (2020) 1713.
155. J. Li, Z. Li, R. Li, Y. Shi, H. Wang, Y. Wang and G. Jin, *RSC Adv.* **8** (2018) 36512.
156. C. Prakash, S. Singh, K. Verma, S. S. Sidhu and S. Singh, *Vacuum* **155** (2018) 578.
157. M. Heiden, E. Nauman and L. Stanciu, *Adv. Healthc. Mater.* **6** (2017) 1700120.
158. W. Liu, J. Wang, G. Jiang, J. Guo, Q. Li, B. Li, Q. Wang, M. Cheng, G. He and X. Zhang, *J. Mater. Chem. B* **5** (2017) 7661.
159. J. Li, Z. Li, Q. Wang, Y. Shi, W. Li, Y. Fu and G. Jin, *RSC Adv.* **9** (2019) 1541.
160. J. He, H. Ye, Y. Li, J. Fang, Q. Mei, X. Lu and F. Ren, *ACS Biomater. Sci. Eng.* **5** (2019) 509.
161. J. Zhang, C. Xu, Y. Jing, S. Lv, S. Liu, D. Fang, J. Zhuang, M. Zhang and R. Wu, *Sci. Rep.* **5** (2015) 13933.

162. A. A. Aliyu, A. M. Abdul-Rani, T. L. Ginta, C. Prakash, E. Axinte, M. A. Razak and S. Ali, *Adv. Mater. Sci. Eng.* **2017** (2017) 1.
163. C. Eriksson, J. Lausmaa and H. Nygren, *Biomaterials* **22** (2001) 1987.
164. Q. Chen and G. A. Thouas, *Mater. Sci. Eng. R Rep.* **87** (2015) 1.
165. S. B. Goodman, Z. Yao, M. Keeney and F. Yang, *Biomaterials* **34** (2013) 3174.
166. Z. A. Uwais, M. A. Hussein, M. A. Samad and N. Al-Aqeeli, *Arab. J. Sci. Eng.* **42** (2017) 4493.
167. M. Kulkarni, A. Mazare, P. Schmuki and A. Iglič, *Nanomedicine* (One Central Press, 2014).
168. A. Kurella and N. B. Dahotre, *J. Biomater. Appl.* **20** (2005) 5.
169. B. D. Ratner, A. S. Hoffman and S. L. McArthur, *Biomaterials Science* (Elsevier, 2020).
170. Z.-Y. Qiu, C. Chen, X.-M. Wang and I.-S. Lee, *Regen. Biomater.* **1** (2014) 67.
171. W. Shi and H. S. Dong, *J. Shanghai Univ.* **9** (2005) 164.
172. D. Xiong, Z. Gao and Z. Jin, *Surf. Coat. Technol.* **201** (2007) 6847.
173. Y. Luo, S. Ge, Z. Jin and J. Fisher, *Proc. Inst. Mech. Eng. J, J. Eng. Tribol.* **223** (2009) 311.
174. M. Niinomi, *Sci. Technol. Adv. Mater.* **4** (2003) 445.
175. B. Yang, M. Uchida, H. M. Kim, X. Zhang and T. Kokubo, *Biomaterials* **25** (2004) 1003.
176. L.-H. Li, Y.-M. Kong, H.-W. Kim, Y.-W. Kim, H.-E. Kim, S.-J. Heo and J.-Y. Koak, *Biomaterials* **25** (2004) 2867.
177. C. Balagna, M. G. Faga and S. Spriano, *Surf. Coat. Technol.* **258** (2014) 1159.
178. M. Hoseini, A. Jedenmalm and A. Boldizar, *Wear* **264** (2008) 958.
179. R. Förch, A. N. Chifen, A. Bousquet, H. L. Khor, M. Jungblut, L. Q. Chu, Z. Zhang, I. Osey-Mensah, E. K. Sinner and W. Knoll, *Chem. Vapor. Depos.* **13** (2007) 280.
180. H. Szymanowski, A. Sobczyk, M. Gazicki-Lipman, W. Jakubowski and L. Klimek, *Surf. Coat. Technol.* **200** (2005) 1036.
181. S. Shadanbaz and G. J. Dias, *Acta Biomater.* **8** (2012) 20.
182. S. L. Peterson, A. McDonald, P. L. Gourley and D. Y. Sasaki, *J. Biomed. Mater. Res. A* **72** (2005) 10.
183. G. Wu, P. Li, H. Feng, X. Zhang and P. K. Chu, *J. Mater. Chem. B* **3** (2015) 2024.

184. P. Chu, *Mater. Sci. Eng. R, Rep.* **36** (2002) 143.
185. Y. Zhao, S. M. Wong, H. M. Wong, S. Wu, T. Hu, K. W. K. Yeung and P. K. Chu, *ACS Appl. Mater. Interfaces* **5** (2013) 1510.
186. C. Díaz, J. Lutz, S. Mändl, J. A. García, R. Martínez and R. J. Rodríguez, *Nucl. Instrum. Methods Phys. Res. Sect. B, Beam Interact. Mater. At.* **267** (2009) 1630.
187. P. Wan, L. Tan and K. Yang, *J. Mater. Sci. Technol.* **32** (2016) 827.
188. R. F. Zhang and S. F. Zhang, *Corros. Sci.* **51** (2009) 2820.
189. M. Tarakci, K. Korkmaz, Y. Gencer and M. Usta, *Surf. Coat. Technol.* **199** (2005) 205.
190. L. Wang, X. Shang, Y. Hao, G. Wan, L. Dong, D. Huang, X. Yang, J. Sun, Q. Wang, G. Zha and X. Yang, *RSC Adv.* **9** (2019) 2892.
191. Y. H. Xia, B. P. Zhang, C. X. Lu and L. Geng, *Mater. Sci. Eng. C* **33** (2013) 5044.
192. A. Janković, S. Eraković, M. Mitrić, I. Z. Matić, Z. D. Juranić, G. C. P. Tsui, C. Y. Tang, V. Mišković-Stanković, K. Y. Rhee and S. J. Park, *J. Alloys Compd.* **624** (2015) 148.
193. J. E. Gray and B. Luan, *J. Alloys Compd.* **336** (2002) 88.
194. H. R. Zheng, Z. Li, C. You, D. B. Liu and M. F. Chen, *Bioact. Mater.* **2** (2017) 1.
195. A. I. Rezk, A. R. K. A. Sasikala, A. G. Nejad, H. M. Mousa, Y. M. Oh, C. H. Park and C. S. Kim, *Sci. Rep.* **9** (2019) 117.
196. Y. Chen, S. Zhao, B. Liu, M. Chen, J. Mao, H. He, Y. Zhao, N. Huang and G. Wan, *ACS Appl. Mater. Interfaces* **6** (2014) 19531.
197. S. S. A. Abidi and Q. Murtaza, *J. Mater. Sci. Technol.* **30** (2014) 307.
198. C. Shuai, P. Li, J. Liu and S. Peng, *Mater. Charact.* **77** (2013) 23.
199. H. A. Ching, D. Choudhury, M. J. Nine and N. A. A. Osman, *Sci. Technol. Adv. Mater.* **15** (2014) 014402.
200. N. Levandowski, N. H. A. Camargo, D. F. Silva, G. M. L. Dalmônico and P. F. Franczak, *Adv. Mater. Res.* **936** (2014) 695.
201. Z. Li, P. Munroe, Z. T. Jiang, X. Zhao, J. Xu, Z. F. Zhou, J. Q. Jiang, F. Fang and Z. H. Xie, *Acta Mater.* **60** (2012) 5735.
202. M. Li, P. Wan, W. Wang, K. Yang, Y. Zhang and Y. Han, *Sci. Rep.* **9** (2019) 933.
203. Y. Su, D. Li, Y. Su, C. Lu, L. Niu, J. Lian and G. Li, *ACS Biomater. Sci. Eng.* **2** (2016) 818.
204. X. Gu, W. Lin, D. Li, H. Guo, P. Li and Y. Fan, *RSC Adv.* **9** (2019) 15013.

205. M. Cheng, T. Wahafu, G. Jiang, W. Liu, Y. Qiao, X. Peng, T. Cheng, X. Zhang, G. He and X. Liu, *Sci. Rep.* **6** (2016) 24134.
206. M. Yazdimamaghani, M. Razavi, D. Vashaee and L. Tayebi, *Mater. Sci. Eng. C* **49** (2015) 436.
207. A. Abdal-hay, N. A. M. Barakat and J. K. Lim, *Ceram. Int.* **39** (2013) 183.
208. Y.-K. Kim, K.-B. Lee, S.-Y. Kim, Y.-S. Jang, J. H. Kim and M.-H. Lee, *Sci. Rep.* **8** (2018) 13264.
209. J. X. Yang, F. Z. Cui, I. S. Lee, Y. Zhang, Q. S. Yin, H. Xia and S. X. Yang, *J. Biomater. Appl.* **27** (2012) 153.
210. M. Mozafari, E. Salahinejad, S. Sharifi-Asl, D. D. Macdonald, D. Vashaee and L. Tayebi, *Surf. Eng.* **30** (2014) 688.

Effect of loads and bio-lubricants on tribological study of zirconia and zirconia toughened alumina against Ti6al4v for hip prosthesis*

S. Shankar[†,‖], R. Nithyaprakash[†], R. Naveen Kumar[‡], R. Aravinthan[†], Alokesh Pramanik[§] and Animesh Kumar Basak[¶]

[†]*Department of Mechatronics Engineering,*
Kongu Engineering College,
Erode, Tamil Nadu, India
[‡]*Department of Mechanical Engineering,*
Kongu Engineering College,
Erode, Tamil Nadu, India
[§]*School of Civil and Mechanical Engineering,*
Curtin University,
Bentley, WA 6102, Australia
[¶]*Adelaide Microscopy,*
The University of Adelaide,
Adelaide, SA 5005, Australia
[‖]*shankariitm@gmail.com*

Zirconia and zirconia toughened alumina (ZTA) are widely used for biomedical applications. Owing to the superior mechanical and biocompatible properties of ZTA, it is now gaining more importance over zirconia, as it has the profound strength of zirconia and alumina. This study examines the tribological behavior of Zirconia and ZTA balls

---

*To cite this article, please refer to its earlier version published in Surface Review and Letters, Vol. 30, No. 1 (2023) 2141006 (10 pages) DOI: 10.1142/S0218625X21410067
[‖]Corresponding author.

sliding against a Ti6Al4V disc using the Ball-on-disc (BOD) Tribometer. The effect of loads and bio-lubricants influencing the tribological behavior is investigated for a sliding distance of 2 km using BoD Tribometer. The friction and wear coefficients are examined under these loading and lubrication conditions. Five bio-lubricants considered for the tribological study of biomaterials include ringer's solution, phosphate buffer saline (PBS), distilled water, NaCl 0.9% saline solution and sesame oil. The result showed that ZTA had a lower coefficient of friction (CoF) value of 0.34 for NaCl bio-lubricant. The ZTA and zirconia exhibited the least wear coefficient values under sesame oil lubrication. Overall, ZTA had better CoF under high loads for PBS, distilled water and sesame oil bio-lubricants. However, zirconia exhibited a better wear coefficient under all loads and lubrication conditions.

Keywords: Zirconia; zirconia toughened alumina; tribometer; biolubricants; wear coefficient.

1. Introduction

Artificial hip implantation mainly consists of acetabulum cup and femoral head replaced by metal, polyethylene, and ceramic bearings. Mostly, bearing couples are classified based on material types such as metal-on-metal (M-o-M), metal-on-plastic (M-o-P), ceramic-on-metal (C-o-M), ceramic-on-ceramic (C-o-C), and ceramic-on-plastic (C-o-P). These biomaterials should be biocompatible, have better wear/corrosion resistance and possess good mechanical properties both *in-vivo* and *in-vitro* conditions to increase the stability of implants.[1] Many *in-vitro* studies analyzed the friction and wear behavior of various biomaterials under dry and lubrication conditions using hip simulators and pin-on-disc Tribometer (PoD).[2-4] Alumina and zirconia ceramics were widely used for joint replacements. However, hydro-thermal ageing and fracture of zirconia ceramics led to use of ZTA for joint replacements.[5,6] The use of ZTA has many advantages in particular, combining the superior mechanical properties of alumina and zirconia in resisting corrosion and crack.[7] In general, synovial fluid is essential in orthopedic implants. *In-vitro* studies were carried out to explore the tribological behavior of hip joint biomaterials using bovine serum instead of synovial fluid.[8] Tribological properties of Ti6Al4V alloy under three bio-lubricants were studied and found that low friction and wear rate can be achieved in a bovine serum environment.[9] The use of bio-lubricants like distilled water, saline solution, ringer's solution, sesame oil and PBS solution in analyzing the wear of implants for hip joint replacement was reported.[9-11] The use of ceramics in joint replacement had several advantages over metallic and polyethylene implants, which include less wear and prevention of metal ions being released into blood stream.

Ball-on-disc Tribometer (BOD) study of analyzing the effect of zirconia and alumina combination under saline bio-lubricant was investigated for joint replacements.[12] Results revealed that high wear rate was observed for zirconia and hence it could not be used for joint replacement. Similarly, the tribological behavior of ZTA against various metal oxides under dry sliding for 50 N load was carried out and resulted with lower friction and wear rate for TiO_2 and MnO_2 metal oxide additives.[13] Another study investigated the dry sliding of ZTA with oxide additives against alumina disc for 30 N load and results revealed that ZTA containing CeO_2 and TiO_2 showed minimum wear rate.[14] The wear rate of ZTA reinforced with alumina/silicon carbide was investigated and lesser wear rate was observed under erosive wear test.[15] Another study investigated the wear of ZTA against steel plate and found very low wear under dry sliding.[16] Recently, tribological study of ceria stabilized zirconia containing doped Samaria was investigated and the results revealed that doped material exhibited better tribological properties.[17] The tribological study plays a crucial role in investigating the wear of composite materials in selecting the better suitable materials for different applications.[18] None of these studies analyzed tribological behavior of ZTA for joint replacement perspectives as friction and wear coefficients plays a major role in analyzing the wear of biomaterials using in-silico approach.[19,20] The *in-silico* submodeling technique will be effective to analyze the wear of the biomaterials using obtained friction and wear coefficients.[21]

This study focuses on ZTA and Zirconia biomaterials sliding against Ti6Al4V alloy (grade 5) for a sliding distance of 2 km using BOD Tribometer under five different bio-lubricants. The bio-lubricants include distilled water, NaCl 0.9% saline solution, PBS, ringer's solution and sesame oil. The friction and wear coefficient are investigated under four loading conditions of 15 N, 20 N, 25 N, and 30 N to replicate hip joint loads.

2. Materials and Methods

The zirconia and ZTA balls with a diameter of 10 mm were used as biomaterials in this study. Ti6Al4V is a counter disc specimen with 160 mm diameter and 4 mm thickness. During the experiment, the friction and wear behavior of biomaterials were examined under five bio-lubricants as mentioned earlier, which converted the hip gait loads to equivalent BoD loads.[11,12] The density of zirconia and ZTA was 6.02 g/cm^3 and 3.65 g/cm^3 as indicated by the supplier. The ZTA had hardness value of 15.9 GPa while zirconia showed hardness value of 13.24 GPa. The friction and wear rate were estimated at room temperature using a BoD tribometer, as shown in Fig. 1. The sliding velocity of disc was kept at 0.7 m/s for a total sliding distance of 2000 m. Each load of 15 N, 20 N, 25 N, and 30 N corresponds to 19 gait activities ranging from normal walking to load transfer of 50 kg gait activities which was considered based on previous literatures.[11,12] After the completion of the wear test, the mass loss of the specimen was determined with the help of weighing balance with an accuracy of 0.01 mg. The tests were repeated thrice to get the optimum friction and wear coefficients. The wear coefficient (mm^3/Nm) is calculated using the following formula:

$$\text{Wear rate} = \frac{\text{Volume Loss}}{\text{Load} \times \text{Sliding Distance}} (\text{mm}^3/\text{N} \cdot \text{m}). \qquad (1)$$

Volume loss of material is determined by the ratio of mass difference of spherical ball material to the density of a material. The BoD Tribometer is shown in Fig. 1.

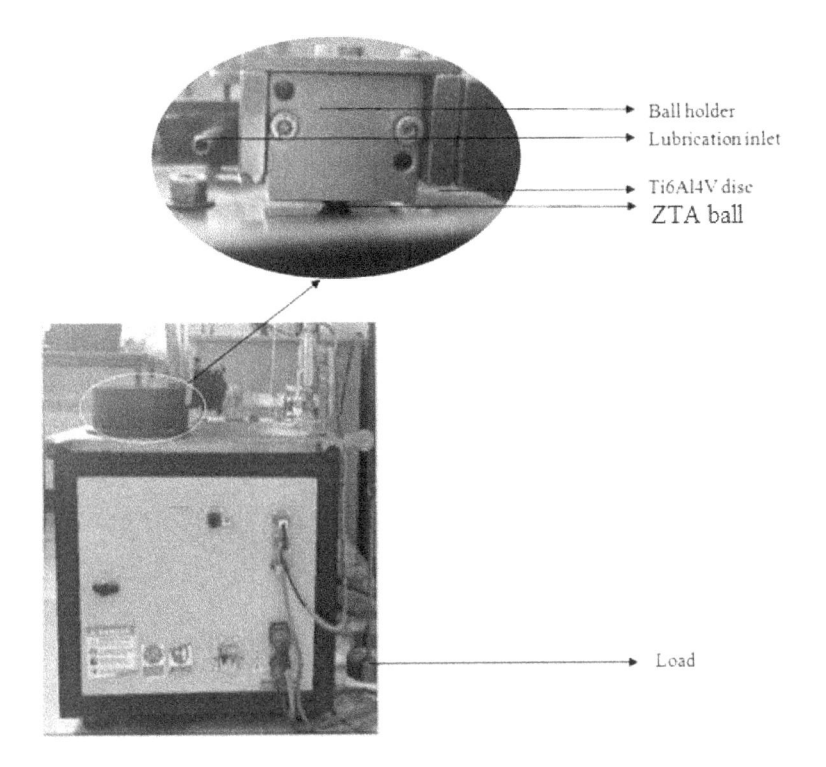

Fig. 1. Ball on Disc Tribometer (Color online).

3. Results and Discussion

3.1. *Friction behavior of zirconia and ZTA*

The coefficient of friction (CoF) of zirconia and ZTA obtained from this work is shown in Figs. 2 and 3. The CoF of zirconia ranges from maximum of 0.45 obtained for distilled water bio-lubricant under 15 N load to minimum of 0.38 for PBS against 25 N load. For ZTA, the CoF ranges from maximum of 0.44 to minimum of 0.34 obtained for saline solution bio-lubricant under 20 N and 15 N load. For sesame oil biolubricant, the zirconia showed CoF value in the range of 0.41–0.43. For distilled water, CoF ranges from 0.42 to 0.45. For PBS solution, the CoF values range from 0.44 to 0.38. For saline solution bio-lubricant, the CoF values range from 0.40 to

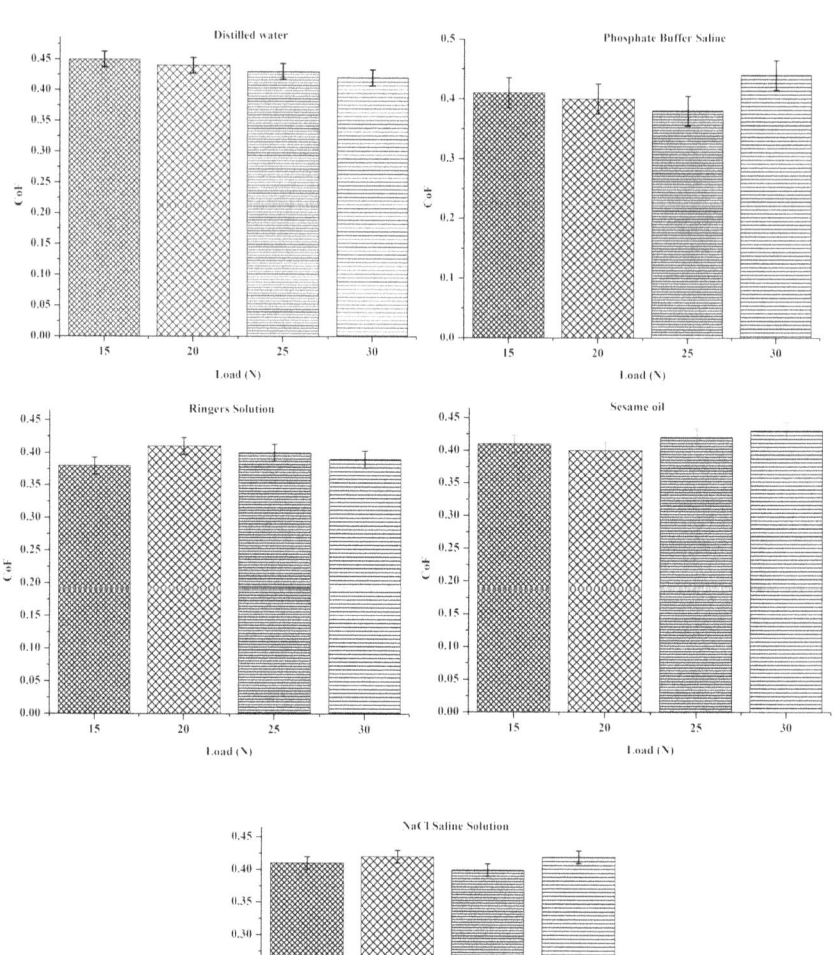

Fig. 2. CoF of Zirconia for different bio-lubricants.

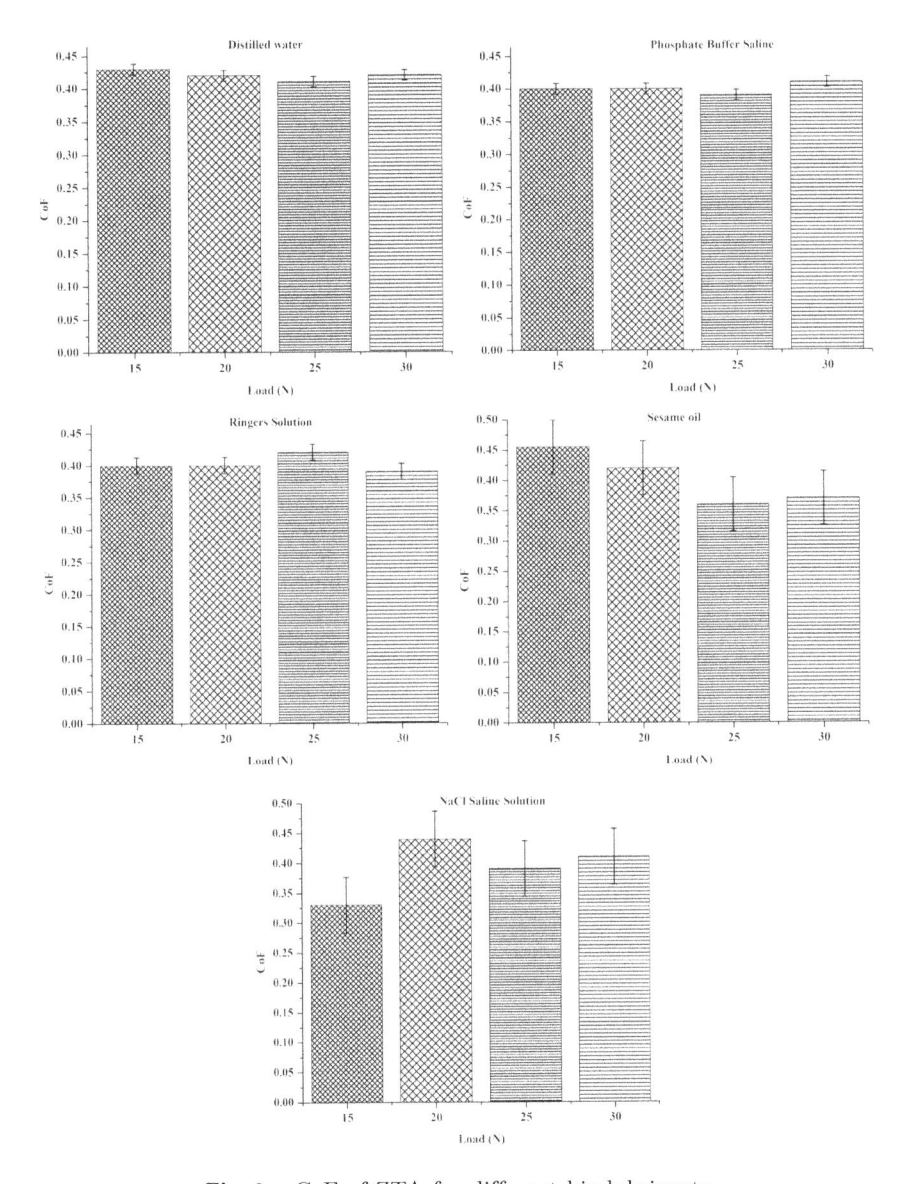

Fig. 3. CoF of ZTA for different bio-lubricants.

0.42. For ringer's solution bio-lubricant, CoF values range from 0.38 to 0.40. For ZTA, CoF values range from 0.36 to 0.44 for sesame oil bio-lubricant. For ringer's solution and PBS, CoF values range from 0.39 to 0.41. For saline solution bio-lubricant, CoF values range from 0.33 to 0.41. For distilled water bio-lubricant, CoF values range from 0.41 to 0.43.

For most of the bio-lubricants, the CoF of ZTA was found to be minimum for higher loads while zirconia showed maximum CoF values for remaining loads. The higher CoF values obtained for both zirconia and ZTA were mainly due to short test run of 2000 m. One of the recent studies showed CoF value of Si_3N_4-Ti6Al4V as 0.22 for sesame oil bio-lubricant while in which the test run was carried out for a sliding distance of 20 km against 20 N load.[11] For this study, 0.40 and 0.42 were the CoF values reported for zirconia and ZTA for the sliding distance of 2000 m. Almost an increase of 81.81% and 90.90% in CoF values was observed in this study when compared with previous study. The dry sliding of ZTA with various metal oxide additives was investigated for 1.8 km sliding distance under 50 N load and the results revealed that the CoF varied from 0.35 to 0.7 and ZTA composite with TiO_2 and MnO_2 achieved least CoF value.[13] The CoF of ZTA against alumina disc with 30 N load for a sliding distance of 2.5 km was investigated and the CoF values obtained were in the range of 0.54–0.57.[14] For this study, none of the biolubricants showed higher CoF values.

3.2. Wear rate behavior

The wear coefficient of zirconia ranges from maximum of $5.64 \times 10^{-5} mm^3/Nm$ for distilled water against 30 N load to minimum of $1.10 \times 10^{-6} mm^3/Nm$ for sesame oil against 15 N load. For ZTA, wear coefficient ranges from maximum of $5.66 \times 10^{-5} mm^3/Nm$ for ringer's solution bio-lubricant under 15 N load to minimum of $3.65 \times 10^{-6} mm^3/Nm$ for sesame oil bio-lubricant with 15 N load. For both ceramic materials, minimum wear coefficient values were observed for sesame oil bio-lubricant. The wear coefficient plot of zirconia and ZTA is shown in Figs. 4 and 5. For zirconia

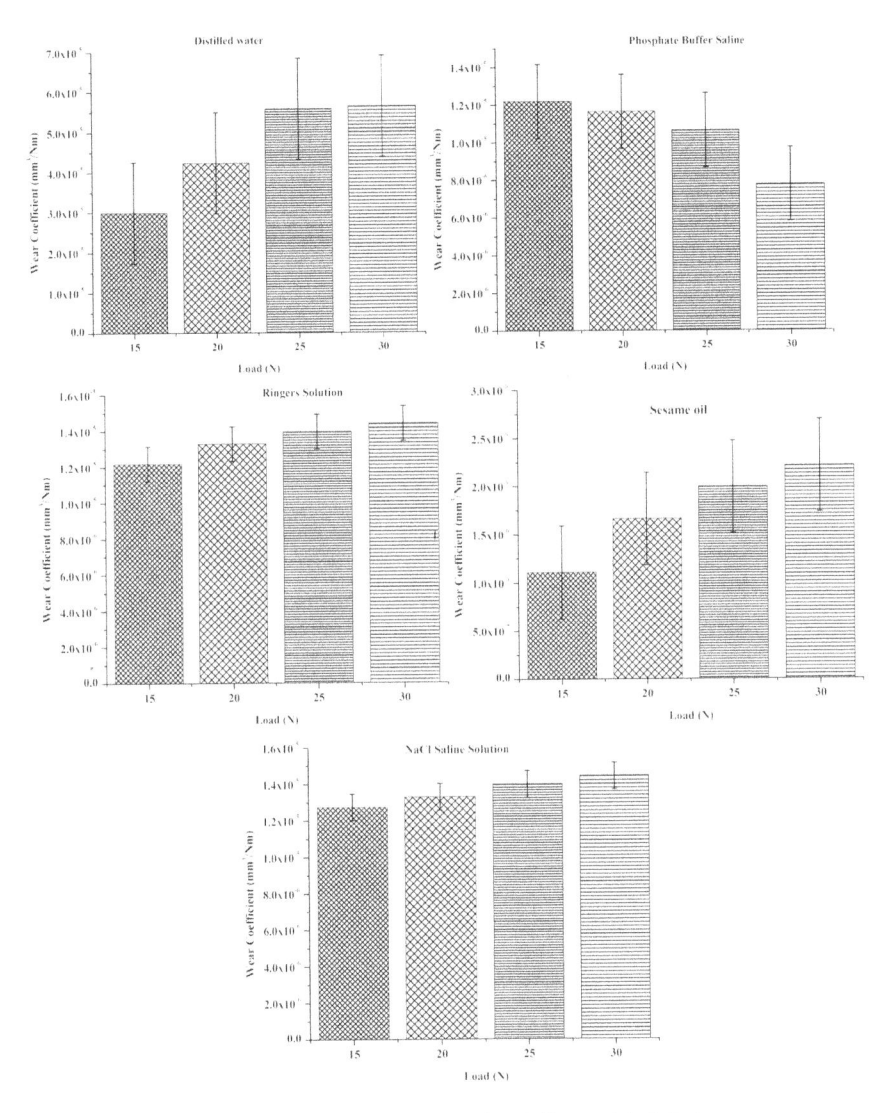

Fig. 4. Zirconia wear coefficient for different bio-lubricants.

ceramic, under sesame oil bio-lubricant, obtained wear coefficient values range from 1.10×10^{-6} mm^3/Nm to 2.21×10^{-6} mm^3/Nm. For distilled water bio-lubricant, wear coefficient values range from 2.99×10^{-5} mm^3/Nm to 5.64×10^{-5} mm^3/Nm. For PBS, wear coefficient values range from 7.75×10^{-6} mm^3/Nm to 1.21×10^{-5} mm^3/Nm.

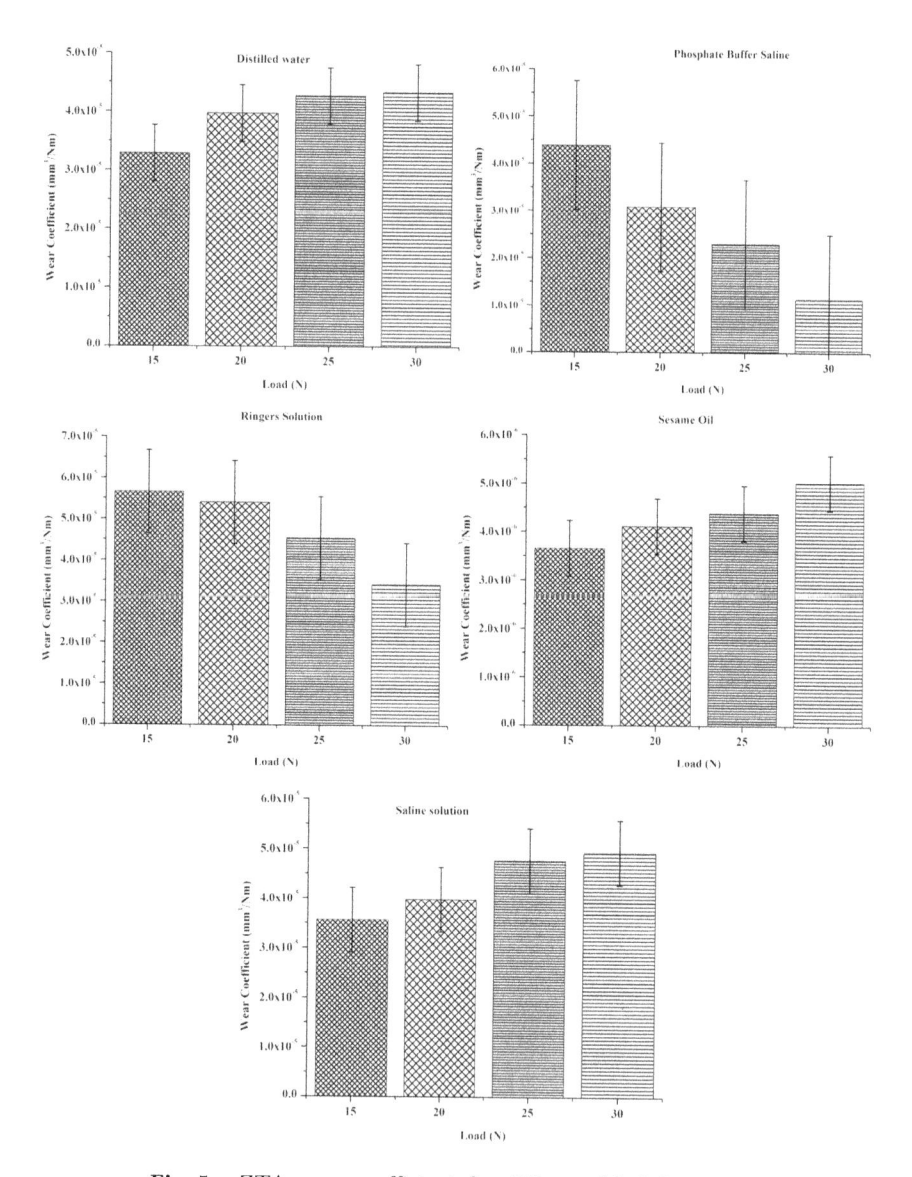

Fig. 5. ZTA wear coefficient for different bio-lubricants.

For saline solution bio-lubricant, wear coefficient values range from 1.27×10^{-5} mm^3/Nm to 1.43×10^{-5} mm^3/Nm. For ringer's solution bio-lubricant, wear coefficient values range from 1.21×10^{-5} mm^3/Nm to 1.43×10^{-5} mm^3/Nm. For ZTA ceramics, for sesame oil bio-lubricant, wear co-efficient values range from 5.02×10^{-6} mm^3/Nm to 3.65×10^{-6} mm^3/Nm. For ringer's solution biolubricant, wear coefficient values range from 5.66×10^{-5} mm^3/Nm to 3.42×10^{-5} mm^3/Nm. For PBS, wear coefficient values range from 4.38×10^{-5} mm^3/Nm to 1.14×10^{-5} mm^3/Nm. For saline solution bio-lubricant, wear coefficient values range from 4.91×10^{-5} mm^3/Nm to 3.56×10^{-5} mm^3/Nm. For distilled water bio-lubricant, wear coefficient values range from 4.33×10^{-5} mm^3/Nm to 3.28×10^{-5} mm^3/Nm.

For PBS bio-lubricant, both zirconia and ZTA showed reduced wear rate for higher loads. For Si_3N_4-Ti6Al4V combination, PBS bio-lubricant showed a minimum wear coefficient value of 3.33×10^{-6} mm^3/Nm. In this study, ZTA wear coefficient magnitude was quite closer to the range of Si_3N_4-Ti6Al4V combination, though the test run was conducted for 2 km.

Higher to lesser specific wear rate of 3.00×10^{-5} mm^3/Nm to 9×10^{-7} mm^3/Nm was observed for ZTA against alumina disc with different additives.[14] For this study, the specific wear rate obtained for the bio-lubricants was closer to that of the previous study as the counter surface disc for this study was Ti6Al4V. Another literature showed a wear rate in the range of 9.2×10^{-5} mm^3/Nm against SiC grit paper for 50 N load.[13] However, this study showed no such higher wear rate for these bio-lubricants.

3.3. *Wear morphology and surface behavior*

The wear behavior of zirconia and ZTA ceramics for maximum and minimum wear coefficient is shown in SEM micrograph (Figs. 6 and 7) for various biolubricants. From the SEM micrograph, it is quite clear that the wear track on the surface of zirconia could be seen clearly while ZTA surface had lesser wear track though the wear coefficient values obtained for ZTA were higher. This indicates that ZTA had high hardness and fracture toughness compared with zirconia as reported earlier.[7] Moreover, from the SEM image of zirconia, the worn out surface of the specimen indicates that it could not be used for higher loads as it might lead to fracture if implanted when humans undergo risky gait activities.

(a) Distilled water

(b) Sesame oil

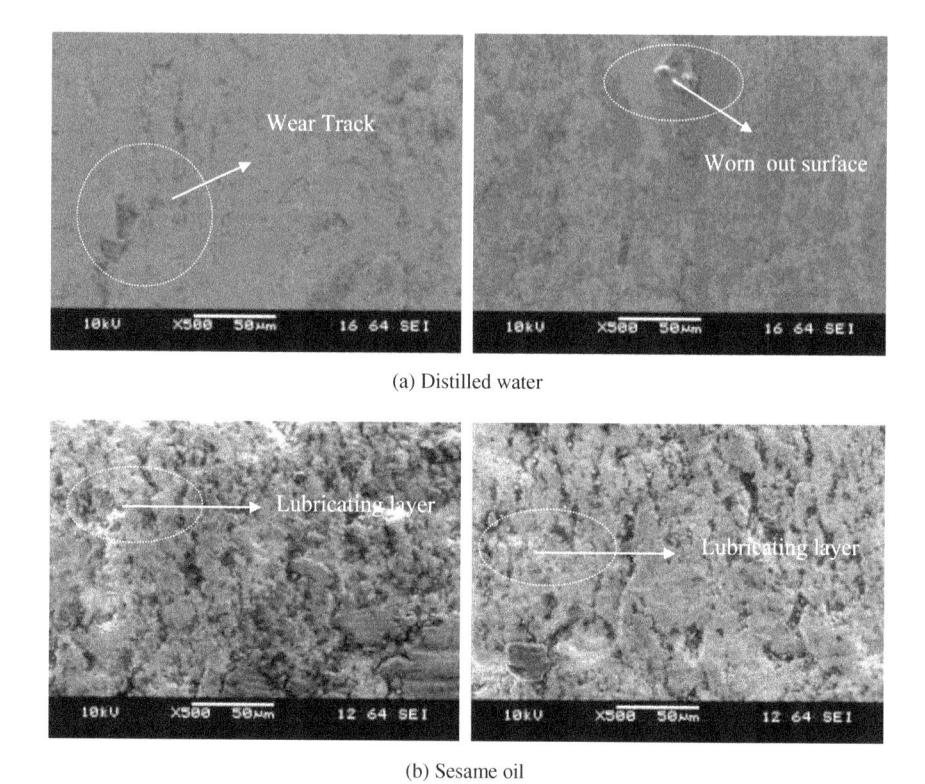

Fig. 6. Wear track of zirconia ball (Color online).

Though this study showed zirconia had better wear coefficient behavior in many bio-lubricants, it is quite evident from the SEM micrograph that lesser wear track in ZTA showed it had better hardness than zirconia. The lubricating layer formed on the surface of the biomaterials under sesame oil bio-lubricant showed better reduction in wear. The *in-vitro* study reported severe wear and fracture for zirconia used in hip joint replacements after few million cycles as reported.[22] The running-in and steady-state wear coefficients have a greater effect in estimating the wear of biomaterials.[23] The submodeling approach could be used to estimate the running-in wear using the obtained friction and wear coefficients.[21] Using these wear coefficients, different parameters like radial clearance,

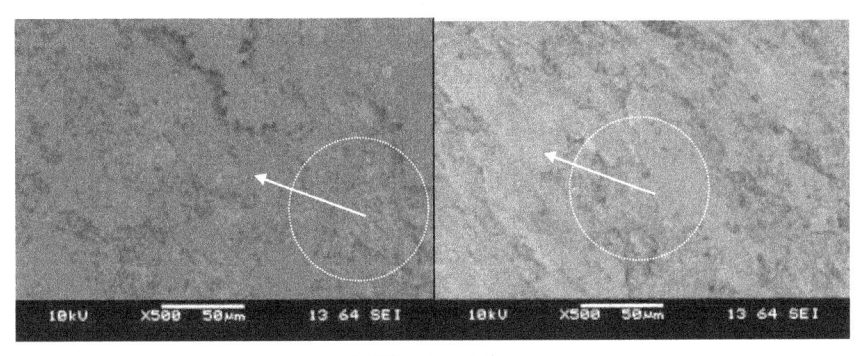

(a) Ringer's solution

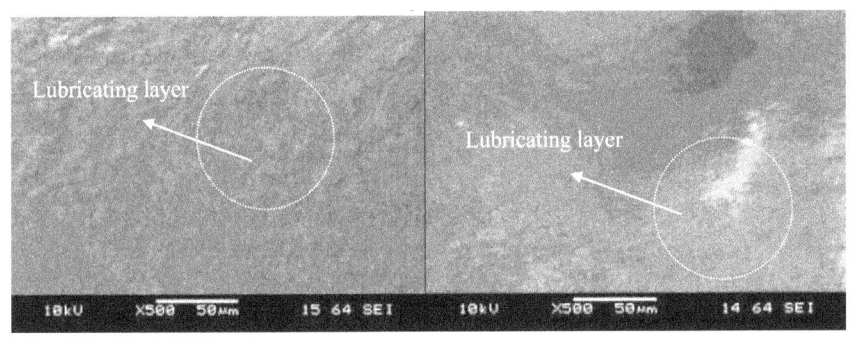

(b) Sesame oil

Fig. 7. Wear track of ZTA ball (Color online).

head diameter and micro-separation influencing wear rate could be investigated.

4. Conclusions

This study analyzed the tribological behavior of Zirconia and ZTA against titanium (Ti6Al4V) alloy for five different bio-lubricants under four loading conditions and the results revealed the following aspects:

- The maximum CoF and minimum CoF of Zirconia are observed for PBS bio-lubricant under 30 N and ringer's solution bio-lubricant under 15 N load.

- The maximum CoF and minimum CoF of ZTA are observed for sesame oil bio-lubricant under 15 N load and NaCl bio-lubricant under 15 N load.
- The maximum wear rate and minimum wear rate of Zirconia are observed for distilled water bio-lubricant under 30 N load and sesame oil bio-lubricant under 15 N load.
- The maximum wear rate and minimum wear rate of ZTA are observed for ringer's solution bio-lubricant under 15 N load and sesame oil bio-lubricant under 15 N load.
- Under high loads i.e. loads corresponding to risky gait activities ZTA showed lesser CoF values compared to zirconia.
- The surface worn out behavior of zirconia under higher load of 30 N clearly indicates that it could lead to fracture if implanted in humans when they undergo risky gait activities.

The zirconia toughened alumina (ZTA) could be used as an alternative for implants in humans who are frequently involved in risky gait activities as it exhibited better surface wear resistance characteristics. Overall, sesame oil showed better tribological behavior in terms of wear coefficient for both ZTA and zirconia ceramic biomaterials.

References

1. G. Shen, F. Fang and C. Kang, *Nanotechnol. Precis. Eng.* **1** (2018) 107.
2. S. Mojumder, S. Sikdar and S. K. Ghosh, *Indus. Lubric. Tribol* **69** (2017) 828.
3. S. B. Wang, S. R. Ge, H. T. Liu and X. L. Huang, *J. Biomimetics Biomater. Tissue Eng.* (Trans Tech Publ.) (2010).
4. J. Park, G. Pekkan and A. Ozturk, *Surf. Rev. Lett.* **16** (2009) 653.
5. I. Clarke, M. Manaka, D. Green, P. Williams, G. Pezzotti, Y.-H. Kim, M. Ries, N. Sugano, L. Sedel and C. Delauney, *JBJS* **85** (2003) 73.
6. S. M. Kurtz, S. Kocagöz, C. Arnholt, R. Huet, M. Ueno and W. L. Walter, *J. Mech. Behav. Biomed. Mater.* **31** (2014) 107.

7. A. De Aza, J. Chevalier, G. Fantozzi, M. Schehl and R. Torrecillas, *Biomaterials* **23** (2002) 937.
8. S. S. Brown and I. C. Clarke, *Tribology Trans.* **49** (2006) 72.
9. Y. Luo, L. Yang and M. Tian, *J. Bionic Eng.* **10** (2013) 84.
10. M. Guezmil, W. Bensalah and S. Mezlini, *Tribology Int.* **94** (2016) 550.
11. S. Shankar, R. Nithyaprakash, B. Santhosh, A. K. Gur and A. Pramanik, *Tribology Int.* **151** (2020) 106529.
12. S. Shankar, R. Nithyaprakash, P. Sugunesh, M. Uddin and A. Pramanik, *J. Bionic Eng.* **17** (2020) 1045.
13. A. K. Dey and K. Biswas, *Ceram. Int.* **35** (2009) 997.
14. A. K. Dey, S. Chatterjee and K. Biswas, *J. Mater. Eng. Perform.* **26** (2017) 6107.
15. H. Kamiya, M. Sakakibara, Y. Sakurai, G. Jimbo and S. Wada, *J. Am. Ceram. Soc.* **77** (1994) 666.
16. C. He, Y. Wang, J. Wallace and S. Hsu, *Wear* **162** (1993) 314.
17. S. Bejugama, S. Chameettachal, F. Pati and A. K. Pandey, *Ceram. Int.* **47** (2021) 17580.
18. S. Agrawal, N. K. Singh, R. K. Upadhyay, G. Singh, Y. Singh, S. Singh and C. I. Pruncu, *Materials* **14** (2021) 2965.
19. R. Nithyaprakash, S. Shankar and M. Uddin, *Med. Biol. Eng. Comput.* **56** (2018) 899.
20. S. Shankar and R. Nithyaprakash, *Proc. Inst. Mech. Eng. J J. Eng. Tribol.* **228** (2014) 717.
21. S. Shankar, R. Nithyaprakash, B. Santhosh, M. Uddin and A. Pramanik, *Comput. Methods Biomech. Biomed. Eng.* **23** (2020) 422.
22. T. D. Stewart, J. L. Tipper, G. Insley, R. M. Streicher, E. Ingham and J. Fisher, *J. Arthroplasty* **18** (2003) 726.
23. S. Shankar, R. Nithyaprakash, A. Sugunesh and M. Uddin, *Int. J. Surface Sci. Eng.* **14** (2020) 192.

Morphological, micro-mechanical, corrosion and *in vitro* bioactivity investigation of superficial layer formation during wire electrical discharge machining of Ti-6Al-4V alloy for biomedical application[*]

Sandeep Malik[†] and Vineet Kumar

Department of Mechanical Engineering, UIET,
Maharshi Dayanand University,
Rohtak, Haryana 124001, India
[†] *sandeep.uietmdu@gmail.com*

In this work, the experimental investigation of the surface integrity and biomechanical properties of the superficial layer obtained by wire electrical discharge machining (W-EDM) of Ti-6Al-4V alloy for biomedical application has been carried out. The surface morphology and elemental composition of the superficial layer have been investigated by field-emission scanning electron microscope (FE–SEM) and energy dispersive X-ray spectroscopy (EDS) techniques. The micro-mechanical behavior in terms of compressive strength and surface hardness was studied using the micro-pillar and nanoindentation technique. The corrosion resistance and *in vitro* bioactivity have been investigated using electrochemical and immersion test. Morphological analysis showed that surface morphology and superficial layer thickness were affected by peak current, pulse-duration and pulse-interval. The niobium (Nb)-rich layer was developed in superficial layer zone. The low peak current (3–6 A), low pulse-duration (5–10 μs) and high pulse-interval (> 45 μs) have been

---

[*]To cite this article, please refer to its earlier version published in Surface Review and Letters, Vol. 30, No. 1 (2023) 2141007 (11 pages) DOI: 10.1142/S0218625X21410079
[†]Corresponding author.

recommended for better surface morphology and thin superficial layer (ranging from 4–6 μm) free from surface defects. The micro-pillar and nano-indentation results showed that the superficial layer comprised of a brittle structure that improved the mechanical properties of the layer and the compressive strength was measured to be 1198 MPa. The corrosion resistance analysis revealed that the Nb-rich layer in the superficial layer improved the corrosion resistance and bioactivity. Excellent apatite growth has been found in the W-EDM-processed zone. The W-EDM can be used for the biomedical industry as a potential surface engineering technique.

Keywords: Ti-6Al-4V alloy; wire EDM; superficial layer; compressive strength; corrosion resistance; *in vitro* bioactivity.

1. Introduction

From the last few decades, electric discharge machining (EDM), a nonconventional machining process, has been the most widely used for processing hard-to-cut materials for die-making, aerospace, and automobile industries.[1–4] EDM has number of variants, like wire EDM (W-EDM),[5–7] powder-mixed EDM (PM-EDM),[8,9] rotary-tool EDM,[10] cryogenic-assisted EDM,[11] magnetic field-assisted EDM,[12] ultrasonic-assisted EDM[13] and ball-burnishing assisted EDM,[14–16] for processing functional materials for various industrial application. Recently, these EDM variants found application for processing and surface modification of biomaterials for medical applications, such as orthopedic accessories (plates and screws), stents and hip implants and knee implants.[6,9] In 1984, Rubeling and Kreilos reported the application of EDM in dentistry.[17] It has been reported that EDM can be used for drilling, shaping and deforming complex geometries of smooth surfaces with a high degree of accuracy. Sillard filed a patent on the application of EDM for developing a fixed removable dental implant system.[18] This technique is mostly utilized in denture cases for implants. Literature reported a short background on the EDM process and its use in dentistry and addressed the many benefits given by the EDM method. Since it is a thermal process, the EDM method is not impacted by metal hardness. EDM may be used without distortion for machining tiny, thin items. The adhesive nature of the workpiece is not altered, because the electrode does not touch the workpiece. EDM also offers a smooth burr-free surface with a precision of less than 0.0001.[19,20] Ntasi et al. investigated the corrosion potential of Co–Cr alloy and Ti-alloy after processing via EDM and compared it with the conventionally finished process. It has been reported that the corrosion resistance of Co–Cr alloy and Ti-alloy was reduced after processing by EDM.[21] Huang et al. reported the potential of ultrasonic vibration-assisted micro-EDM for developing micro-hole in nitinol alloy.[22] Theisen and Schuermann reported that EDM can be used as a surface modification technique for shape memory nitinol (SMA-NiTi) alloy.[23] A TiC-rich layer was synthesized on SMA-NiTi alloy that improved the biomechanical properties. Guo reported that the fatigue and corrosion performance of NiTi-based alloy depends upon the superficial layer generated by any process. It has been reported that the superficial layer comprised of brittle structure, controlled low surface roughness, and rich-layer of oxides and carbides, which improved the mechanical properties and resulted in enhancing the fatigue and corrosion performance of NiTi-based alloy. On the other hand, the conventionally machined surface comprised of high roughness and burrs that reduced the fatigue and corrosion performance.[24]

Peng et al. developed a nano-porous layer of TiO_2 and nano-γ-TiH-(γ-hydride)-phase on Ti-6Al-4V (Ti64) alloy using EDM process, which has been proved as biocompatible for the cell growth and osteointegration-type activities.[25] Furthermore, Yang et al. and Lee et al. confirmed that nano-porous TiO_2-layer was bioactive and promotes various biological activities of Mg-63 cell line on the modified surface.[26,27] However, Harcuba et al. examined the effect of surface roughness on cell attachment and fatigue performance.[28] It was observed that cell growth was increased with the increase in roughness and fatigue performance was reduced. In continuity, Strasky et al. reported that low surface roughness was beneficial for high fatigue performance and bioactivity.[29] Strasky et al. and Havlikova et al. used postheat treatments techniques, like shot blasting, chemical-etching and chemical milling to remove surface defects to improve the biological and fatigue performance.[30,31] Bin et al. developed a TiC-rich layer by EDM process that improved the tribological and mechanical properties.[32] Prakash et al. used the PM-EDM for the synthesis of nano-porous biocompatible-rich layer.[33] It has been reported that the layer developed by PM-EDM increased the surface hardness, corrosion resistance, and bioactivity of Ti-alloy. It is very important to optimize the process parameters of PM-EDM to generate a biocompatible layer and the NSGA-II algorithm has been used to determine a single optimal condition to obtain low roughness and high hardness.[34] Moreover, finite element simulation has been carried out to understand the phenomena of crater formation during the PM-EDM process and the bone-implant interface strength has been studied.[35,36] Prakash et al. reported that HA-rich can be easily deposited by the EDM process, which improved the biomechanical properties, corrosion resistance and in vitro bioactivity.[37,38]

From the literature, it can be seen that EDM/PM-EDM has the potential to develop biocompatible surfaces but the superficial layer comprises surface defects that may affect the performance of the implant. Most of the researchers studied surface roughness, hardness and cracks; there is an urgent need to investigate the micro-mechanical behavior of superficial layers, such as micro-structure, nano-indentation and micro-piler compression tests along with corrosion and in vitro bioactivity analysis. Moreover, very few studies are available on the micromechanical behavior of the superficial layer of implants processed by the W-EDM process. So, in this work, an attempt has been made to investigate the micro-structure, morphology, micro-mechanical performance, and corrosion resistance of superficial layer formed during W-EDM of Ti-Al6-V4 alloy for biomedical application.

2. Materials and Methods

In this section, details about the materials, experimentation, testing and characterization have been discussed.

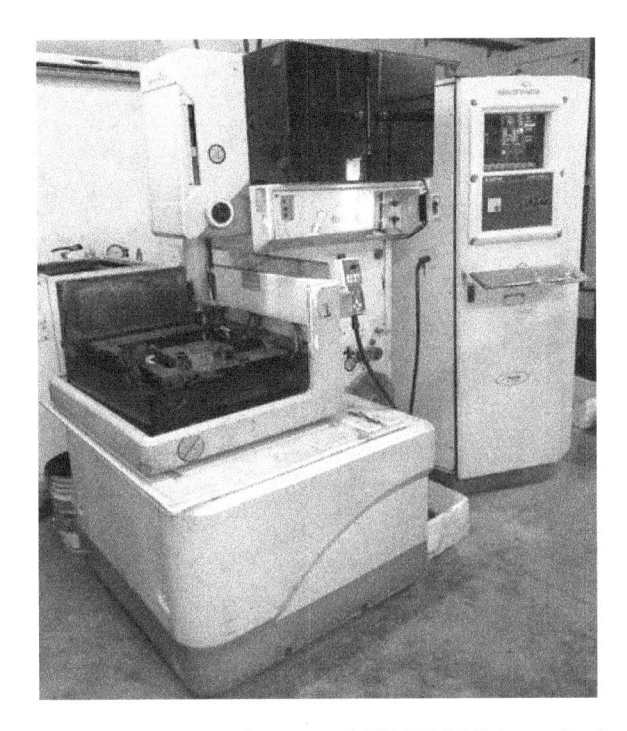

Fig. 1. Experimental set-up of W-EDM (Color online).

2.1. Experimentation

Commercially available Ti-6Al-4V (Ti64) alloy has been chosen as the substrate. The Ti64 alloy was used for preparing orthopedic accessories (plate and screws). The W-EDM (Model: EUROCUT MARK-2, Made: Electronica Machine Tools Ltd) has been utilized for processing Ti64 alloy. The brass wire-coated with niobium (Nb) of diameter 0.25 mm was used as electrode tools in W-EDM. The experimental set-up of W-EDM is shown in Fig. 1. The W-EDM process parameters are presented in Table 1 and chosen as per the literature.

2.2. Morphology, cryosection and mechanical characterization

The test specimens of dimension (Ø 15 mm with 10-mm height) were produced by W-EDM. Then, the specimens were mounted on

Table 1. Process parameters of W-EDM.

Process parameters	Symbol	Unit	Range
Peak current	I_p	A	3, 6, 9
Pulse-duration	T_{on}	µs	5, 10, 20
Pulse-interval	T_{off}	µs	15, 30, 45
Voltage	—	V	80
Speed	S	m/min	10
Wire tension	Wt	gf	1400

resin for detailed characterization and testing. The surface morphology and superficial layer were measured using field-emission scanning electron microscopy (FE-SEM, JEOL: 7600F) coupled with energy-dispersive X-ray spectroscopy (EDS) analysis. The surface roughness was measured in terms of average roughness (R_a) using a noncontact three-dimensional (3D) surface profilometer (Talysurf CCI lite). To check the thickness of the superficial layer, the cross-section of the process zone was selected and the surface was well polished using metallographic polishing. The grade series of emery paper 220, 600, 800, 1200 and 2000 were used to flatten the surface. After that, the surface was subjected to the diamond slurry for nano-finishing.

The micro-mechanical behavior of the superficial layer was investigated in terms of elastic modulus, nano-hardness and compressive strength. The elastic modulus and nano-hardness were measured using HyistronPI-88nano-indentation system with a diamond indenter using 10,000 µN load with 2s dwell time. The Oliver–Pharr method was used to plot the depth versus load graph.[39,40] The compressive strength of untreated and W-EDM-processed Ti64 alloy was investigated by micro-pillar testing analysis using a nano-indentation system. The 3-µm diameter pillar was developed in the superficial zone using nano-milling. The compressive force was applied using a diamond punch at 3 nm/s loading rate and $10^{-3}s^{-1}$ strain rate.

2.3. *In vitro corrosion and bioactivity analysis*

The corrosion resistance of W-EDM-processed Ti64 alloy was accessed using an electrochemical workstation in simulated body fluid (SBF) at room temperature condition (37 ± 1°C). The superficial layer chosen for corrosion analysis, graphite/carbon rod and Ag/AgCl were utilized as a working cathode, anode and reference electrode, respectively, as reported in the previous study.[39,40] The corrosion test has been performed by subjecting an electric potential from –0.8 V to 0.5 V at a scan rate of 1mV/s. The corrosion performance of the W-EDM-processed Ti64 alloy was determined in terms of parameters, such as corrosion potential (E_{corr}), current density (I_{corr}) and corrosion rate (R) using the Tafel extrapolation technique. The bioactivity of W-EDM-processed Ti64 alloy was accessed using the immersion method by measuring the apatite growth on the surface, as reported in previous studies.[39]

3. Results and Discussion

In this section, results and discussion on the top surface morphology, surface characteristics and properties have been presented.

3.1. *Surface morphology, cross-section, and roughness analysis*

Figure 2 shows the top-surface morphology and EDS spectrum of untreated and W-EDM-processed Ti64 alloy. The untreated surface of Ti64 alloy shows scratch marks of machining (Fig. 2(a)). The associated EDS spectrum confirmed the presence of Ti, V, Al and O elements (Fig. 2(b)). The surface morphology and elemental composition of Ti64 alloy have been altered after the W-EDM process. The W-EDM surface comprised of discharge craters and redeposited molten metal (Fig. 2(c)). The associated EDS spectrum confirmed the presence of Nb element along with base material elements (Ti, V, Al and O elements) (Fig. 2(d)). The peaks corresponding to Nb

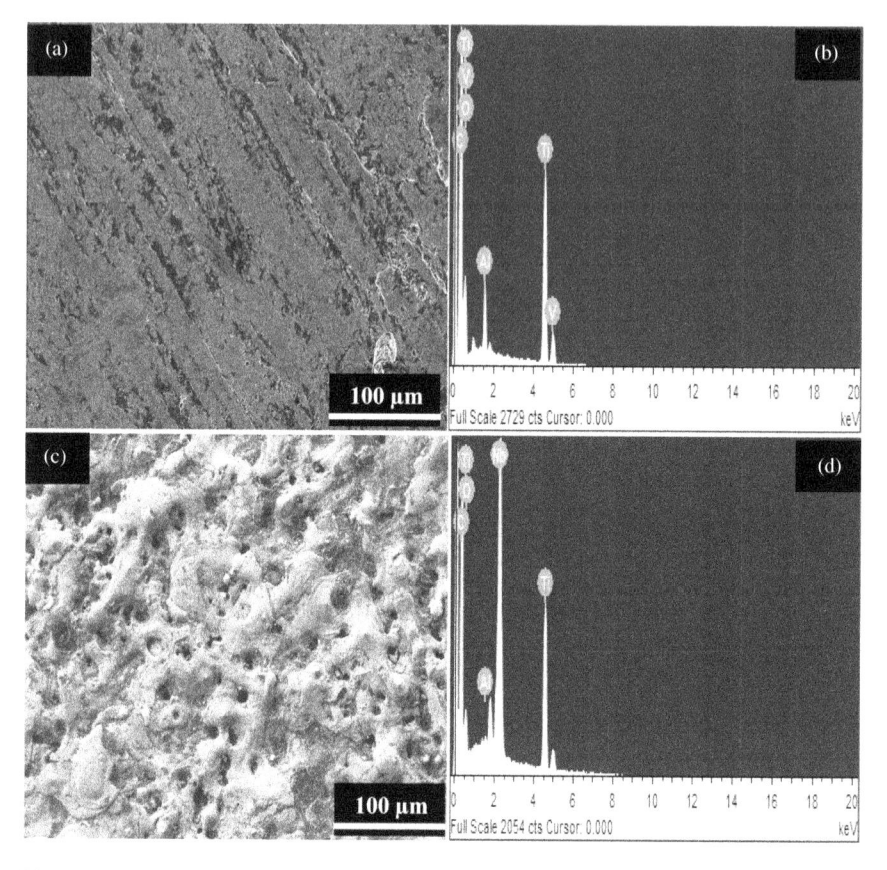

Fig. 2. Top-surface morphology and EDS spectrum of (a) and (b) untreated and (c) and (d) W-EDM processed Ti-6Al-4V alloy (Color online).

appeared because the Cu-wire was Nb coated, and the Nb element was transferred and deposited on the workpiece surface during the process. The surface morphology of the processed surface was dependent upon the W-EDM process parameters. The effect of peak current, pulse-on and pulse-interval on the surface morphology and thickness of the superficial layer was studied in detail. Figure 3 shows the effect of peak current, pulse-on and pulse-interval on the surface morphology and various surface features, such as discharge craters, micro-cracks, globules and ridges of re-deposited molten

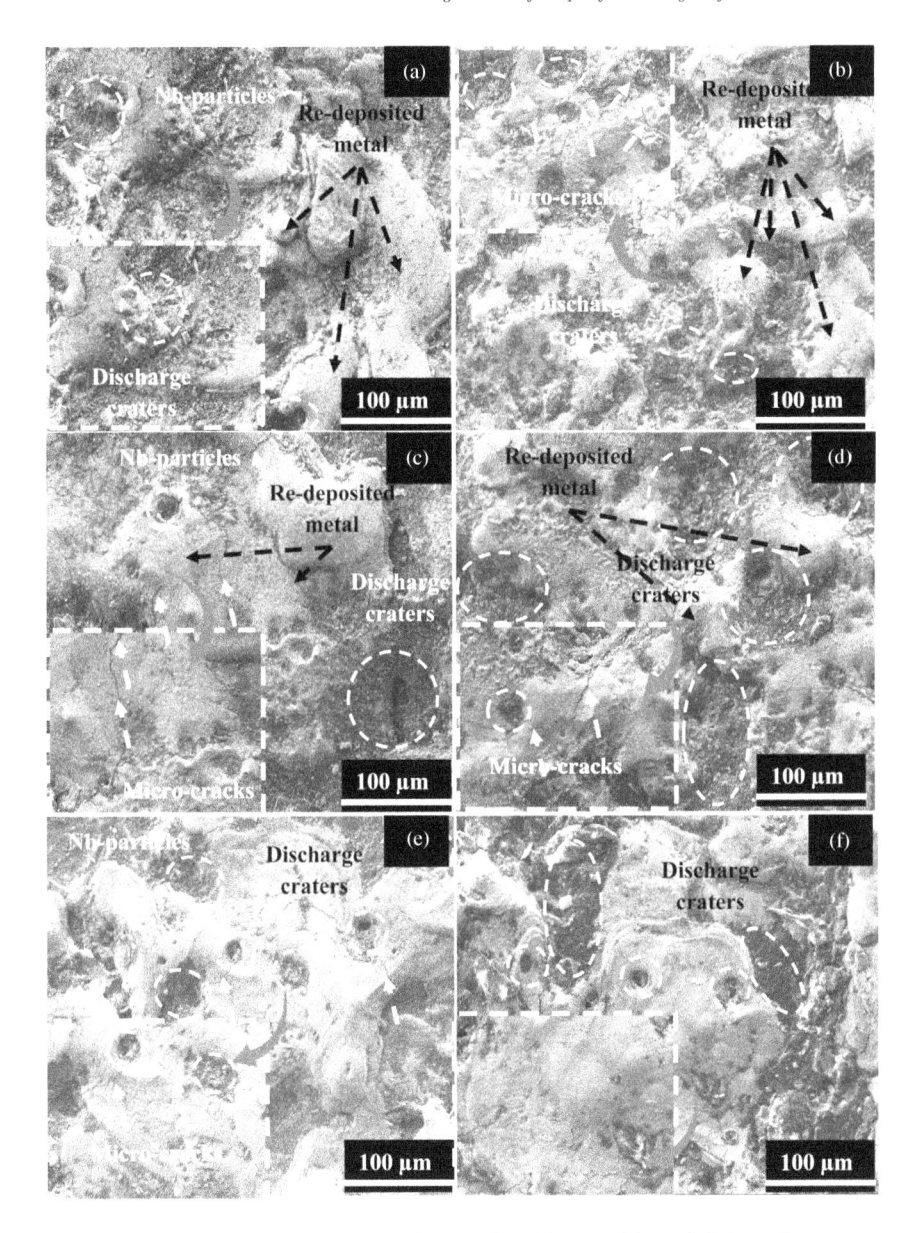

Investigation of superficial layer formation 101

Fig. 3. Top surface morphology of superficial layer: (a) and (b) peak current 3 and 9 A, (c) and (d) pulse-on 5 and 20 µs, and (e) and (f) pulse-interval 15 and 45 µs (Color online).

metal can be seen. Figures 3(a) and 3(b) show the morphology (at different magnification levels) of the machined surface after the W-EDM process at different levels of peak current keeping pulse-duration and pulse-interval constant. At 3-A peak current, the surface comprised of a small number of discharge craters and deposited Nb-particles, which is less defective (Fig. 3(a)). At large magnification (1800×), the deposition of Nb particles can be seen, which confirmed the deposition of Nb and is expected to improve the biomechanical characteristics and properties required for biomedical applications.[6] As the peak current increases, the discharge energy generated by electrical sparks was increased, as a result, more thermal energy sinks into the workpiece surface; thus, more materials removed from the workpiece surface, and various surface defects developed.[8] At 9-A peak current, the surface comprised of a large number of discharge craters and deposited Nb particles, which were less defective (Fig. 3(b)). At large magnification (1800×), deep and wide craters can be seen which was due to high thermal energy generation between wire and Ti64 alloy. Figures 3(c) and 3(d) show the morphology (at different magnification levels) of the machined surface after the W-EDM process at different levels of pulse-duration, keeping peak current and pulse-interval constant. The discharge energy is also a function of pulse-duration, which also affects the surface morphology of the machined surface. At 5-μs pulse-duration, the surface was less defective and comprised of discharge craters, micro-cracks and the deposited Nb particles (Fig. 3(c)). At large magnification (1800×), micro-cracks can be seen. This is because, at low pulse-duration, the discharge energy is low which removed material from the workpiece surface in the form of micro-scale debris.[8,33] As the pulse-duration increases, the radius and depth of discharge energy increase, which develops large-sized craters and micro-cacks on the surface. Figure 3(d) shows the machined surface at 20-μs pulse-duration and from the FE-SEM high ridges of re-deposited metal and micro-cracks can be seen on the surface. At large magnification (1800×), deep and wide craters along with micro-cracks can be seen which was due to high thermal energy generation between the wire and Ti64 alloy. Figures 3(e) and 3(f) show the morphology (at different magnification levels) of the machined surface after the W-EDM process at different levels of pulse-interval keeping peak current and pulse-duration constant. The high pulse-interval provides sufficient time to settle the discharge products and surface characteristics improved. At low pulse-interval (15 μs), the surface morphology was uneven (Fig. 3(e)). At large magnification (1800×), it can be seen that the surface is comparatively flat but microcracks can be seen. However, at high pulse-interval (45μs), the machined surface is flat and comprises fewer defects (Fig. 3(f)). At large magnification (1800×), a flat surface can be seen. From these observations, it can be concluded that low peak current (3–6 A), low pulse-duration (5–10 μs), and high pulse-interval (>45 μs) have been recommended for better surface morphology.

Figure 4 shows the cross-section morphology of the superficial layer (at different magnification levels) of the machined surface after the W-EDM process at different levels of peak current keeping pulse-duration and pulse-interval constant. From the cross-section images, it can be seen that the superficial layer comprises a brittle structure in comparison with the base surface. Moreover, microcracks can be seen in all the results. This is because, during W-EDM process, rapid heating and quenching take place simultaneously, which is responsible for the brittle structure formation in the superficial layer. This improved the mechanical properties of the layer. The superficial layer ranges from 4–12 μm, which depends upon various process parameters. Figures 4(a) and 4(b) show the cross-section morphology of the superficial layer of the machined surface after the W-EDM process at different levels of peak current keeping pulse-duration and pulse-interval constant. At 3-A peak current, the layer thickness was measured around 8–10 μm. The micro-cracks were also measured in the range of 3–5 μm in length (Fig. 4(a)). As the peak current increases, the layer thickness increases. This is because, with the increase in peak-current, the energy generated by electrical sparks was increased, as a result,

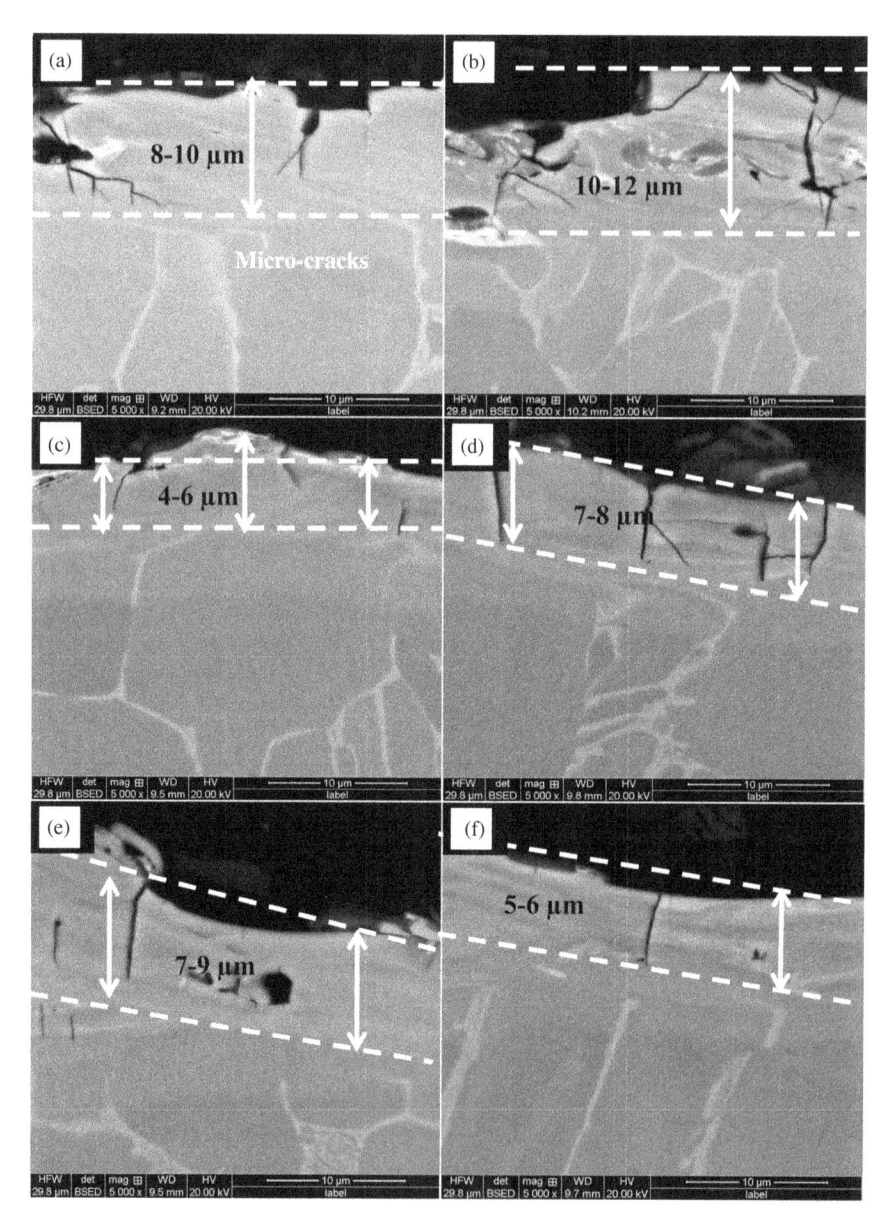

Fig. 4. Cross-section morphology of superficial layer: (a) and (b) peak current 3 and 9 A, (c) and (d) pulse-on 5 and 20 μs, and (e) and (f) pulse-interval 15 and 45 μs (Color online).

a large amount of molten pool was generated, and owing to quenching action, the same was re-deposited on the workpiece surface causing the increase in the thickness of the superficial layer. Moreover, the intensity and scale of micro-cracks also increased. At 9-A peak current, the layer thickness was measured around 10–12 μm. The micro-cracks were also measured in the range of 8–10 μm in length (Fig. 4(b)). Figures 4(c) and 4(d) show the cross-section morphology of the superficial layer of the machined surface after the W-EDM process at different levels of pulse-duration keeping peak current and pulse-interval constant. At 5-μs pulse-duration, the layer thickness was measured around 4–6 μm. The micro-cracks were also measured in the range of 2–4 μm in length (Fig. 4(c)). The layer thickness increases with the increase in pulse-duration. This is because, the discharge energy is also a function of pulse-duration, as a result, more melting and re-deposition of the molten pool. At 20-μs pulse-duration, the layer thickness was measured around 7–8 μm. The micro-cracks were also measured in the range of 7–8 μm in length (Fig. 4(d)). Figures 4(e) and 4(f) show the cross-section morphology of the superficial layer of the machined surface after the W-EDM process at different levels of pulse-interval keeping peak current and pulse-interval constant. At 15-μs pulse-interval, the layer thickness was measured around 7–9 μm. The micro-cracks were also measured in the range of 5–6 μm in length (Fig. 4(e)). This is because, at low pulse-interval, the discharge did not get sufficient time to flush out, as a result before subsequent spark large amount of molten pool deposited on the workpiece surface. Apart from the surface micro-cracks, micro-scale porosities were also identified. At 45-μs pulse-interval, the layer thickness was measured around 5–6 μm. The micro-cracks were also measured in the range of 4–5 μm in length (Fig. 4(f)). This is because, at high pulse-interval, the discharge gets sufficient time to flush out, as a result before the subsequent spark, less amount of molten pool deposited on the workpiece surface. For the thin layer, low peak current (3–6 A), low pulse-duration (5–10 μs), and high pulse-interval (>45 μs) have been recommended.

3.2. Micro-mechanical behavior analysis

Figure 5 shows the engineering stress–strain curves of the recast layer and untreated base Ti64 alloy subjected to *in situ* micro-pillar compression. From the graph, it can be seen that the untreated Ti64 alloy exhibits 870 MPa compressive strength. However, the compressive strength of the superficial zone (recast layer) of the W-EDM-processed surface exhibits 1198 MPa. The formation of brittle structure and deposition of Nb-element was assumed for the improvement in mechanical properties of the modified surface by the W-EDM process. From the micro-pillar compression test, it has been revealed that un-treated Ti64 alloy failed within 3% strain, however, the W-EDM-processed sample failed within 7–8% strain. The hardness of the untreated Ti64 alloy and W-EDM-processed

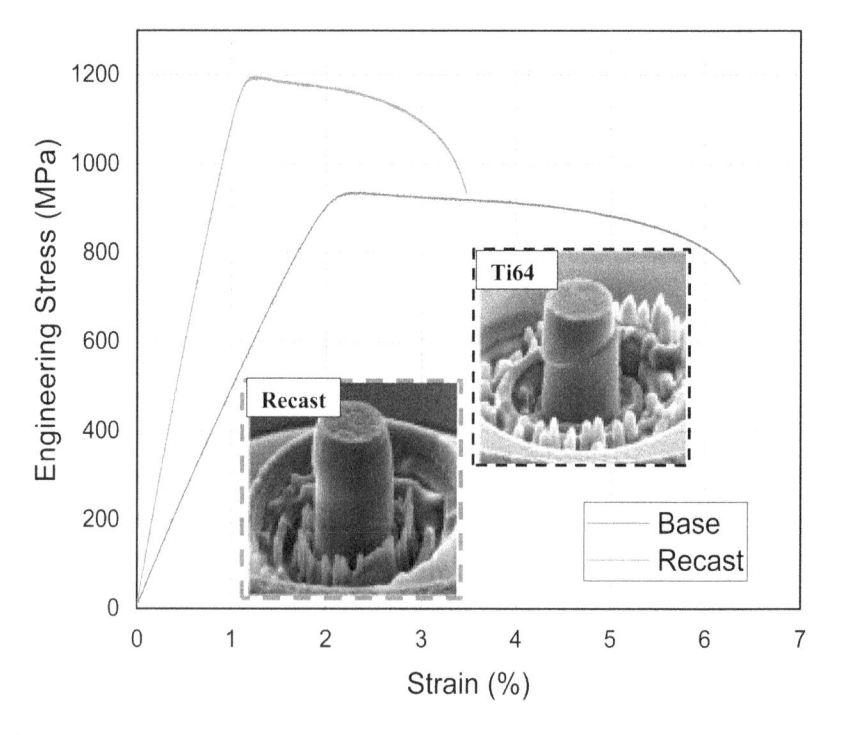

Fig. 5. Engineering stress-strain curves of recast layer and untreated base. Ti-6Al-4V alloy subjected to *in situ* micro-pillar compression (Color online).

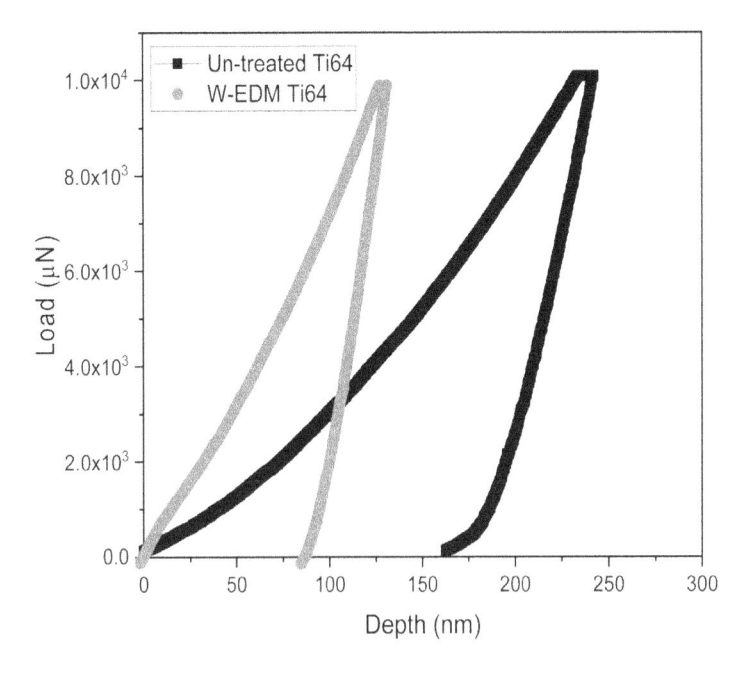

Fig. 6. Typical load–displacement curves on recast layer and base-Ti-6Al-4V alloy (Color online).

superficial zone has been determined by the nano-indentation technique. W-EDM-processed superficial zone had high surface hardness (745 HV) as compared to untreated Ti64 alloy. The indentation depth was high (250 nm) in the case of untreated Ti64 alloy (380 HV), whereas in the case of W-EDM-processed superficial zone, the indentation depth was high (125 nm).

3.3. *In vitro corrosion and bioactivity analysis*

Figure 7 shows the Tafel exploration curve for the untreated and W-EDM-processed Ti64 alloy. The superficial zone was chosen as an exposed area to SBF medium (corrosion environment). The corrosion current density (I_{corr}) and corrosion potential (E_{corr}) of untreated and W-EDM-processed Ti64 alloy were 6.28 μA/cm^2 and –270 mV, and 8.15 μA/cm^2 and –365 mV, respectively. It has been reported that the surface holding the low value of I_{corr} and high E_{corr}

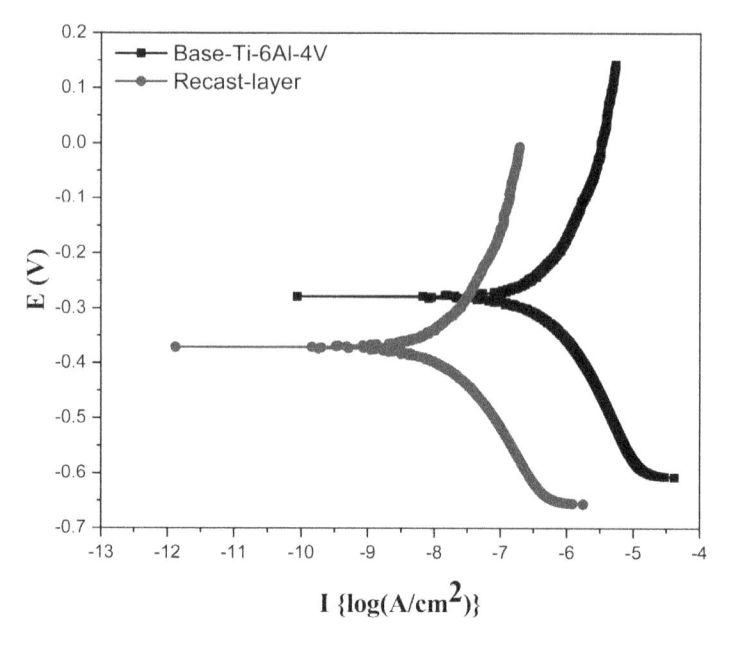

Fig. 7. Tafel exploration curve of base-Ti-6Al-4V alloy and recast layer obtained by W-EDM (Color online).

offered high corrosion resistance.[33,39,40] From the I_{corr} and E_{corr} observations, the untreated Ti64 alloy surface possessed the least corrosion resistance. This is because the untreated Ti64 alloy comprised of Al and V elements, which were released in the corrosive medium. This causes a high degree of micro-cracks and deep and wide craters, which acted as an active site for pitting corrosion. This is because the Ti64 alloy exhibits a very thin layer of TiO_2, which failed under corrosion environment and promote pitting corrosion. This results in severe corrosion on the surface and failed the specimen. Figure 8 shows the morphology and EDS spectrum of corroded samples. Figure 8(a) shows the formation of pits and shredded regions and apatite growth. The corresponding EDS spectrum shows the presence of Ca, P and O elements along with base substrate elements (Ti, Al and V) confirmed the growth of apatite layer (Fig. 8(b)). It has been reported in previous studies that Ca, P and O element's presence was attributed to the apatite layer

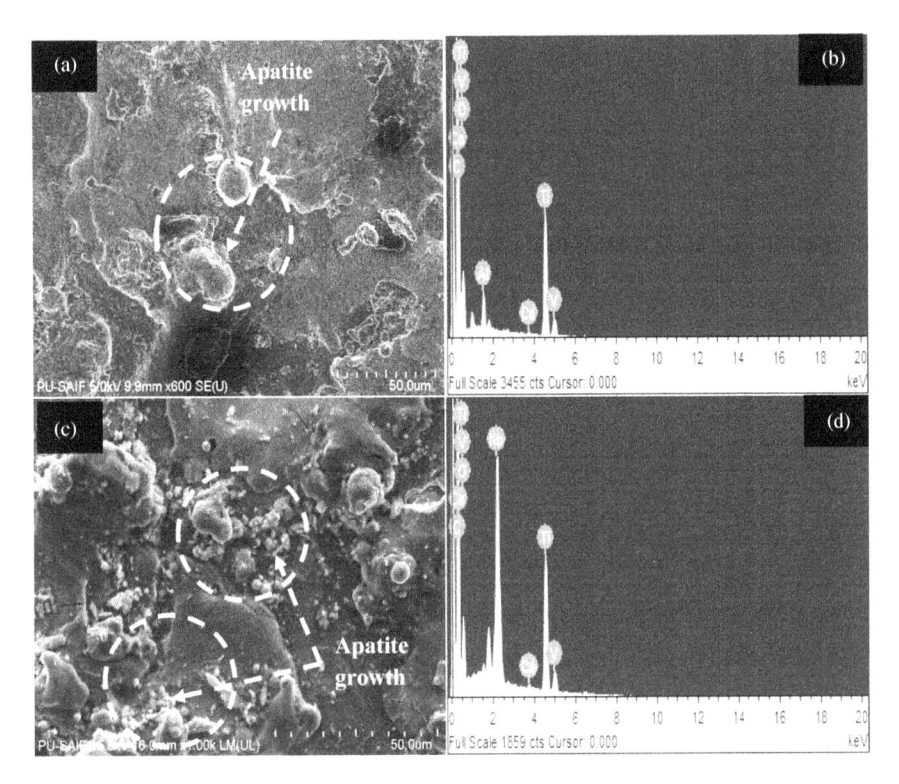

Fig. 8. Bioactivity analysis using apatite growth and EDS analysis: (a) and (b) base Ti-6Al-4V alloy; (c) and (d) recast layer obtained by W-EDM (Color online).

growth. The results are corroborative with previous research works.[34,37] On the other hand, the W-EDM-processed surface offered higher corrosion resistance as compared to the untreated surface. The Tafel exploration curve re-positioned toward the left side and offered a low value of I_{corr} as compared to the untreated specimens. This is owing to the presence of the Nb-rich layer on the superficial layer of W-EDM-processed samples, which helps to develop an apatitelike layer that offers a barrier to corrosion attack. Figure 8(c) shows the formation of the apatite layer in the craters. The rough surface of the superficial layer of W-EDM-processed samples and Nb element promotes the growth of apatite at a faster pace compared to the untreated Ti64 alloy. The corresponding EDS spectrum shows the presence of Nb, Ca, P and O elements along with base substrate elements (Ti, Al and V), which confirmed the growth of apatite layer (Fig. 8(d)). The peaks intensity of Ca, P and O elements was very high as compared to the apatite growth sample of untreated Ti64 alloy. The W-EDM-processed surface exhibits excellent bioactivity and higher corrosion resistance despite surface defects, which is due to the brittle structure and Nb element presence.

4. Conclusion

In this work, the morphological, micro-mechanical, corrosion, and *in vitro* bioactivity behavior of superficial layer formed by W-EDM of Ti-Al6-V4 alloy for biomedical application has been investigated. The low peak current (3–6 A), low pulse-duration (5–10 µs), and high pulse-interval (>45 µs) have been recommended for better surface morphology and thin superficial layer (ranging from 4–6 µm) free from surface defects. The superficial layer comprised of a brittle structure that improved the mechanical properties of the layer. The superficial zone (recast layer) of the W-EDM-processed surface exhibits a compressive strength 1198 MPa. The Nb-rich layer in the superficial layer of W-EDM-processed samples improved the corrosion resistance and bioactivity. Excellent apatite growth has been found in the W-EDM-processed zone. In the light of all observations, it can be concluded that W-EDM-modified specimens exhibit excellent biomechanical integrity, which could be helpful for the bone-ingrowth process in the host body and recommended for the biomedical industry as a potential surface engineering technique.

References

1. N. M. Abbas, D. G. Solomon and M. F. Bahari, *Int. J. Mach. Tools Manuf.* **47** (2007) 1214.
2. U. S. Yadav and V. Yadava, *Proc. Inst. Mech. Eng. B, J. Eng. Manuf.* **229** (2015) 1764.

3. E. Uhlmann, S. Piltz and U. Doll, *J. Mater. Process. Technol.* **167** (2005) 488.
4. C. Prakash, S. Singh, C. I. Pruncu, V. Mishra, G. Królczyk, D. Y. Pimenov and A. Pramanik, *Materials* **12** (2019) 1006.
5. N. K. Gupta, N. Somani, C. Prakash, R. Singh, A. S. Walia, C. Singh and C. I. Pruncu, *Materials* **14** (2021) 2292.
6. C. Prakash, H. K. Kansal, B. S. Pabla, S. Puri and A. Aggarwal, *Proc. Inst. Mech. Eng. B, J. Eng. Manuf.* **230** (2016) 331.
7. K. H. Ho and S. T. Newman, *Int. J. Mach. Tools Manuf.* **43** (2003) 1287.
8. C. Prakash, H. K. Kansal, B. S. Pabla and S. Puri, *Mater. Manuf. Process.* **32** (2017) 274.
9. A. A. A. Aliyu, A. M. Abdul-Rani, T. L. Ginta, C. Prakash, E. Axinte, M. A. Razak and S. Ali, *Adv. Mater. Sci. Eng.* **2017** (2017) 8723239.
10. R. Teimouri and H. Baseri, *Int. J. Adv. Manuf. Technol.* **67** (2013) 1371.
11. S. V. Kumar and M. P. Kumar, *J. Manuf. Process.* **20** (2015) 70.
12. R. Teimouri and H. Baseri, *Appl. Soft Comput.* **14** (2014) 381.
13. M. R. Shabgard and H. Alenabi, *Mater. Manuf. Process.* **30** (2015) 991.
14. C. Prakash, R. Wandra, S. Singh, A. Pramanik, A. Basak, A. Aggarwal and N. Yadaiah, *Mater. Lett.* **301** (2021) 130282.
15. R. Wandra, C. Prakash and S. Singh, *Mater. Today, Proc.* (2021).
16. R. Wandra, C. Prakash and S. Singh, *Mater. Today, Proc.* (2021).
17. G. Rubeling and H. A. Kreilos, *Quintessence Dental Technol.* **8** (1984) 649.
18. R. Sillard, Fixed removable dental implant system, US patent: US4931016A (1990).
19. R. Sillard, *Implant Soc.* **3** (1992) 13.
20. E. F. R. Contreras, G. E. P. Henriques, S. R. Giolo and M. A. A. Nobilo, *The Journal of Prosthetic Dentistry* **88** (1992) 467.
21. A. Ntasia, W. D. Muellerb, G. Eliadesa and S. Zinelis, *Dental Mater.* **26** (2010) 237.
22. H. Huang, H. Zhang, L. Zhou and H. Y. Zheng, *J. Micromech. Microeng.* **13** (2003) 693.
23. W. Theisen and A. Schuermann, *Mater. Sci. Eng. A* **378** (2004) 200.

24. Y. Guo, Electrical discharge machining of biomedical nitinol alloys and the resulting fundamental relationship of microstructure-property-function, https://www.nsf.gov/awardsearch/showAward?AWDJD=1234696.
25. P. W. Peng, K. L. Ou, H. C. Lin, Y. N. Pan and C. H. Wang, *J. Alloys Compd.* **492** (2010) 625.
26. T. S. Yang, M. S. Huang, M. S. Wang, M. H. Lin, M. Y. Tsai and P. Y. W. Wang, *Implant Dent.* **22** (2013) 374.
27. W. F. Lee, T. S. Yang, Y. C. Wu and P. W. Peng, *J. Exp. Clin. Med.* **5** (2013) 92.
28. P. Harcuba, L. Bacakova, J. Strasky, M. Bacakova, K. Novotna and M. Janecek, *J. Mech. Behav. Biomed. Mater.* **7** (2012) 96.
29. J. Strasky, M. Janecek, P. Harcuba, M. Bukovina and L. Wagner, *J. Mech. Behav. Biomed. Mater.* **4** (2011) 1955.
30. J. Strasky, J. Havlikova, L. Bacakova, P. Harcuba, M. Mhaede and M. Janecek, *Appl. Surf. Sci.* **281** (2013) 73.
31. J. Havlikova, J. Strasky, M. Vandrovcova, P. Harcuba, M. Mhaede, M. Janecek and L. Bacakova, *Mater. Sci. Eng. C, Mater. Biol. Appl.* **39** (2014) 371.
32. T. C. Bin, L. D. Xin, W. Zhan and G. Yang, *Appl. Surf. Sci.* **257** (2011) 6364.
33. C. Prakash, H. K. Kansal, B. S. Pabla and S. Puri, *J. Mater. Eng. Perform.* **24** (2015) 3622.
34. C. Prakash, H. K. Kansal, B. S. Pabla and S. Puri, *J. Mech. Sci. Technol.* **30** (2016) 4195.
35. C. Prakash, H. K. Kansal, B. S. Pabla and S. Puri, *Nanosci. Nanotechnol. Lett.* **8** (2016) 815.
36. C. Prakash, H. K. Kansal, B. S. Pabla and S. Puri, *J. Comput. Inf. Sci. Eng.* **14** (2015) 1.
37. C. Prakash and M. S. Uddin, *Surf. Coat. Technol.* **326** (2017) 134.
38. C. Prakash, S. Singh, L. Y. Wu, H. Y. Zheng and G. Królczyk, *J. Braz. Soc. Mech. Sci. Eng.* **43** (2021) 1.
39. H. Singh, C. Prakash and S. Singh, *J. Bionic Eng.* **17** (2020) 1029.
40. S. Singh, C. Prakash and H. Singh, *Surf. Coat. Technol.* **398** (2020) 126072.

Preparation and characterization of Sr-doped HAp biomedical coatings on polydopamine-treated Ti6Al4V substrates*

Gurmohan Singh[†,¶], Abhineet Saini[‡] and B. S. Pabla[§]

[†]*Chitkara University School of Engineering and Technology, Chitkara University, Himachal Pradesh, India*
[‡]*Chitkara University Institute of Engineering and Technology, Chitkara University, Punjab, India*
[§]*Department of Mechanical Engineering, National Institute of Technical Teacher Training and Research, Chandigarh, India*
[¶]*gurmohan.singh@chitkarauniversity.edu.in*

Ti6Al4V alloy of titanium is a significant biomaterial due to its biocompatible nature, but it lacks required bioactivity to make it mimic properties to a human bone. Thus, hydroxyl-apatite (HAp), an inorganic compound found in human bones, is generally coated onto Ti6Al4V substrates to improve their bio-characteristics. But, HAp itself lacks certain bio-functionalities, such as allowing tissue bone regeneration and poor binding to the Ti6Al4V substrate, which results in osteoporosis and reduced bioactivity of the bio-implant, respectively. The proposed way out for this is the further doping of HAp with Strontium (Sr) for enabling tissue bone regeneration as well as addition of Polydopamine (PDA) for improved adhesion of HAp-based coatings with the substrate. Moreover, PDA results in increased drug delivery area and thus can be used as a material for enhancing resistance to

---

*To cite this article, please refer to its earlier version published in Surface Review and Letters, Vol. 30, No. 1 (2023) 2141009 (7 pages) DOI: 10.1142/S0218625X21410092
*Corresponding author.

bacterial growth. The present study demonstrates an experimental work on deposition of HAp, HAp with PDA and HAp with PDA and Sr coatings deposited onto Ti6Al4V alloy by means of biomimetic coating technique. Initially the pure HAp coatings were deposited using 10 SBF (simulated body fluid) solution and optimized in terms of time duration for desired coating uniformity. Then, for the optimized coating duration, the PDA pretreated Ti6Al4V substrates were coated, utilizing HAp, and Sr (at two different compositions) combinations were deposited through modified 10 SBF solutions. The characterization involving microstructural analysis and phase detection was performed for all these coatings using Scanned Electron Microscopy (SEM), Energy Dispersive Spectroscopy (EDS) and X-Ray Diffraction (XRD) of the coated substrates and adhesion strength was calculated using a standard pull out adhesion test ISO 10779–4. The study showed an effective and comparatively cheap method of depositing organic coatings using biomimetic technique to obtain improved bio-functionalities in metallic implants at low temperatures.

Keywords: Titanium alloy; HAp; polydopamine; Strontium; bioimplants; biomimetic coating, composite coatings.

1. Introduction

Titanium alloys, most commonly Ti6Al4V, have gained significant attention as suitable material for fabrication of bioimplants, owing to their superior mechanical properties such as higher specific strength, excellent corrosion resistance and lower modulus of elasticity when compared to conventional steel or ceramic implants.[1-4] Although being biocompatible and light weight, Ti6Al4V alloy lacks desired bioactivity *in vivo* and depicts minimal integration with the neighboring tissues. Thus, it results in bio-inertness and consequent implant loosening.[5,6] Hence, to improve the bioactivity of Ti alloys, different calcium phosphate (CaP)-based surface modifications have been suggested by researchers.[7] Hydroxyapatite (HAp) is one of the most commonly studied forms of CaP coatings owing to its similarity to the extracellular matrix of the bone.[8,9] Various coating techniques such as dip coating, electrophoretic deposition, sol-gel coatings, sputtering, plasma spray, biomimetic coating, etc., can be used to obtain HAp deposition on the metallic substrates.[10,11]

The plasma spray method, a high-temperature coating method followed by rapid cooling, is the most commonly used method for coating dental and orthopedic implants.[12,13] However, the rapid variation in temperature during the deposition process has resulted in a heterogeneous coating of HAp in some cases.[14,15]

With advancements in coating techniques and shift in emphasis to achieve overall biofunctionality of implants,[16-18] there is a significant change in the focus of researchers to develop alternative coating methods that are not only economical and use minimal energy but also mimic *in vivo* biochemical processes.[19,20] The term biofunctionality is referred to combining the bioactivity of the materials with other important aspects of biocompatibility, such as osseointegration, osteoporosis, enhanced antimicrobial activity and capability of drug delivery.

Strontium (Sr), a natural trace mineral found in human bones, is selected as a doping agent, as it is known to cure osteoporosis by enhancing mineralization and supporting hard tissue formation.[21,22] Mussel proteins have gained prominence in recent years, as they can display strong adhesion to the substrates *in vivo*[23,24] and can be used as an interface to bind HAp with the Ti6Al4V surface. Polydopamine (PDA), having a similar molecular structure as 3,4-dihydroxy-L'phenylalanine proteins, can be used as a binder to be deposited on organic and inorganic surfaces. PDA, with the ability to stick to the substrate surface and inheriting capability to covalently immobilize molecules and enhancing surface adhesion, is identified as a suitable element for surface modifications. The other applications of PDA include cell proliferation, mitigation and cell spreading, thereby enhancing the overall functionality of the current implants.[25-27] As described by Tas and Bhaduri, a rapid biomimetic technique using 10 × Simulated Body Fluid (SBF) forms the basis of the coating technique in the study.[28] This study aims to prepare a Sr-doped HAp coating and use PDA as a binder to Ti6Al4V substrate and hydroxyapatite, with a focus on improving the surface adhesion of the metallic implants by implying a low cost low temperature rapid biomimetic coating technique.

Table 1. PMI report data for procured Ti6Al4V material.

Element	Ti	Al	V	Fe
Chemical composition of Ti6Al4V				
% allowed			3.5–4.5	0.25 max
% identified	90.25	6	4.46	0.11

2. Materials and Methods

2.1. *Preparation of Ti6Al4V substrates*

A Ti6Al4V round bar of diameter 10 mm and length 1000 mm was purchased from, M/s Bhagyashali Metals, Mumbai, India along with Positive Material Identification (PMI) report, as attached in Table 1. The round bar was cut into parts of 8 mm height each and the obtained specimens were abraded with 800 grit SiC paper manually. For the chemical and thermal treatment of the substrates before coating process as per available protocol,[28–30] bars were heat-treated in a muffle furnace at 600°C for 1 h, followed by cleaning in acetone and ethanol for 15 min each and lastly, rinsing in de-ionized water. Specimens were then etched in 5 molar KOH solution for 24 h at room temperature before proceeding with the coating process.

2.2. *Preparation of 10 SBF and pure HAp coating process*

Based on the coating protocol prescribed by Tas and Bhaduri,[28] chemicals were added to 800 mL of de-ionized water in chronological order as per Table 2, to prepare a 2 L SBF solution. It was made sure that the first chemical was completely dissolved before adding the next chemical. The magnetic stirrer stirred at 800rpm to dissolve the chemicals. Water was added to make the solution to 2 L and pH was raised to 6.4 by gradually adding $NaHCO_3$.

Table 2. Chemicals used to make 10 SBF.

S. no.	Name	Quantity (g)	Concentration (mM)	Make
1	NaCl	116.886	1000	Loba-Chemie
2	KCl	0.7456	5	Loba-Chemie
3	$CaCl_2 \cdot 2H_2O$	7.3508	25	Loba-Chemie
4	$MgCl_2 \cdot 6H_2O$	2.0330	5	Loba-Chemie
5	NaH_2PO_4	2.3996	10	Loba-Chemie

The solution was then transferred to glass bottles, and the specimen was suspended in it for 12, 24, and 48h at room temperature. The samples were marked as PHA-12, PHA-24 and PHA-48. After the stated time interval, the specimens were taken out, washed with de-ionized water to remove any unattached chemicals, and dried in a vacuum oven at 35-40°C at pressure 0.03 bars. The characterization was done on the obtained samples to settle the optimum coating time for the other coating process.

2.3. Preparation of dopamine solution in TRIS buffer and pretreatment of substrate

Dopamine Hydrochloride (Sigma-Aldrich) 2mg/mL in TRIS Buffer (10 mM, pH 8.5), was used to form a thin layer of PDA on substrates. The TRIS buffer was prepared in the laboratory by adding 1.21 g of Trizma (R) base (Luba-Chemie) to de-ionized water and adjusting the final volume of solution to 1 L with de-ionized water. The pH was adjusted to 7.4 value by slowly adding approximately 6–7mL concentrated HCl. The solution was allowed to cool down before testing for pH and make further addition of HCl if required. The dopamine solution was then transferred to 250mL sealed glass bottles, and substrates were dipped in the bottles for 24h at room temperature, as prescribed by Zhe et al.[31] After the reaction time, the samples were removed from the solution, cleaned thoroughly in de-ionized water to remove unattached dopamine and dried at room temperature.

2.4. Sr-doped SBF and coating process

The dopamine-treated samples were further coated with pure hydroxyapatite and Sr-doped hydroxyapatite, marked as PDA-HAp and PDA-Sr1 and PDA-Sr2. The protocol as applied to coat PHA-12, 24, and 48 samples was used to coat the pretreated samples. 0.5 nM and 1 mM Sr nitrate was used to replace calcium chloride in 10 SBF solutions and rapid biomimetic coating was done.

2.5. Characterization of obtained coatings

The obtained samples were characterized using Scanned Electron Microscopy (SEM) on a JSM- IT100 In Touch Scope™ SEM from JEOL Ltd., Japan. The equipment operates on an accelerating voltage of 30.0 kV. The Energy Dispersive Spectroscopy (EDS) analysis was done on Oxford Instruments X-act $-10mm^2$ SDD Detector embedded with SEM from JEOL Ltd., Japan. The instrument is in compliance with ISO 1502:2012, and has a premium resolution of 125 eV. It detects elements from beryllium to plutonium, compatible with AZtec® EDS analysis software. For the purpose of X-Ray Diffraction (XRD), D8 ADVANCE ECO, BRUKER XRD instrument was used to analyze crystalinity and phase composition of the obtained coatings. Data were collected in the range of 20–60° with a step size of 0.02° at a scanning rate of 1° per min.

2.6. Adhesion strength of the obtained coatings

The adhesion strength was evaluated by a standard pull out adhesion test ISO 13779-4. To perform the test, a solid titanium cylinder with a diameter of 10 mm was glued with Loctite 907 Hysol epoxy adhesive to the obtained coated surface on the specimen. The pasted samples were rested for 24 h at room temperature for curing purposes. Afterwards, the specimens were subjected to tensile testing on a universal testing machine with a speed of 1 mm/min, tensile load applied perpendicular to the specimen, until coating detached from the surface of the specimen. Three samples per each coating group were tested and arithmetic mean was used to calculate average adhesive strength for each group using the following formula:

$$\sigma_{Adh} = F/A_{Det},$$

where σ_{Adh} is adhesive strength, F is the load at the point of detachment and A_{Det} is the coating area detached from the substrate.

3. Results and Discussions

Images of deposited coatings obtained by SEM for PHA-12, PHA-24 and PHA-48, as well as PDA-HAp, PDA-Sr1 and PDA-Sr2 samples, are shown in Fig. 1. The nucleation of calcium and phosphate was evident in all the cases, even after 12h of immersion, as in PHA-12. Deposition of HAp around the substrate would increase with an increased immersion period, images obtained for PHA-24 and PHA-48 suggested the same and conform to the earlier results obtained by various researchers.[28,32,33]

The EDS data are compiled in Table 3 to include elemental compositions of various possible elements present and stoichiometric ratio (Ca/P). The EDS results are extracted as an average value from the SEM images, shown in Fig. 1, by identifying and quantifying the zones available throughout the surface profile. Ca/P ratio is used to signify the presence of hydroxyapatite and other calcium-based coatings. Ca/P ratio of 1.67 is considered an ideal ratio to confirm the presence of HAp at the investigation site and for adequate mechanical properties.[34] In the case of pure HAp, it is observed that the Ca/P ratio for 12h sample is 2.35 which is very high when compared to the ideal Ca/P ratio, while the value obtained for 24h and 48h sample is well within the ideal ratio as given in Table 3. Based on the above calculations it can be stated that 12h immersion is not sufficient for coating of HAp on substrates. The stoichiometric ratio for samples doped with Sr was calculated taking into account the atomic percentages of calcium and its substitute Sr and dividing it with atomic % of P. The ratio was close to the prescribed ratio of 1.67. The EDS data of the

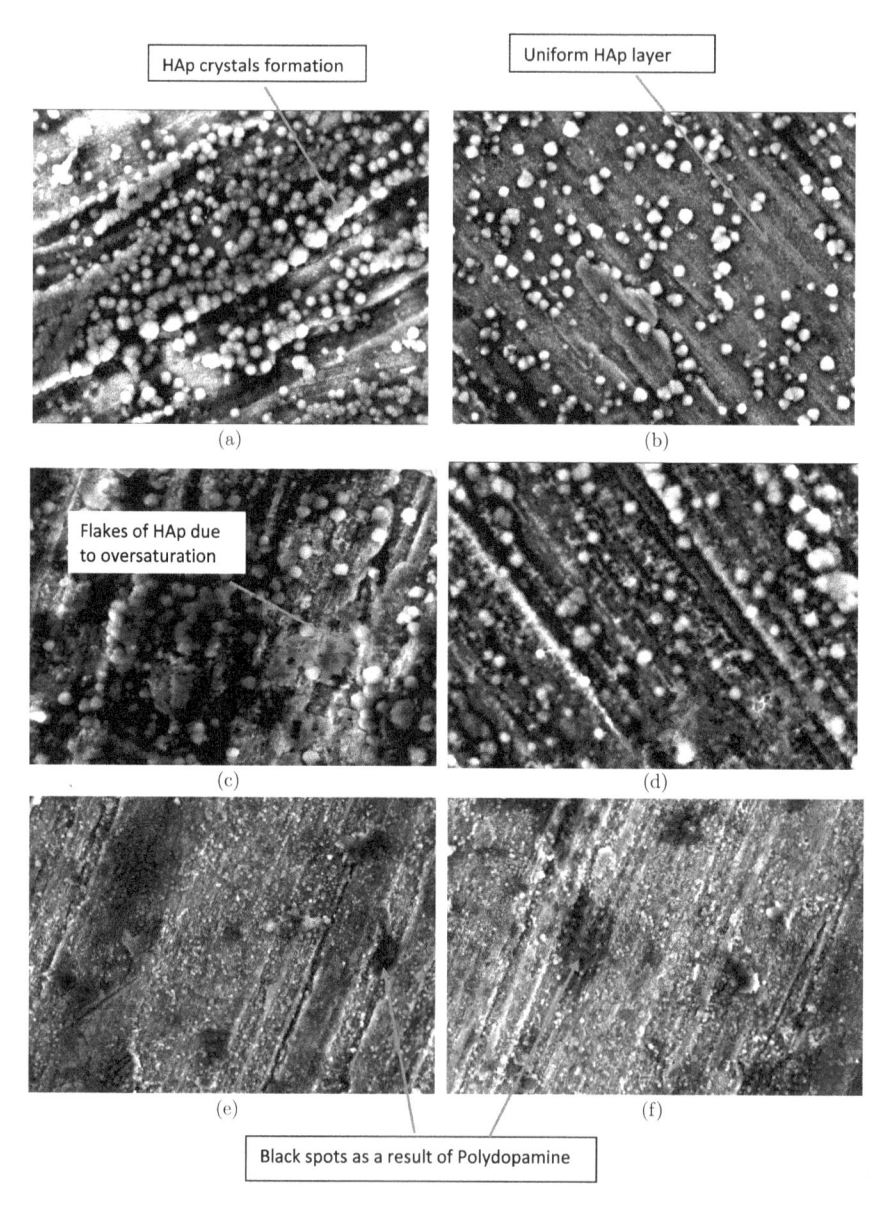

Fig. 1. SEM Images for (a) PHA-12, (b) PHA-24, (c) PHA-48, (d) PDA-HAp, (e) PDA-Sr1 and (f) PDA-Sr2 at 1000 × (Color online).

Table 3. EDS Analysis of the obtained coatings.

Elements	PHA-12 Wt.%	PHA-12 At.%	PHA-24 Wt.%	PHA-24 At.%	PHA-48 Wt.%	PHA-48 At.%	PDA-HAp Wt.%	PDA-HAp At.%	PDA-Sr1 Wt.%	PDA-Sr1 At.%	PDA-Sr2 Wt.%	PDA-Sr2 At.%
Ti	59.83	37.38	37.89	22.69	36.22	20.62	27.51	14.79	32.72	17.48	32.47	17.31
O	27.97	52.32	43.62	64.77	44.94	66.79	47.92	68.82	46.12	67.47	47.18	68.19
Sr	nd	nd	nd	nd	nd	nd	nd	nd	0.38	0.11	0.53	0.19
Ca	7.88	5.89	10.55	6.39	10.84	6.51	15.83	8.87	12.24	7.49	11.97	7.46
P	2.61	2.51	5.35	3.89	5.58	4.01	6.43	5.48	5.74	4.9	5.65	4.78
Al	1.71	1.9	2.59	2.26	2.42	2.07	2.31	2.04	2.8	2.55	2.2	2.07
Total	100.00	100.00	100.00	100.00	100.00	100.00	100.00	100.00	100.00	100.00	100.00	100.00
Ca/P	2.35		1.64		1.62		1.62		1.55		1.60	

PDA-HAp sample showed a notable increase in mass percentages of Ca and P, validating the fact that pretreatment with PDA can enhance the deposition rate. Although the Ca and P percentage decreased with the addition of Sr into the SBF solution, the presence of Sr in EDS analysis signifies that the stated technique can be effectively used to deposit composite HAp coatings on metallic substrates.

Figure 2 presents the XRD patterns of 12, 24, and 48 h coated pure HAp samples. All the coatings revealed typical HAp-XRD peaks as compared with JCPDS-09-0432 and conform to the same. The sharper HAp peaks are observed with an increase in immersion time that can be attributed to the increased concentration. The peaks obtained in developed coatings adhere to standard HAp and no other CaP-based phases were identified in the analysis. The results obtained were in line with the relevant research,[35,36] and it was inferred that phase pure coatings were obtained. XRD peaks

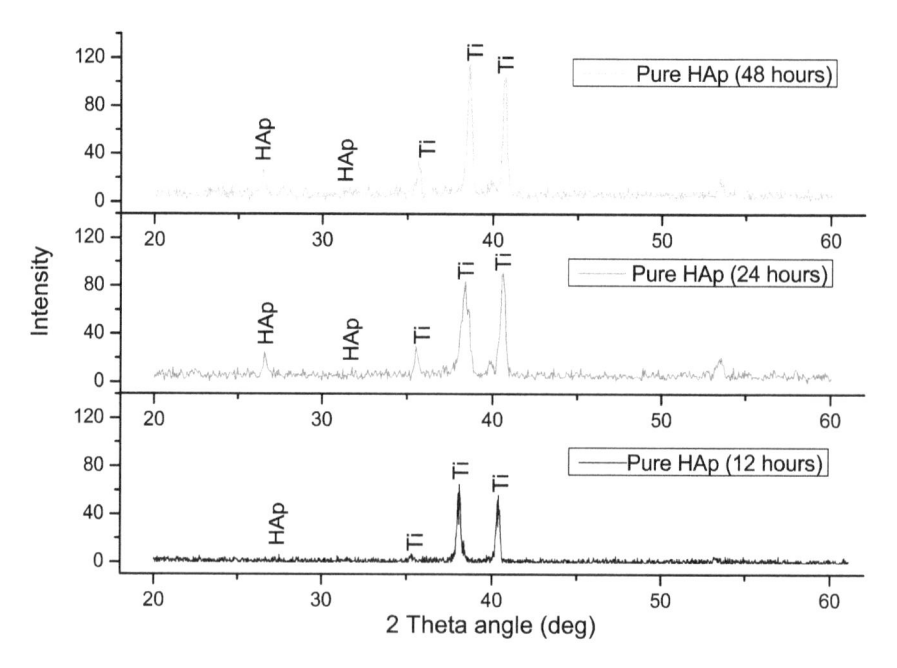

Fig. 2. Comparative chart of XRD Patterns obtained for PHA-12, PHA-24, and PHA-48 (bottom to top).

for 48 h immersion duration showed higher peaks for Ti, which can be possibly caused by higher concentration of Ti on the topographical surface.

The XRD patterns obtained for pretreated samples are compared in Fig. 3, it is observed that Sr- doped HAp specimen depicted typical HAp XRD peaks similar to the one obtained for pure hydroxy-apatite specimen and in accordance with JCPDS-09-0432. Although the presence of Sr is not identified in XRD analysis as peaks are similar to that of pure HAp, as signified by various researchers in their study of Sr-doped HAp.[37,38] However, the presence of Sr in EDS analysis indicated that Ca was effectively substituted by Sr in the HAp structure. The same was inferred about the PDA-treated HAp samples also from diffraction analysis.

The coating adhesion strength for the pure HAp coatings on the untreated samples was calculated to be 6.17 ± 0.2, 8.31 ± 0.4 and 9.28 ± 0.2 MPa, respectively, for the 12, 24, and 48 h immersion

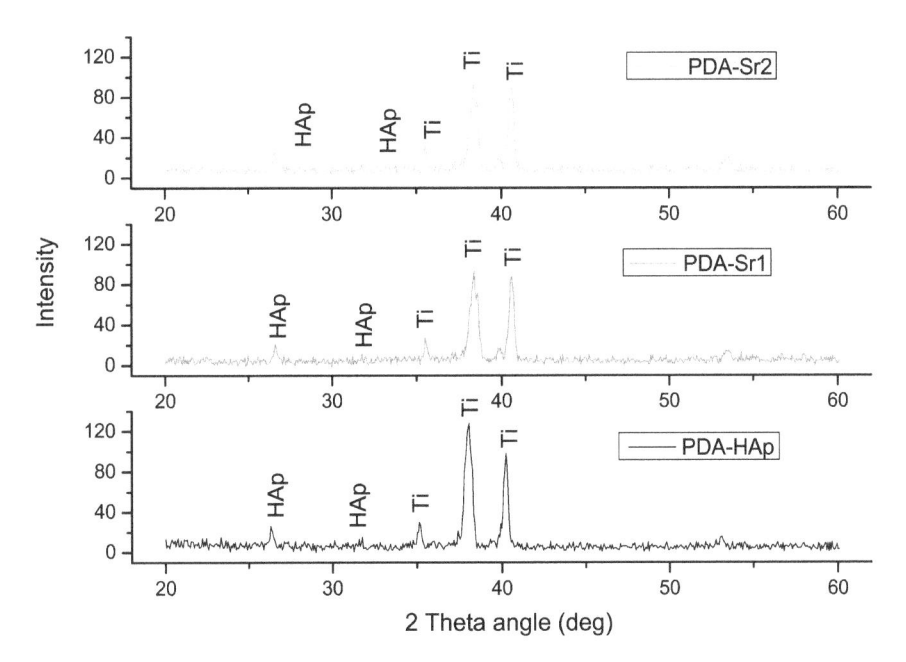

Fig. 3. Comparative chart of XRD Patterns obtained for PDA-HAp, PDA-Sr1, and PDA-Sr2 (bottom-top).

times and that for PDA-treated 24 h specimen was 10.08 ± 0.4 MPa. It can be calculated that pretreatment with PDA has resulted in an improvement of 20% in adhesion strength of pure HAp coatings for 24 h time duration. Furthermore, the adhesion strength values for the composite coating group's PDA-HAp-Sr1 and PDA-HAp-Sr2 were 9.12 ± 0.4 MPa and 8.86 ± 0.2 MPa, respectively, and are still in closer range to the pure HAp samples. So, the results suggest that the adhesion between HAp coating and the titanium surface was improved by the pre-treatment of substrates with the PDA.

4. Conclusion and Future Scope

Rapid biomimetic coating technique was used to obtain Sr-doped HAp composite coatings on PDA-treated Ti6Al4V samples at low temperatures. The untreated samples were coated for a varying duration to assess the optimum period for the coating process. The data obtained from SEM, EDS, and XRD were analyzed and based on Ca/P ratio and XRD graphs which showed relevant peaks for hydroxyapatite formation, 24 h was used as the optimal time for the coating process. Coating pure HAp on pretreated Ti6Al4V samples demonstrated enhanced the calcium deposition rate and improved the Ca/P coefficient when compared to stoichiometric HAp molar ratio of 1.67. Hence, signifying the fact that the presence of mussel-inspired proteins or similar structures enhances the surface. The doping of Sr into hydroxyapatite had no considerable effect on the surface morphology of the obtained coatings and caused a slight decrement in the deposition of calcium on the substrate, but a higher Sr content compensated that, where (Ca+Sr)/P coefficient of 1.60 was closer to a molar ratio of 1.67. The noteworthy improvement in the adhesion strength of the PDA-treated samples further highlights the significance of pretreatment of the substrates. The current study further confirms the ability of Sr to incorporate within hydroxyapatite structure through low temperature rapid biomimetic coating method.

Further studies can be undertaken to establish the overall biofunctionality of the obtained coatings, such as *in vitro* anti-bacterial capabilities, resistance to corrosion, and the coatings' thickness.

References

1. H. J. Rack and J. I. Qazi, *Mater. Sci. Eng.* **26** (2006) 1269.
2. F. Barrere, T. A. Mahmood, K. De Groot and C. Van Blitterswijk, *Mater. Sci. Eng.: R: Rep.* **59** (2008) 38.
3. M. Niinomi, T. Narushima and M. Nakai (eds.), *Advances in Metallic Biomaterials*, Vol. **3** (2015), pp. 179–213.
4. J. I. Lim, B. Yu, K. M. Woo and Y. K. Lee, *Appl. Surf. Sci.* **255** (2008) 2456.
5. M. P. Bajgai, D. C. Parajuli, S. J. Park, K. H. Chu, H. S. Kang and H. Y. Kim, *Bioceram. Dev. Appl.* **1** (2010) 685.
6. G. Singh and A. Saini, Developments in metallic biomaterials and surface coatings for various biomedical applications, in *Advances in Materials Processing* (Springer, Singapore, 2020), pp. 197–206.
7. Y. Su, C. Luo, Z. Zhang, H. Hermawan, D. Zhu, J. Huang, Y. Liang, G. Li and L. Ren, *J. Mech. Behav. Biomed. Mater.* **77** (2018) 90.
8. S. Jaya, T. D. Durance and R. Wang, *J. Compos. Mater.* **43** (2009) 1451.
9. M. C. Chang, C. C. Ko and W. H. Douglas, *Biomaterials* **24** (2003) 2853.
10. E. Mohseni, E. Zalnezhad and A. R. Bushroa, *Int. J. Adhes. Adhes.* **48** (2014) 238.
11. G. M. Rodriguez, J. Bowen, D. Grossin, B. Ben-Nissan and A. Stamboulis, *Colloids Surf. B* **160** (2017) 154.
12. L. Sun, C. C. Berndt, K. A. Gross and A. Kucuk, *J. Biomed. Mater. Res.* **58** (2001) 570.
13. O. Albayrak, O. El-Atwani and S. Altintas, *Surf. Coat. Technol.* **202** (2008) 2482.
14. L. Yan, Y. Leng and L. T. Weng, *Biomaterials* **24** (2003) 2585.
15. Y. P. Lu, G. Y. Xiao, S. T. Li, R. X. Sun and M. S. Li, *Appl. Surf. Sci.* **252** (2006) 2412.
16. L. Cao, I. Ullah, N. Li, S. Niu, R. Sun, D. Xia, R. Yang and X. Zhang, *J. Mater. Sci. Technol.* **35** (2019) 719.

17. T. Hanawa, *J Periodontal Implant Sci* **41** (2011) 263.
18. C. D. Reyes, T. A. Petrie, K. L. Burns, Z. Schwartz and A. J. Garcia, *Biomaterials* **28** (2007) 3228.
19. F. Thammarakcharoen and J. Suwanprateeb, Rapid biomimetic coating of calcium phosphate on titanium: Effect of soaking time, temperature and solution refreshing, in *Key Engineering Materials* (Trans Tech Publications Ltd., Stafa-Zurich, Switzerland, 2016), pp. 81–86.
20. J. Forsgren, F. Svahn, T. Jarmar and H. Engqvist, *Acta Biomater.* **3** (2007) 980.
21. K. J. Joshi, N. M. Shastri and N. M. Shah, *AIP Conf. Proc.* (AIP Publishing Center, Melville, NY, USA, 2020), p. 070005.
22. Q. Wang, P. Li, P. Tang, X. Ge, F. Ren, C. Zhao and K. Duan, *Colloids Surf. B* **182** (2019) 110359.
23. H. Lee, S. M. Dellatore, W. M. Miller and P. B. Messersmith, *Science* **318** (2007) 426.
24. Q. Ye, F. Zhou and W. Liu, *Chem. Soc. Rev.* **40** (2011) 4244.
25. L. Jia, F. Han, H. Wang, C. Zhu, Q. Guo, J. Li and B. Li, *J. Orthop. Trans.* **17** (2019) 82.
26. Z. Y. Xi, Y. Y. Xu, L. P. Zhu, Y. Wang and B. K. Zhu, *J. Membr. Sci.* **327** (2009) 244.
27. J. Yuan, Z. Zhang, M. Yang, F. Guo, X. Men and W. Liu, *Tribol. Int.* **107** (2017) 10.
28. A. C. Tas and S. B. Bhaduri, *J. Mater. Res.* **19** (2004) 2742.
29. L. Jonášová, F. A. Müller, A. Helebrant, J. Strnad and P. Greil, *Biomaterials* **25** (2004) 1187.
30. H. M. Kim, H. Takadama, F. Miyaji, T. Kokubo, S. Nishiguchi and T. Nakamura, *J. Mater. Sci. Mater. Med.* **11** (2000) 555.
31. W. Zhe, C. Dong, Y. Sefei, Z. Dawei, X. Kui and L. Xiaogang, *Appl. Surf. Sci.* **378** (2016) 496.
32. F. Thammarakcharoen, N Hobang and J. Suwanprateeb, Rapid biomimetic coating of biocompatible calcium phosphate on titanium: influence of pretreated NaOH concentration and cleaning method. In *Advanced Materials Research*, Vol. 1119 (Trans Tech Publications Ltd., 2015), pp. 81–86.
33. J. Suwanprateeb, W. Suvannapruk and K. Wasoontararat, *Mater. Med.* **21** (2010) 419.
34. Z. Boukha, M. P. Yeste, M. Á. Cauqui and J. R. González-Velasco, *Appl. Catal. A: Gen.* **580** (2019) 34.

35. M. Avci, B. Yilmaz, A. Tezcaner and Z. Evis, *Ceram. Int.* **43** (2017) 9431.
36. J. Terra, E. R. Dourado, J. G. Eon, D. E. Ellis, G. Gonzalez and A. M. Rossi, *Phys. Chem. Chem. Phys.* **11** (2009) 568.
37. M. Furko, V. Havasi, Z. Kónya, A. Grünewald, R. Detsch, A. R. Boccaccini and C. Balázsi, *Bol. Soc. Esp. Ceram. V.* **57** (2018) 55.
38. O. Kaygili, S. Keser, M. Kom, Y. Eroksuz, S. V. Dorozhkin, T. Ates, I. H. Ozercan, C. Tatar and F. Yakuphanoglu, *Mater. Sci. Eng. C* **55** (2018) 538.

Performance of thermally sprayed hydroxyapatite coatings for biomedical implants: A comprehensive review*

Gaurav Prashar[†,§] and Hitesh Vasudev[‡,¶]

[†]*Rayat Bahra Institute of Engineering and Nano-Technology, Hoshiarpur 146104, India*
[‡]*School of Mechanical Engineering, Lovely Professional University, Phagwara 144411, Punjab, India*
[§]*grvprashar@yahoo.co.in*
[¶]*hiteshvasudev@yahoo.in*

Lalit Thakur

Mechanical Engineering Department, NIT Kurukshetra, Kurukshetra 136119, Haryana, India
lalitthakur@nitkkr.ac.in

Amit Bansal

I.K. Gujral Punjab Technical University, Kapurthala 144603, Punjab, India
amit.bansal978@gmail.com

Metallic bioimplant are widely used now-a-days to replace a part of human body in a physiologically accepted manner. However, the biocompatibility of the metallic bioimplant was mainly achieved by the

---

*To cite this article, please refer to its earlier version published in Surface Review and Letters, Vol. 30, No. 1 (2023) 2241001 (29 pages) DOI: 10.1142/S0218625X22410013
[¶]Corresponding author.

incorporation of a bio-compatible coating on its surface through suitable surface modifications techniques. In surface modifications techniques, the thermal spraying is widely used for modification of metallic bioimplant due to its versatile nature. In thermal spraying, the hydroxyapatite (HAp)-based coatings are mainly preferred because this coating responses to physiochemical environment and adapts itself accordingly. But bulk of the HAp coating give out due to less adhesion strength of the HAp coating and its poor mechanical properties. The properties of HAp-based coatings can be designed as per requirements by reinforcing this coating through hard particulates in suitable proportions. In this paper, the various thermal sprayed (TS) coatings used for performing HAp-based coatings on bioimplant were discussed. The influence of reinforcements on the mechanical and bio-compatible properties of the coatings is also discussed in detail. Lastly, the challenges in the TS HAp coatings alongwith the future perspective of TS coating in fabricating of 3D biomedical implants by using cold spray (CS) has also been summarized.

Keywords: Hydroxyapatite; corrosion, wear; thermal spray; heat treatment; 3D.

1. Introduction

1.1. *Need of biomaterial/Implant*

The use of metallic implants to treat imperfections and fractures of bones comes into picture in the early 19th century, when the industrial revolution metal industry begin to expand.[1] However, no experimental trials of implanting the metallic devices, like bone pins and spinal wires fabricated from silver, iron or gold were successful until 1860. But with the advent of Lister's aseptic surgical technique in 1860, there is an unprecedented boom in this field.[1] After solving the primary problems with this surgical technique, the focus then shifted to another milestone called as biomaterials. The area of biomaterials gained its world-wide recognition in South Carolina (1969) after the 1st meeting held on biomaterials at Clemson. Afterwards, this area continued to gain substantial attention of the researchers. Biomaterials may be artificial or natural, employed in developing implants or structures, to replace the diseased biological or lost structure to restore both form and function. Thus, they provide answers to many problems encountered in the field of medical sciences. Biomaterial helps in enhancing the

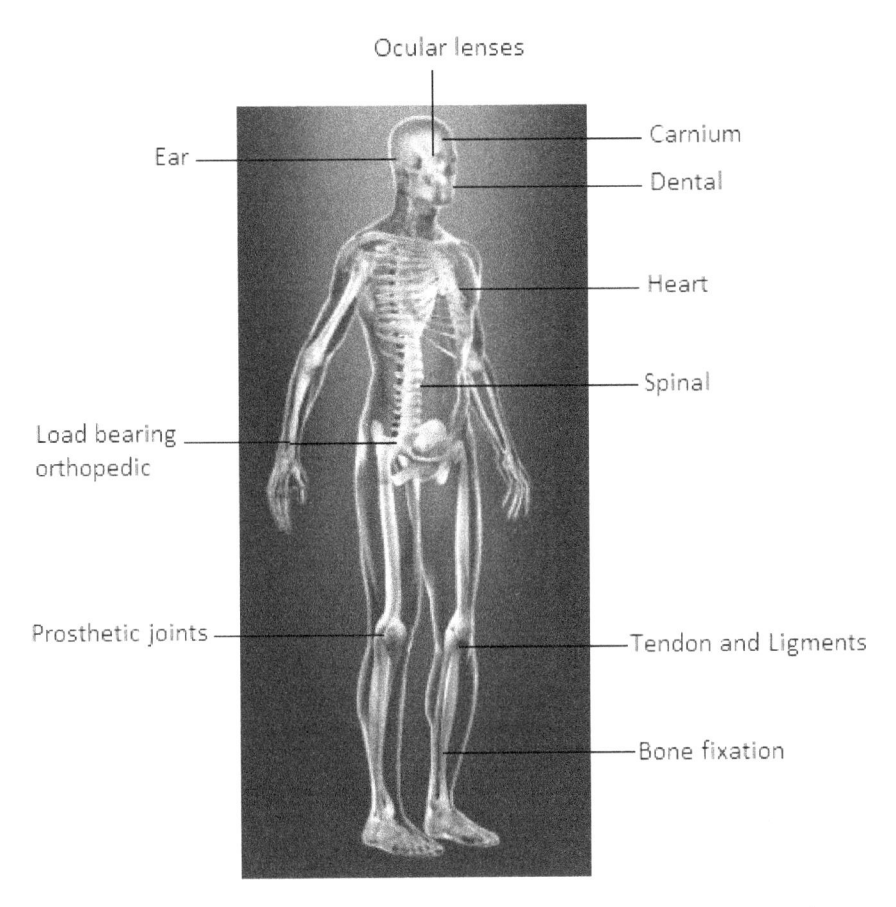

Fig. 1. Various parts of human body where biomaterials can be used (Color online).

longevity and quality of life in human beings. Biomaterials are used in various parts of body like in the heart as artificial valves, in blood vessels as stents, replacement implants in human shoulders, elbows, knees, hips, and orodental structures (dental).[2,3] Figure 1 illustrates the various parts of a human skeleton that can be replaced with biomaterial. The example of implants used in joints of hip and knee is shown in Fig. 2.

As indicated in Fig. 3, the number of primary total knee arthroplasty (TNA) procedures conducted in the United States is predicted to increase by more than 673% by 2030 compared to

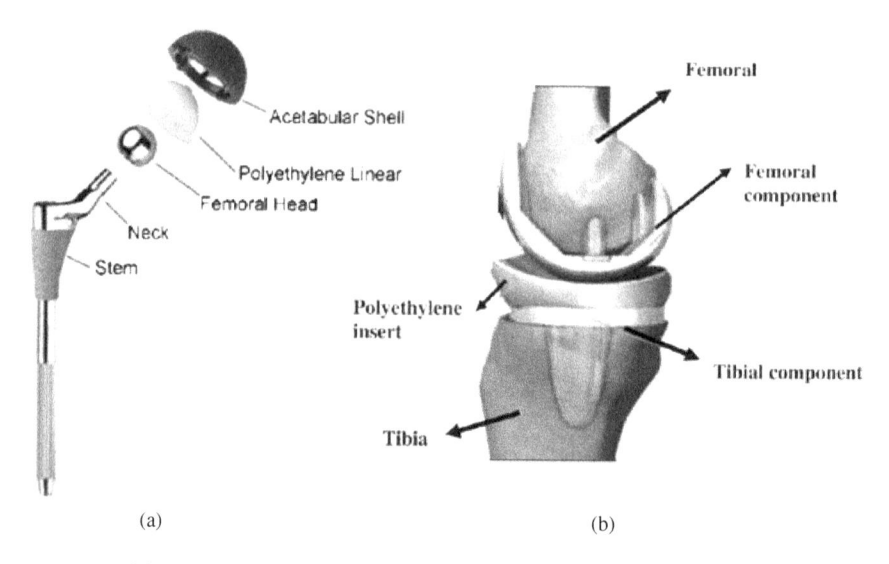

Fig. 2. Total replacements of hip and knee with implants.[4]

Years/ 2005-2030	
THA	TKA
Annual No. of procedures[x 1,000]	
upto 700	upto 3500

Fig. 3. The projected number of primary THA and TKA procedures from 2005 to 2030 in the United States.[5]

2005, while primary total hip arthroplasty (THA) procedures are expected to increase by almost 174% during the same time period.[5] Other factors contributing towards this rise include the truth that baby boomers are now senior citizens, our life expectation keeps on increasing and the obesity pandemic means we are putting more wear and strain on these weight-bearing joints.

From Fig. 3, it is clear that the biomaterials are a multibillion-dollar business that is growing at a rapid pace around the world. Millions of patients' lives have been improved, supported, and sustained because of the use of biomaterials, and this figure is growing every year. As a result, research in this field is critical in order to enhance materials and bring down costs. Improving the fixing time and lifespan of the implant anchoring inside the human body is a major goal in modern day orthopedics and bioengineering. Further, the materials ability to perform its intended function without undergoing any reactions with the host body is measured with the term biocompatibility. The prime and most necessary condition needed for the present-day biomaterials is to achieve this biocompatibility when implanted in the human body.[6] Materials not satisfying this criterion (materials with secondary side or undesirable effects) are no longer required for medical or surgical purposes. Other factors on which the levels of biocompatibility depend are the contact time among biomaterial and biological tissue that may range from few seconds to several decades. Biomaterials can be summarized mainly in four groups depending upon the alloying element as illustrated in Table 1. On the other hand, centered on the human body response, biomaterials can be graded into three groups: (i) bioinert materials (1st generation), (ii) bioresorbable materials (2nd generation) and (iii) bioactive materials (3rd generation).[7]

Table 1. Four groups of metallic biomaterials.[7]

Type	Primary utilizations	Application
Stainless steels	Total hip replacements	Applied routinely
(Fe-based biomaterials)	Temporary devices	
Co-based alloys	Total joint replacement Dentistry castings	
Ti-based alloys	Nails, pacemakers Stem and cup of total hip replacements	
Others	Vascular stents	Approved by FDA
NiTi	Orthopedic staples Catheter guide wires Intracranial aneurysm clips	
Mg	Biodegradable orthopedic implants	
Ta	A radiographic marker Wire sutures to be used in plastic and neurosurgery	

Table 2. The impacts of corrosion in the host body as a result of various metallic elements present in different biomaterials.[11]

Various elements in biomaterials	Effects on the human body
Ni	Skin effects like dermatitis
Co	Anemia B restricting Fe from being absorbed into the stream of blood
Cr	Leads to ulcers and disturbances in central nervous system
Al	Results in epileptic impact and Alzheimer's disease
Va	Toxic in its elementary state

2. Major Implant Failure Causes

2.1. *Corrosion of implants*

Inside the body of human beings, the prevailing environment is chemically and physically different from the outside conditions. For instance, a metal that serves well in ambient conditions may undergo a severe corrosion in the human body environment. In fact, the most corrosion-resistant stainless steel (SS) commonly produce toxic reactions inside the body, which are only discovered after a considerable length of time has passed after implantation.[8–10] The pH and oxygen concentrations in different areas of the body differ. Due to oxidation and acidic erosion, an implant that operates well in one part of the body may experience unacceptably high levels of corrosion in another part of body. Table 2 depicts the impact of corrosion in the host body as a result of diverse biomaterials, while Table 3 depicts the types of corrosion in traditional materials used for biomedical implant.[11] Theoretically, resistance to corrosion should be increased if the release of metal ions will be minimized from a metallic implant in the body under harshest

Table 3. Corrosion classification in traditional materials employed for biomedical implants in clinical applications.[11]

Corrosion type	Implant material	Location of implant	Implant shape
Pitting corrosion	SS-304 and Co-based alloy	Orthopedic/dental alloy	
Crevice corrosion	316 L SS	Screws and bone plates	
Stress corrosion cracking	CoCrMo and 316 L SS	Only *in vivo*	

(*Continued*)

Table 3. (Continued)

Corrosion type	Implant material	Location of implant	Implant shape
Corrosion fatigue	CoCrNiFe and 316 SS	Bone cement	
Fretting corrosion	CoCrSS and Ti6Al4V	Ball joints	

Galvanic	304SS/316SS, CoCr + Ti6Al4V, 316SS/Ti6Al4V or CoCrMo	Screws, nuts and oral implants
Selective leaching	Mercury from gold	Oral implants

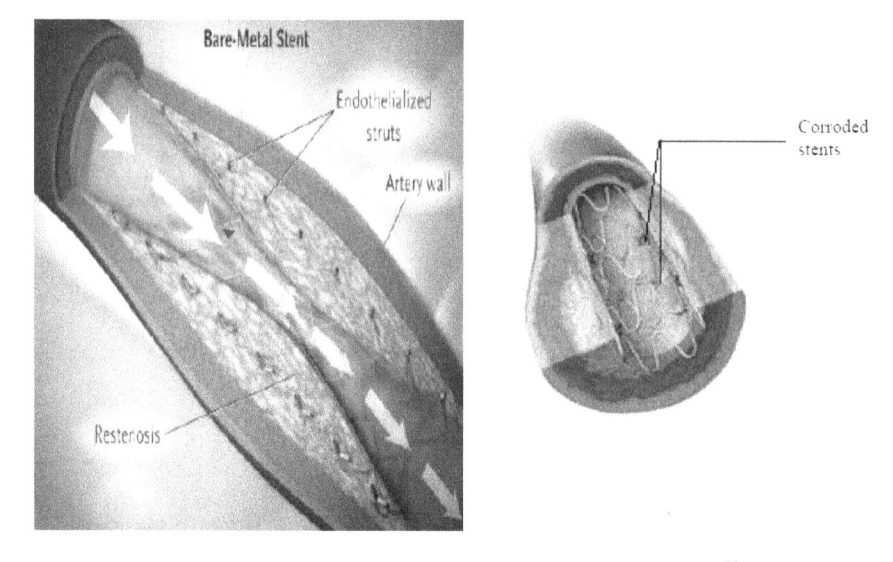

Fig. 4. Stent corrosion in bare metal (Color online).[11]

conditions. Figure 4 shows the corrosion of a stent implanted in the heart. The most popular metal for stents is 316L SS, which can be used with or without a coating. But the allergic reactions caused by the element Ni release from the SS implant was discovered, when the SS implant was used in human body. As a result, it is critical that a material be screened for corrosion behavior in all severe conditions in order to ensure that it does not fail in real applications.

2.2. *Wear of implants*

Wear is an unavoidable issue in every joint replacement, regardless of the materials utilized. The type of joint determines the material that should be used for it. Joints can be of two types: (i) mobile (joints among long bones such as hip, elbow, shoulder and ankle) and (ii) static (like tooth, skull and wrist). Mobile joints may be congruent and incongruent. A ballshaped head fits tightly to a cup-like socket in the congruent joints, like the hip and shoulder joints, distributing stress uniformly (Fig. 5). Any strong material,

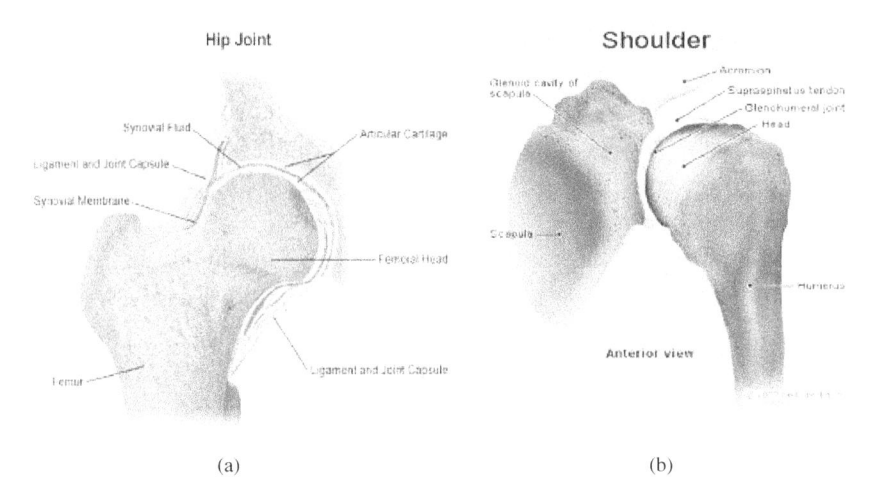

Fig. 5. Congruent joints in human body. (a) Hip and (b) shoulder (Color online).[7]

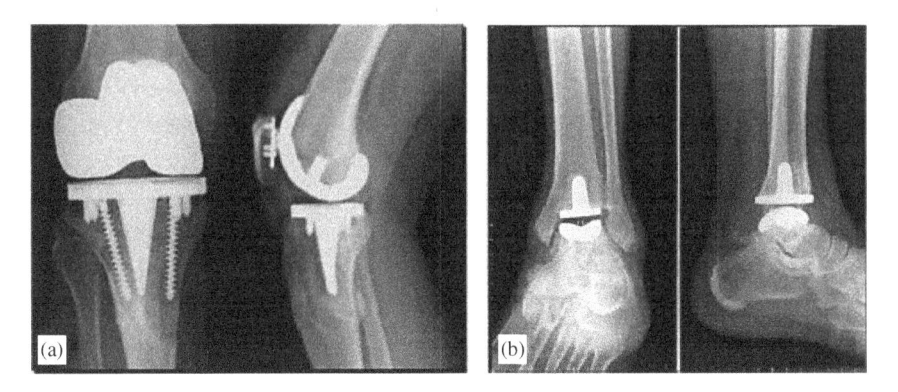

Fig. 6. Artificial incongruent joints in human body. (a) Knee and (b) ankle.[7]

including brittle ceramic materials, can withstand such mechanical loads. For incongruent type (like joints of knee and ankle shown in Fig. 6), the heterogeneous stresses were developed due to the contact of two hard incongruent surfaces. Because brittle ceramic materials are unable to withstand such stresses, metallic and rigid polymeric materials are favored in these situations.[7]

Concerns have also been raised about the wear debris' biological impact and long-term implications. In THA, wear has been identified as a primary cause of osteolysis. Submicron particles travel into the effective joint space, triggering a foreign-body reaction that leads to bone loss.[12] Wear debris driven osteolysis in the implant causes the bone around the implant to wear away, resulting in impaired joint function in many people.[13] Wear debris is created by the acetabular cup against the femoral head in a typical metal-on-polymer bearing combination, and it can cause substantial harm to living tissues. Wear debris, in a nutshell, produces severe adverse reactions, prompting revision surgery. Patients, doctors, and health-care systems all bear a significant burden as a result of revision surgeries. The replacement surgeries in the United States are believed to be over 2 million each year.[14] Metal debris formed during use, on the other hand, can become ionized and travel throughout the body. While there is no direct correlation among metal ions and health problems, but people with metal-on-metal implants have demonstrated elevated metal levels in their blood and organs, raising concerns about systemic health implications.[13] As a result, there has been enhanced focus on the fabrication of alternative materials or enhancement of the wear resistance of present materials for the purpose of total joint replacements so to avoid the costly revision surgeries triggered by the wear debris. As a result high corrosion resistance and wear resistance are crucial properties that define the longevity of joint implants. A flow chart showing relation among types of wear and operating conditions is shown in Fig. 7.

3. Preventive Measures to Minimize the Implant Failure

3.1. *Protective coatings*

The interaction of all implants with the host tissue, which is most typically assessed in terms of biocompatibility, is a major concern. The failure or success of the bioengineered system among the host and prostheses is determined by the tissue–implant interaction. Therefore, the modification of prosthesis surface is recommended

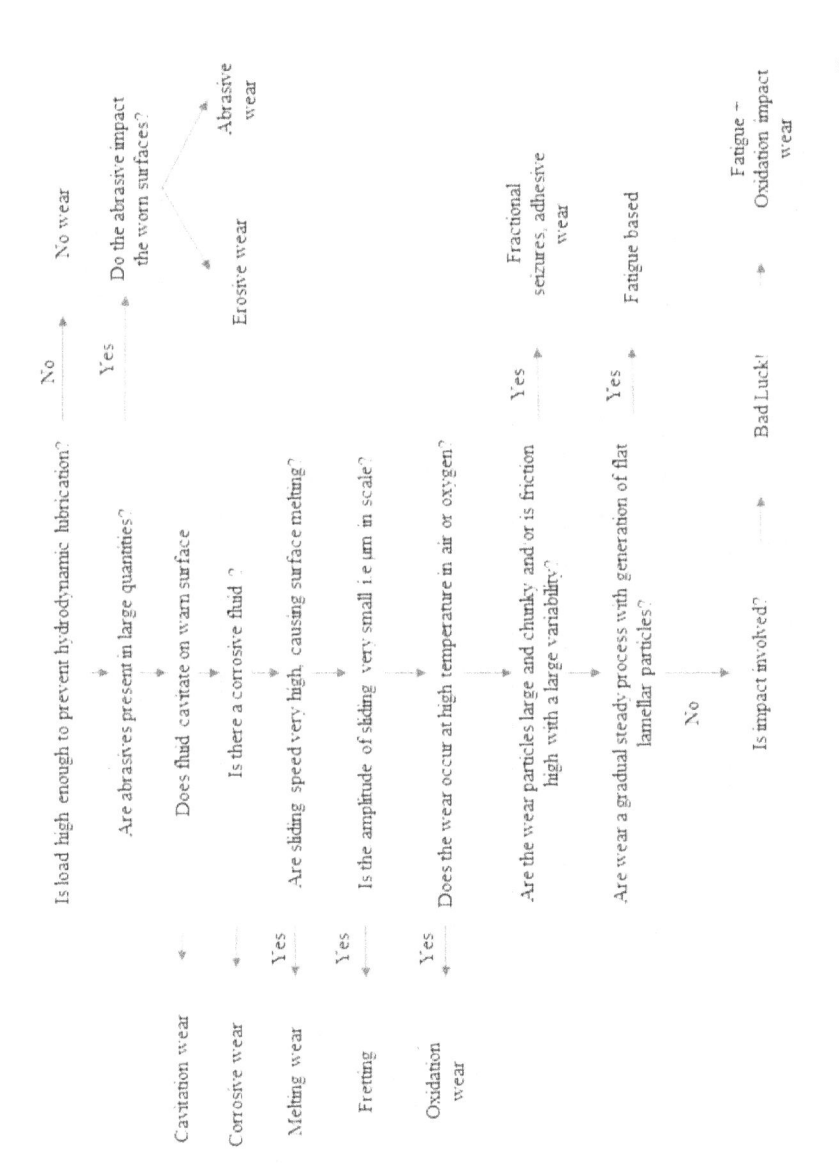

Fig. 7. Flow chart showing relation among types of wear and operating conditions (Color online).[15]

```
┌─────────────────┐   ┌─────────────────┐   ┌─────────────────┐   ┌─────────────────┐
│ Generation-1    │   │ Generation-2    │   │ Generation-3    │   │ Generation-4    │
│ Pre-peg coatings│ ⇨ │ Passive coatings│ ⇨ │ Active coatings │ ⇨ │ New generation  │
│ applied to hip  │   │ consisting of   │   │ based on        │   │ coatings        │
│ stems to improve│   │ thermal sprayed │   │ hydroxyapatite  │   │                 │
│ stabilization   │   │ porous Ti alloys│   │                 │   │                 │
│ and fixation of │   │                 │   │                 │   │                 │
│ implants        │   │                 │   │                 │   │                 │
└─────────────────┘   └─────────────────┘   └─────────────────┘   └─────────────────┘
```

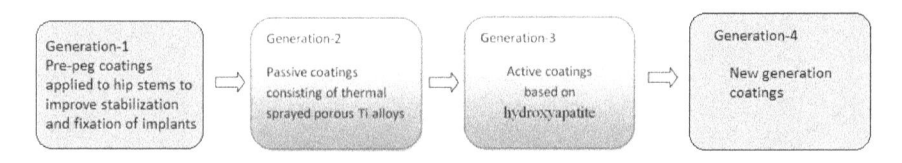

Fig. 8. The generations of coating for surface modification of the prostheses.

as a promising method to achieve good outcomes in patients. Prostheses surface modification has a well-known history and can be divided into mainly four generations as shown in Fig. 8. However, researchers and clinicians are interested in "New generation coating" that either leads to a higher level of bone formation or has bio-function than thermally sprayed hydroxyapatite (HAp).

However, as Table 4 summarizes variety of chemistries of powders available which can be deposited on surface via different thermal spray (TS) techniques. The key parameter is the thickness of the coating deposited on implant surface since the thickness has main bio-mechanical considerations with regard to stress distributions at the interface of implant-coating.

Generally metallic biomaterials are used widely for implants all-round the globe due to their good mechanical strength, biocompatibility and better corrosion resistant properties. To name some of them, such as Ti, SS316L and Ti alloys are recommended in biomedical and orthopedic applications.[31] To perform better inside the host body for longer time span, no toxic effect should be noticed even when the medical instrument is in contact with the living tissue. Moreover, during prolonged exposure, medical instruments inside the patient body should not cause carcinogenesis, mutagenesis or cytotoxicity. But due to corrosion, metals or their alloys are not suitable for direct implantation in the human body. The secretion of certain metallic ions (such as Fe^{+2}) in the vicinity of these organs may cause fibrosis, which is not good for the body. As a result, these metals or alloys should be coated with materials that are biocompatible and have a composition that is comparable to that of bones, allowing for continued bone growth and development.

Table 4. Protective coatings which can be deposited by TS techniques.

Coating material	Coating technique	Coating thickness (μm)	Ref.
Bioactive glass K_2O–CaO–P_2O_5–SiO_2	HVSFS	10–15	16
HAp $Ca_{10}(PO_4)_6(OH)_2$	HVSFS	14–23	17
Hardystonite ($Ca_2ZnSi_2O_7$)	APS	15–18	18
Bioactive glass K_2O–Na_2O–CaO–P_2O_5–SiO_2	HVSFS	20–50	19
Tricalcium phosphate (TCP)	HVSFS	20–25	16
HAp $Ca_{10}(PO_4)_6(OH)_2$	HVSFS	27–37	20
HAp $Ca_{10}(PO_4)_6(OH)_2$	APS	30–40	21
Bioactive glass K_2O–Na_2O–CaO–P_2O_5–SiO_2	SPS	<50	19
HAp $Ca_{10}(PO_4)_6(OH)_2$	APS	50	22
HAP/titanium (HAp/Ti: 1/1) composite	APS	80	23
HAp/β TCP composite	APS	80–90	24
Wollastonite $CaSiO_3$	FS	100–150	25
Baghdadite ($Ca_3ZrSi_2O_9$)	APS	120	26
Diopside ($CaMgSi_2O_6$)	APS	200–300	27
Dicalcium silicate (Ca_2SiO_4)	APS	380	28
HAp and HAp/TiO_2	HVFS	135 for HA 140 for HA/TiO_2	29
HAp + 10 wt.% Aluminum oxide			30
HAp + 10 wt.% Zirconia oxide	VPS	—	

The main objective in the modern bioengineering and orthopedics is to improve the longevity of the implant anchorage and fixation time in the human body. These outcomes can be fulfilled by using bioactive and bioresorbable coating materials on the metallic implant. HAp is extensively employed as a coating material due to its good bioactivity, bioresorbility, biocompatibility and osteoinduction.[32]

HAp coating on metallic alloys improves bone bonding ability, as well as biocompatibility and decreases bioimplant toxicity in living organisms. Furthermore, these coatings are biocompatible and provide a local source of calcium and phosphate ions, which are necessary for bone cell growth. Figure 9 represents the crystal structure which is related commonly to HAp. In Fig. 9, the Ca, PO_4 and OH ions are tightly packed in the unit cell to depict the apatite structure.[33]

However, there are numerous techniques introduced to deposit corrosion resistant coatings (mainly HAp-based coatings) on the metallic implant due to the availability of the broad combinations of coatings materials. The selection of a particular technique

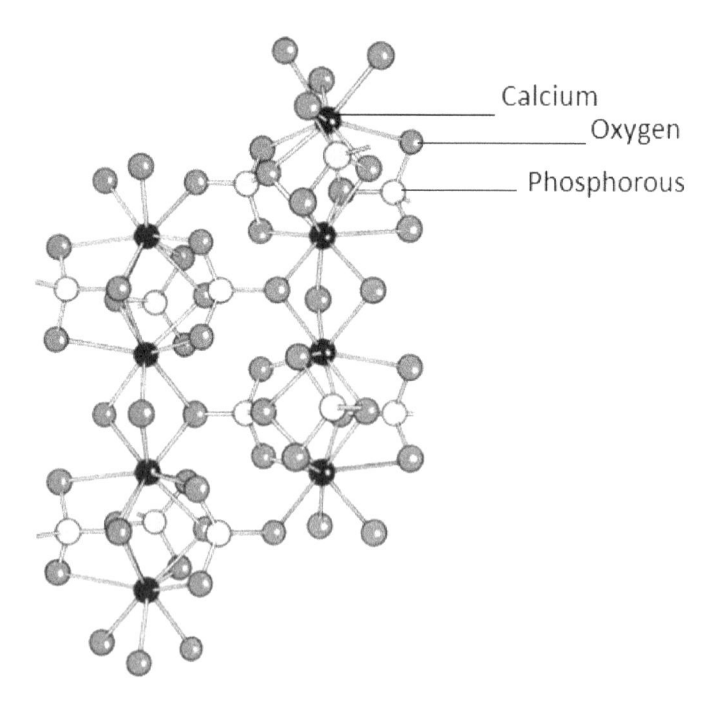

Fig. 9. HAp fluorapatite crystal structure (Color online).[33]

depends upon the factors like substrate material, coated material applications and thickness of coating layer as discussed in Table 4. There are several coating technologies available, each with unique capabilities; however, only a handful of these approaches are sufficiently trust worthy to be used in biomedical applications.[6] These techniques improve the substrate's corrosion resistance and biocompatibility at the same time. Amongst the various coatings techniques available, the TS is most widely applied in biomedical applications. The principle of this technique along with its various variants is explained in the following sections.

3.2. TS coating methods

TS fabricates mechanically robust and high-performance biomedical implants like replacement joints that are stronger than identical materials fabricated with conventional manufacturing techniques. In this context, during the recent year's, high-performance TS coatings has been examined widely over other manufacturing techniques. A number of industry giants use TS coatings for a variety of biomedical industry applications. TS coatings have been shown to improve biocompatibility, corrosion resistance and wear resistance in biomedical equipment which is valuable to the field of biomedical engineering. For example, the TS process is integral in fabrication of top quality medical implant prosthetics. These implants were made compatible for human use by applying dense HAp spray-based coatings by utilizing a TS process. The HAp-based coatings are preferred because of its resemblance with the main phase present in the bone and it forms strong implant–bone interfacial bond to improve prosthesis fixation.

The standard DIN EN 657 describes the basic concept and the meaning of the word thermal spraying.[34] Thermal spraying is a process in which the spray material in the form of powder, wire or rod is fed into the heat source. The spraying material may be fed inside or outside the spraying gun, and due to the high temperature, the feedstock material is melted thoroughly or superficially

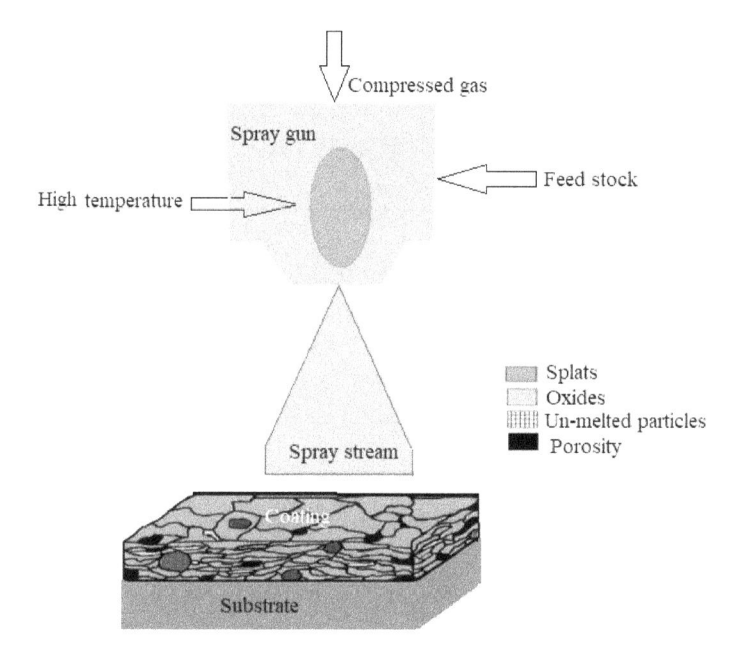

Fig. 10. Schematic of the TS coating process (Color online).

heated until it becomes soft. Then a stream of gas propels or accelerates the melted feedstock particles towards the substrate's surface, whereon impact with the surface, the particles deformed plastically in the shape of splats to form a coating. The basic principle of thermal spraying is shown in Fig. 10. During the process of coating, the substrate surface is not melted. However, it is subjected to medium thermal stresses. The splat's final shape developed by the molten particle impact on the solid substrate depends on fluid and heat transfer flow. During impact, a splat can fragment or remain intact, and solidifying in the form of a disc.

The inherent attributes associated with TS technologies like flexibility and versatility enable them to be used in a wide range of biomedical applications. Differed by the technique applied and materials used both for the coating and substrate, these techniques presently span the complete spectrum from support to repair to replacement applications. This topical review focused primarily on TS techniques and coating materials to combat corrosion and wear

Performance of thermally sprayed HAp coatings for biomedical implants 147

of metallic implants. In TS, precursors in solution or suspension form are used for coating the surface of the substrate to achieve desired properties. TS is classified into flame, electrical arc and plasma arc sprays. Different approaches like vacuum plasma spray (VPS), atmospheric plasma spraying (APS), suspension PS (SPS), HVOF and high velocity suspension flame spray (HVSFS) can be used to deposit coatings on biomedical implants. The different variants of the TS coatings employed to deposit coatings on biomedical implant are explained in the following sections.

3.2.1. *HVSFS*

HVSFS technology can produce coatings that are both thin and dense (10–70 μm). HVSFS is a refined version of HVOF coating method that uses suspensions to deposits a layer of coating on substrates with the required composition.[35] A schematic diagram showing the basic principle of HVSFS is shown in Fig. 11. In this

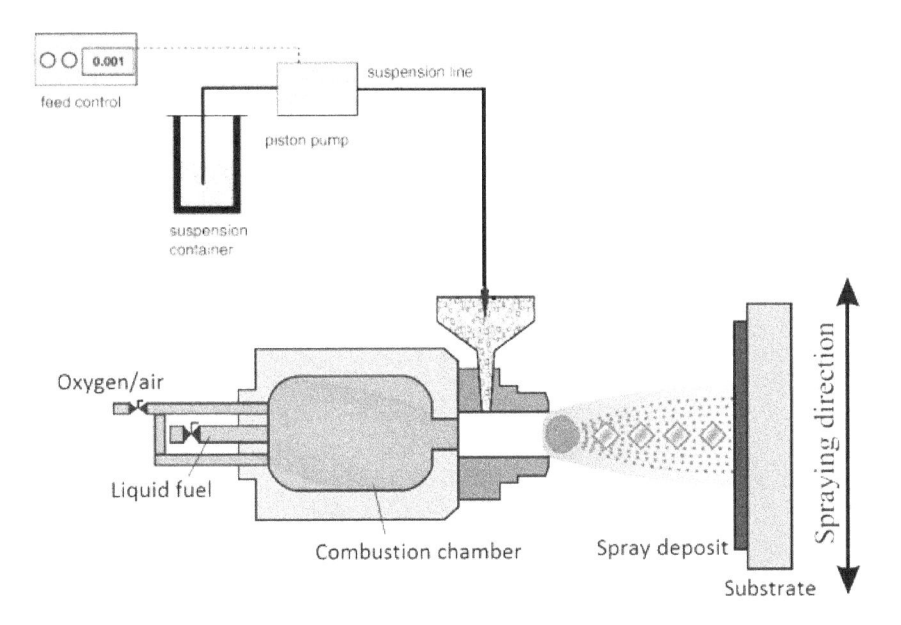

Fig. 11. Schematic diagram showing basic principle of HVSFS spraying (Color online).

technique, an inlet propels the material into the hot flame stream and accelerates the coating material towards the surface of target. The heated material is subsequently deposited on the substrate surface in the form of splats. The equipment employed in this technique is simple and economical as compared to the equipment used in electron beam machine (EBM) and ion beam machine (IBM).

3.2.2. *Plasma spraying*

The plasma spraying can be effectively applied for coating of complicated implant shapes with good chemical and microstructural control. The coatings with required thickness can be deposited by using PS technique. The gun employed for PS consists of a copper anode and a tungsten cathode. The heated gaseous mixture becomes a high-temperature plasma of about 15,000 K (ionized gas) and exits the nozzle with velocities approximately 800 ms^{-1} or above.[36] A feedstock material for coating in the powder form is discharged into the high-temperature plasma jet, mainly in a radial direction, as shown in Fig. 12.

Further, the PS techniques can be categorized depending upon the environment where it is performed.

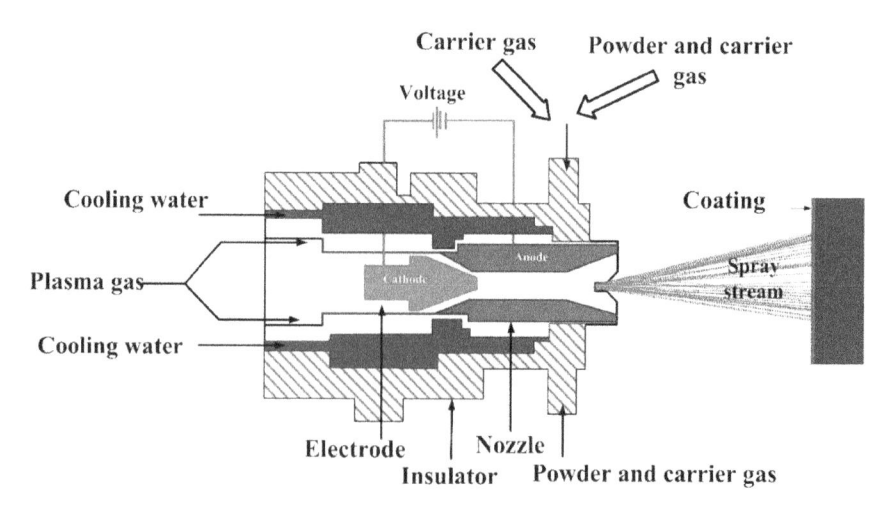

Fig. 12. Schematic of coating deposited by PS technique (Color online).[37]

Atmospheric PS: The atmospheric PS technique uses a direct current (DC) arc source to produce heat through a noncarburizing or nonoxidizing gas flow to deposit HAp coatings mainly. The power output for commonly used PS guns ranges from 20 kW to 250 kW, and various configurations of spray guns are available according to various spraying requirements. However, APS coatings present good resistance against abrasive, adhesive and sliding wear. APS is the most widely used technology for the development of HAp-based coatings on metallic implants which is also approved by food and drug administration (FDA).[38,39]

Vacuum or low-pressure PS (VPS/LPPS): As compared to the inert atmospheric spray process, this method is also carried out in a controlled atmosphere but at a pressure in the range 10–50 kPa which is slightly below the atmospheric pressure. The equipment used is costlier than that used in inert atmospheric spraying. Transferred arc etching is used to remove the oxide layer from the substrate surface before the coating deposition, and the coating process is conducted under the reversed polarity. During the deposition operation, the component is heated to a significantly high temperature to achieve a strong diffusion bonding between the component and the coating material. This method is mainly used for the deposition of coatings over the turbine blades. Several advantages of this technique are enhanced bonding, high coating density, better control over the coating thickness (even with a surface of an irregular shape), and excellent deposition efficiency.[30] Moreover, during low pressures, the diameter and length of the plasma become significant, and by the use of nozzles (convergent/divergent), a high-speed plasma jet can be generated. The absence of oxygen and the ability to operate at high temperature results in dense and more adherent coatings with lower oxide content.

3.2.3. *Flame spraying*

Flame spraying, according to chronology, is the first spraying technique invented by Swiss engineer MU Schoop in 1912, used initially for spraying of metals having low melting points such as tin and lead. Later on, it was extended for the spraying of refractory metals and ceramics.[40] The flame spray (FS) process deposits quality coatings by utilizing a heat source to melt the feedstock materials, usually available in powder, wire and rod forms. The flame spraying is performed using a heat source developing from the chemical reaction between the hydrocarbon gaseous fuel and oxygen. This heat source is in the form of a hot gas stream made up of combustion products that come out of the spray torch at high velocity. After that, spray material in powder form is supplied into the hot flame using the compressed air that gets melted and propelled towards the substrate's surface. If a wire is used as the spraying material, it is melted and atomized before it is propelled towards the substrate surface. Schematic for the techniques utilized is shown in Fig. 13. From the discussion, it has been found that there are a number of TS techniques available for performing biomedical coatings on the metallic implants. Table 5 summarizes some of the advantages and drawbacks of various TS techniques employed for depositing HAp coatings on metallic implants.

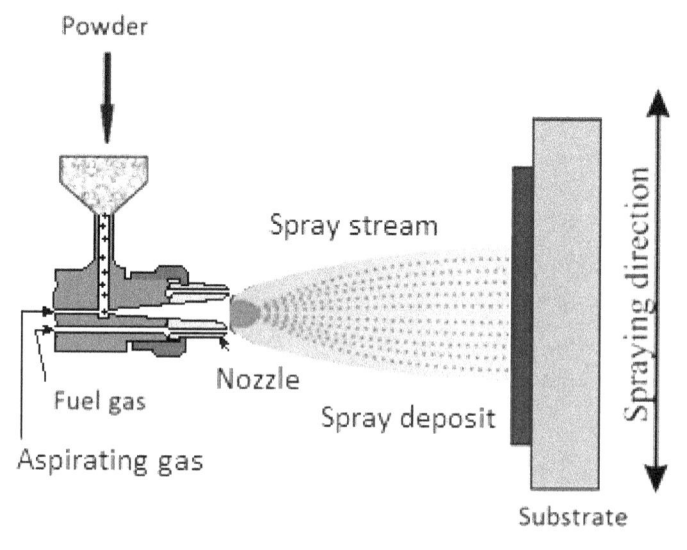

Fig. 13. Schematic of powder flame spraying technique (Color online).

Table 5. Advantages and drawbacks of different TS methods employed for depositing HAp coatings on metallic implants.

Technique	Thickness of coating layer	Advantages	Disadvantages	Ref.
PS	<20 μm	• Economical • Deposited smooth coating layer • Fast process	• Poor bonding among HAp film and surface of metal • Uniformity of HAp film density may effected	41–45
HVSFS	<50 μm	• No post treatment needed • Economical • Uniform coating layer • Nanometric porosity	• Needs higher temperatures	16, 19 and 46
FS	100–250 μm	• Most economical TS method • Easily adaptable • Porous coating	• Needs post treatments • Cracks may develop at lower temperatures	47

4. Performance of the Various HAp (Pure and Reinforced) Coatings Deposited by TS Techniques

In 1786, Werner was the first to designate HAp as a mineral. As a naturally occurring phosphatic compound, HAp is abundant on the planet. In terms of chemical formula and composition, HAp is remarkably similar to natural human bone.[48] On the other hand, there are concerns about HAp coatings' poor mechanical properties, such as excessive brittleness, wear and low fracture toughness, which limit their clinical utility in the orthopedic applications. High wear resistance of coated implants is an important property which guarantees its long-term success especially in load bearing applications. Corrosion is another important element to consider when choosing an implant material since the corrosive environment of the body, which is caused by blood, water, Na, Cl, proteins, plasma and amino acids, can adversely impair the mechanical and biological qualities of the implant. Corrosion can degrade an implant to the point where the metal can no longer bear typical loads, leading to the implant failure.[11] Therefore, to avoid the dissolving of surface oxide, which introduced more ions into the human body environment and caused several chronic diseases, an implant material should have a strong corrosion resistance. As a result, without compromising implant performance, it is necessary to improve the wear and corrosion performance of HAp coatings.

High hardness and fracture toughness are the two important mechanical properties that will improve the wear resistance of coated implants. HAp alone cannot possess high hardness and fracture toughness. However, its hardness and fracture toughness can be improved by reinforcing it by hard particles like ceramics in it. The improvement in properties by reinforcing the HAp with hard particles has been reported in literature earlier. The effect of reinforcing different hard particles on the various properties (microhardness, fracture toughness and tribological properties) of the HAp coatings has been illustrated in Table 6. The detailed study regarding the effect of each individual hard reinforcement on the HAp coating has been explained in the following sections.

4.1. Role of zirconia/YSZ reinforcement in performance of HAp coatings

Because of its good strength and stress-induced phase transformation toughening nature, yttria-stabilized zirconia (YSZ) reinforcement in suitable proportion can greatly improve the mechanical properties of HAp coating.[52] It was observed that YSZ reinforcement (10, 20 and 30 wt.%) enhances adhesive strength, hardness, sliding wear rates and corrosion resistance of HAp coatings as illustrated in Fig. 14.

Similar findings were also noticed on the improvement in mechanical properties of YSZ reinforced coatings in literature. Lei et al.[57] observed 113% growth in the bond strength of 10 wt.% YSZ

Table 6. Performance of HAp reinforced coatings.

Feedstock material	Metallic implant	Technique	Findings	Ref.
• HAp/polymer composite coating	MS	FS	• Composite coating shows good fracture toughness, reasonable dissolution behavior, better adhesion strength and increased young modulus	49
• HAp • HAp + CNT	Ti6Al4V	PS	• CNT reinforced HAp coating shows enhanced wear resistance in comparison with pure HAp coating • CNT reinforcement enhances fracture toughness and biocompatibility • Bridging and stretching of CNT assist in wear debris pinning resulting in less weight and volume loss of reinforced coating	39
• HAp • HAp + SiO_2 (10 and 20 wt.%)	SUS 304SS	PS gas tunnel type	• With rise in silica content, the coating becomes dense and hard • Highest abrasive wear resistance is shown by 20 wt.% SiO_2 reinforced coating • Presence of SiO_2 enhances the adhesive strength due to improvement in coating bonding strength at interface	50
• HAp • HAp + 10 wt.% (80 Al_2O_3 and 20 TiO_2)	Ti6Al4V	PS	• Reinforcement leads to dense and crack free coating • Tensile strength increases due to reinforcement • Reinforced coatings exhibit better corrosion resistance than pure HAp coating	51
• HAp • HAp + YSZ (10, 20 and 30 wt.%)	SS 316L	PS gas tunnel type	• YSZ reinforcement remarkably increases the hardness, adhesive strength and sliding wear resistance of HAp coatings	52

(Continued)

Table 6. (Continued)

Feedstock material	Metallic implant	Technique	Findings	Ref.
• HAp	Ti	HVSFS	• YSZ reinforcement also improves cell attachment and cell adhesion during cell culture test • Reinforced coatings show better corrosion resistance in simulated body fluid (SBF) than pure HAp coating	53
• Zn doped HAp coating	Ti	FS	• Di-ethylene glycol coatings are more stable in solution of SBF than water suspension coatings. • Di-ethylene glycol coatings are more reliable and crystalline • Zn doped HAp coating has good biocompatibility and better antibacterial properties	54
• HAp • HAp + TiO_2	SS 316L	HVSFS	• Both wear resistance and tensile strength of reinforced coating increases in comparison with single layer pure HAp coating	55
• HAp • HAp + 10 wt.% Al_2O_3 • HAp + 10 wt%ZrO_2	SS 316L and Ti6Al4V	VPS	• As-sprayed coatings shows better wear performance in comparison to that of the heat-treated coatings • Microcracks and pores reduce after heat treatment (HT) due to diffusion process	30
• HAp • HAp + TiO_2	Ti6Al4V	FS (high-velocity)	• Both coatings were helpful in reducing the corrosion rates • The reinforced coatings were observed to retain its identity even after corrosion testing • HAp coating showed good in vitro corrosion resistance	29
• HAp • HAp + graphene nanoplatelets (1 and 2 wt.%)	Ti6Al4V	PS	• Graphene reinforcement results in crystallinity improvement and cell proliferation rate of composite coatings • With increase in plasma power, the crystallinity decreases but hardness, elastic modulus and fracture toughness increases	56

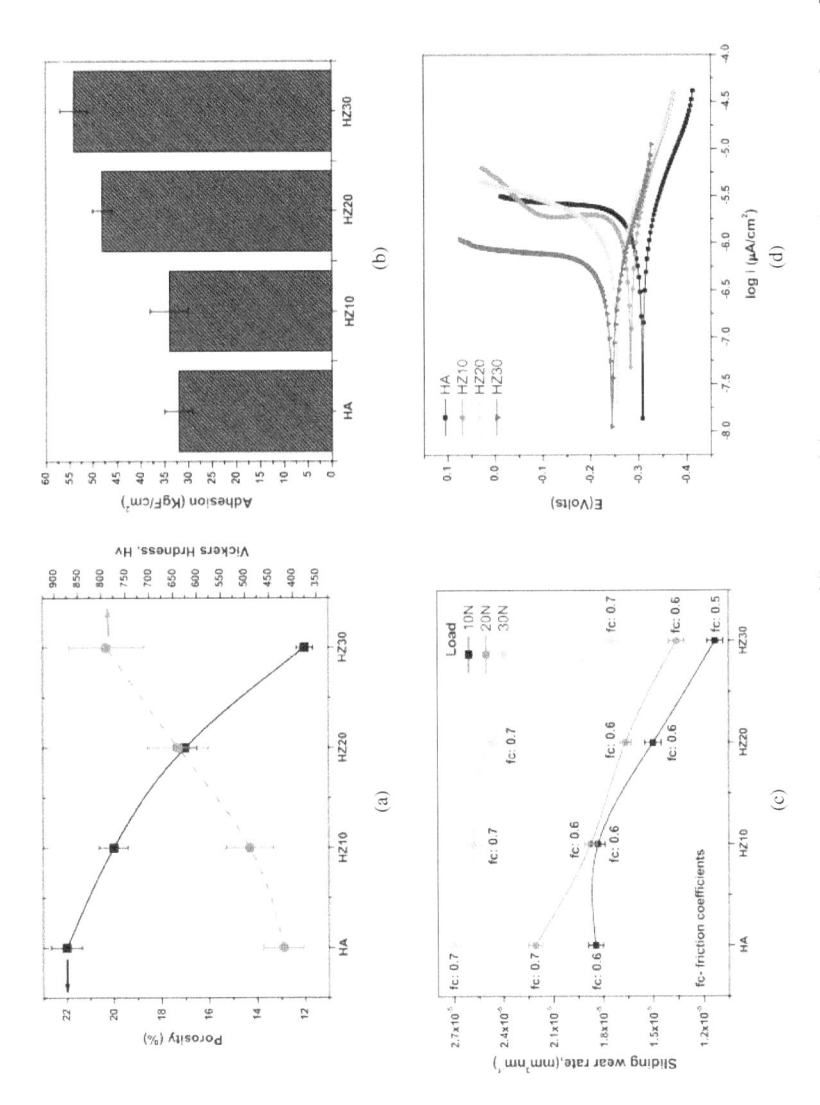

Fig. 14. Effect of addition of YSZ (10, 20 and 30 wt.%) on the (a) microhardness and porosity values of coatings, (b) adhesion strength of coatings, (c) friction coefficients and sliding wear rate readings and (d) potentiodynamic polarization curves of YSZ reinforced and pure HAp coatings (Color online).[52]

reinforced coating in comparison with pure HAp coating. Balani et al.[58] recorded a comparable increase in bond strength, of about 110%, with YSZ reinforcement in pure HAp powder. Gu et al.[59] observed that reinforcement of YSZ in pure HAp results in positive influence on the mechanical properties such as bond strength, Young's modulus (YM) and Knoop hardness of composite coating as shown in Fig. 15. They compared PS HAp/Ti6Al4V, and HAp/Ti6Al4V/YSZ composite coatings with pure HAp coating.

Reports on PS YSZ reinforced HAp coatings revealed that composite coatings produced good mechanical properties in comparison with bulk HAp alone owning to the addition of second phase particle.[60]

4.2. Role of CNTs reinforcement on the HAp coatings

To improve the mechanical strength of HAp-based composites, a few researchers have tried carbon nanotubes (CNTs) as reinforcement in HAp coatings.[39,58,61–63] First to report such an attempt was Chen et al.[63] They demonstrated that by adding upto 20% CNTs into HAp, a composite coating formed which resulted in a significant improved wear resistance. Meng et al.[61] found that by adding 10% CNT to an HAp, its flexural strength increased as compared to pure HAp coatings with the addition of 2% CNT to HAp. Kaya[62] found an enhancement in the bond strength of composite coating deposited on Ti–6Al–4V implant substrate. The elastic modulus, fracture toughness, and hardness of these coatings were also showing similar trends.[62,63] Balani et al.[58] found that a PS composite coating of HAp reinforced with 4 wt.% CNT increased fracture toughness by 56% and reduced wear volume by 60%.[39] Bridging and stretching of CNTs assist in wear debris pinning, which results in a less weight and volume loss of reinforced coating as shown in Figs. 16 and 17, respectively.

In another study Tercero et al.[38] compared the three PS coatings: pure HAp coating, HAp + 20 wt.% alumina, and HAp + 18.4% Al_2O_3 + 1.6 wt.% CNTs reinforced coatings. They found that by simply using alumina as a reinforcement resulted in the increased

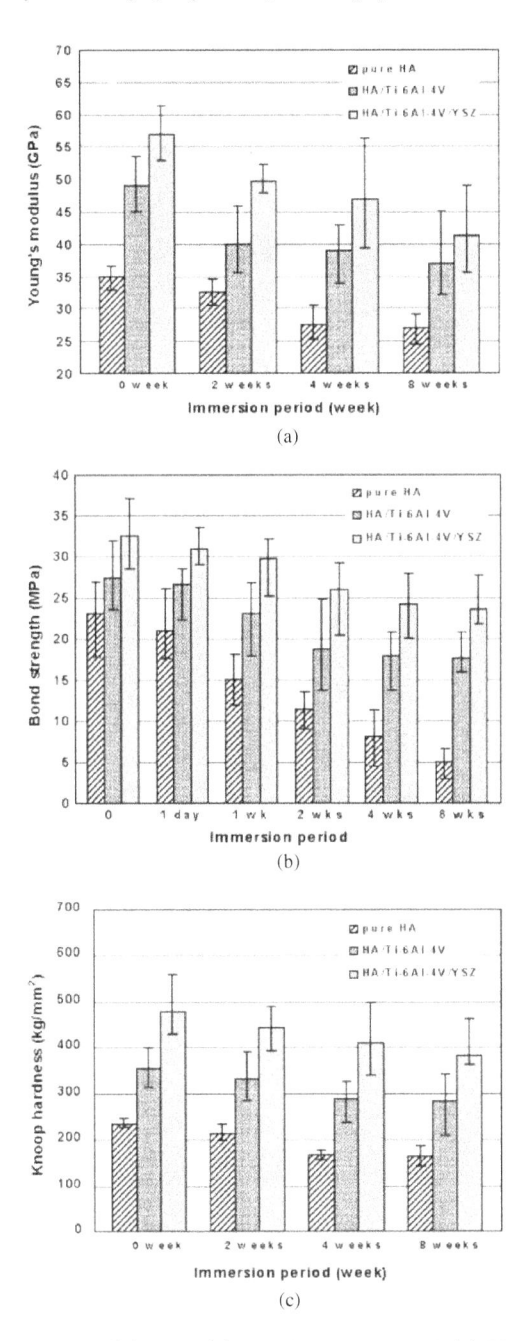

Fig. 15. Comparison of (a) YM, (b) bond strength and (c) Knoop hardness of different combinations of composite coatings.[59]

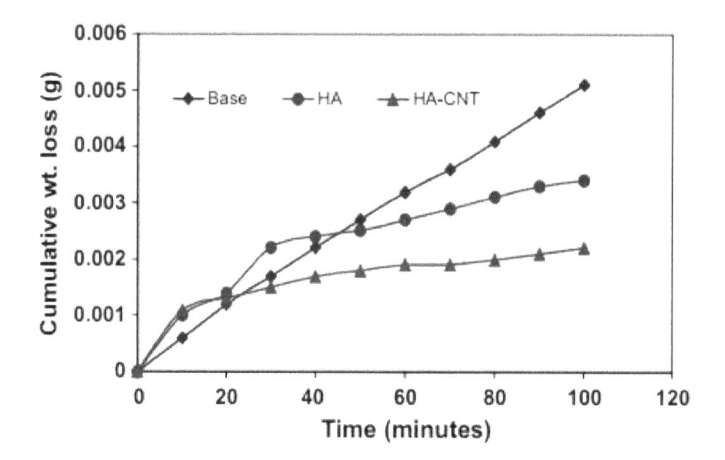

Fig. 16. Comparison showing wear rate of base metal, HAp coated, and HA-CNT composite coating (Color online).[39]

fracture toughness by 158%. On the other hand, by using CNTs with alumina causes increased in fracture toughness by 300%, as illustrated in Fig. 18. Authors after comparing the growth rate of cells in 3 coatings in SBF condition finds that after one day of immersion, pure HAp had higher cell growth rate (23.75 cm^2/cell) in comparison with hap reinforced with Al_2O_3 (14.2 cm^2/cell) and HAp reinforced with Al_2O_3-CNTs (15.5 cm^2/cell) coatings. Whereas after seven days of immersion in SBF, the results of pure HAp (52.5 cm^2/cell) and Al_2O_3-CNTs reinforced hap (54.2 cm^2/cell) coatings were comparable but for Al_2O_3 reinforced hap coating it was less (47 cm^2/cell).

4.3. *Role of silica reinforcement on the HAp coatings*

According to some of the findings,[64-66] silicon is an important bioactive material for bone and muscle bonding, as well as a cross-linking agent in the connective tissue. Silica is believed to perform three functions in HAp coatings. The first is that it aids in gene expression and cellular growth. Secondly, it serves a chemical function as the bonding nature to the bone of bio-glasses is related to the

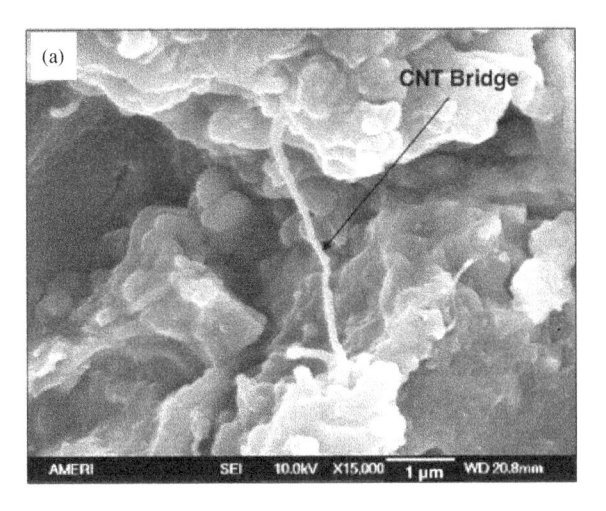

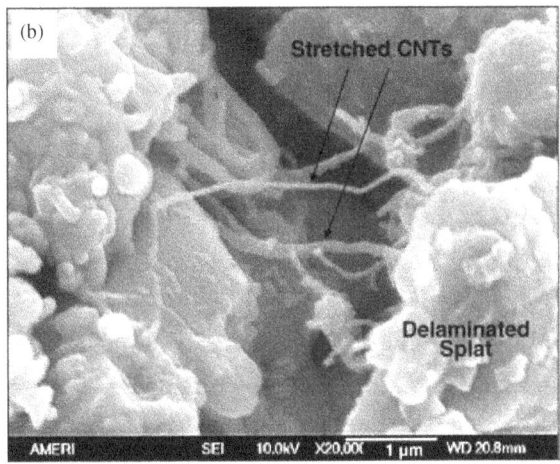

Fig. 17. High magnification scanning electron microscope images showing (a) CNTs bridging of splats, and (b) wear debris anchoring by stretched CNTs.[39]

in vivo solubility of these glasses, which is a function of their silica content. Third, silica particles appear to boost the strength of an HAp coating by particle mediated reinforcing, resulting in crack arrest or its deflection. Fused silica (i.e. amorphous SiO_2) is a promising choice for improving HAp coating mechanical properties.[49] Morks[50] observed that with a rise in the silica content in HAp, the coating becomes dense and hard. Highest abrasive wear resistance is

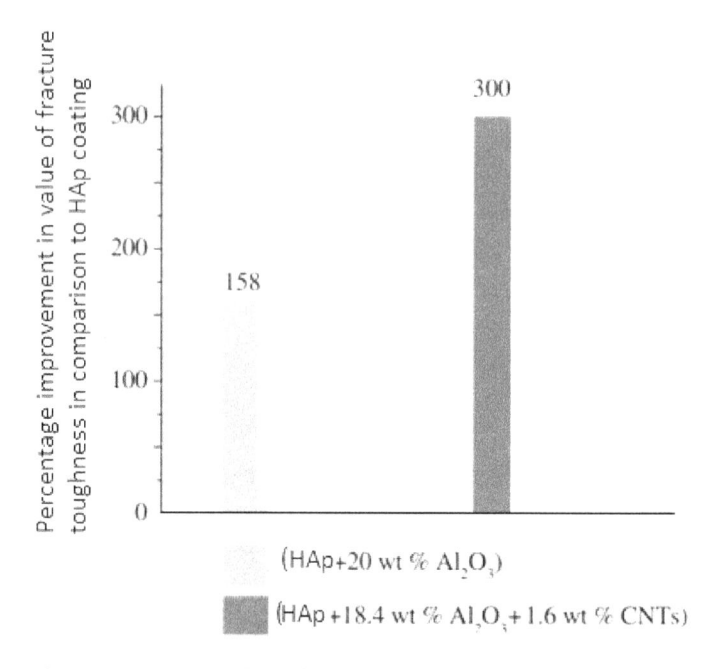

Fig. 18. Improvement in value of fracture toughness with reinforcing alumina and CNTs in the HAp powder (Color online).[38]

shown by 20 wt.% SiO_2 reinforced coating. Presence of SiO_2 enhances the adhesive strength due to improvement in coating bonding strength at interface. The microhardness and wear resistance of coatings showed upward trend with addition of 20 wt.% SiO_2 reinforcement in the hap coatings. Microhardness and wear resistance of 20 wt.% reinforced coating also improves. The percentage changes in the various coatings properties with the addition of 10 and 20 wt.% SiO_2 reinforced coating are shown in Figs. 19(a)–19(c).

4.4. Role of titania reinforcement on the HAp coatings

Previous study has found that Ti could be effective as a porous cell carrier material with qualities including strong permeability and excellent biocompatibility that help to boost cell vitality.[67] To investigate the influence of TiO_2, Li et al.[68] used an HVOF spray method to create pure HAp and TiO_2 reinforced (10 and 20 wt.%) HAp

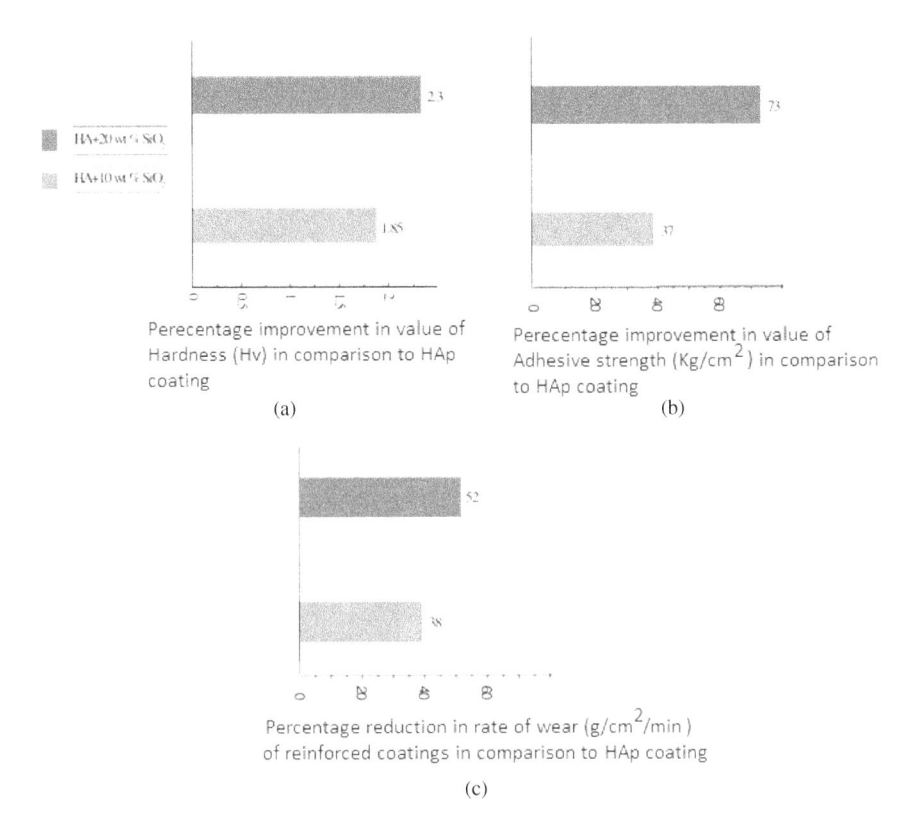

Fig. 19. Percentage changes in the properties with 10 and 20 wt.% SiO$_2$ reinforced in HAp coating (Color online).[50]

composite coatings and thereafter various properties of coating were compared like shear strength, bond strength, YM and fracture strength. The influence of TiO$_2$ on various coating properties is shown in Figs. 20(a) and 20(b).

They found that when TiO$_2$ content increased, the shear strength and fracture strength improved, but bond strength decreased. When TiO$_2$ content is reinforced by 10 wt.%, the YM increases. Whereas, when TiO$_2$ content is increased from 10 wt.% to 20 wt.%, this characteristic decreased. The HAp + 10 wt.% (80 Al$_2$O$_3$ and 20 TiO$_2$) coatings were prepared by Singh et al.[51] on Ti6Al4V alloy by PS coatings. Reinforcement leads to dense and crack-free coating. Tensile strength and microhardness increases

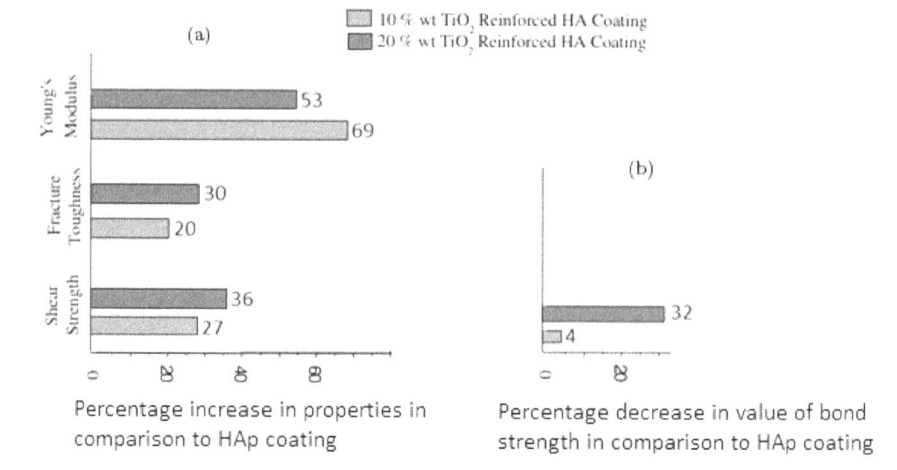

Fig. 20. Influence of TiO$_2$ on various coating properties (Color online).[68]

due to reinforcement. Reinforced coatings exhibit better corrosion resistance than pure HAp coating. The effect of TiO$_2$ reinforcement on various coatings properties like tensile strength, microhardness and corrosion resistance is shown in Figs. 21(a)–21(c).

4.5. Role of calcium silicate reinforcement on the HAp coatings

Previous studies have shown that adding reinforcement to pure HAp generally lowers the percentage of amorphous phases.[69-71] Calcium silicate (CS; CaSiO$_3$) is a bioceramic material that is extremely bioactive as well as degradable in nature. This material can be utilized to regenerate bone tissue.[72] Apatite of CS (wollastonite) forms at a faster rate than other bioactive glass and glass-ceramics. Because of its high rate of degradation, CS cannot be employed alone as a coating material, which can raise the surrounding environment pH value and disrupt cell vitality.[73] This drawback may be overcome by using CS in a composite form. CS increases the mechanical properties of the composite systems, making CS an excellent material for use as reinforcement filler in composite form.[74] Recent research on the HAp–CS composite suggests that the HAp–CS composite could be used for load-bearing applications.

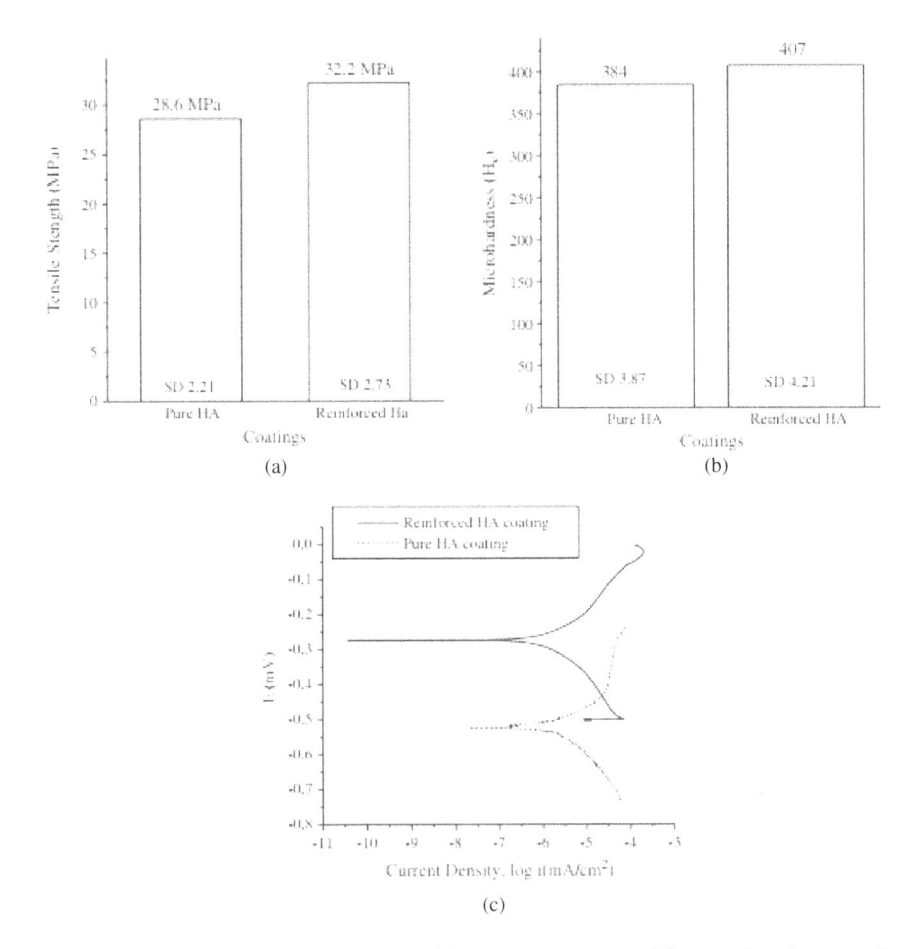

Fig. 21. Effect of reinforcement on (a) tensile strength, (b) microhardness and (c) corrosion resistance of pure HAp and reinforced coatings.[51]

Beheri et al.[75] developed the HAp–CS composite and discovered that the addition of CS to HAp resulted in improving the mechanical properties. Lin et al.[76] produced the HAp–CS composite with different wt.% ratios in another investigation. The researchers discovered that increasing the CS concentration in HAp boosted the proliferation rate of bone marrow mesenchymal stem cells.

Recently, Singh et al.[77] reported the improvement in corrosion performance of the HAp reinforced CS (10 and 20 wt.%) coatings deposited on Ti alloy by utilizing PS technique. The corrosion

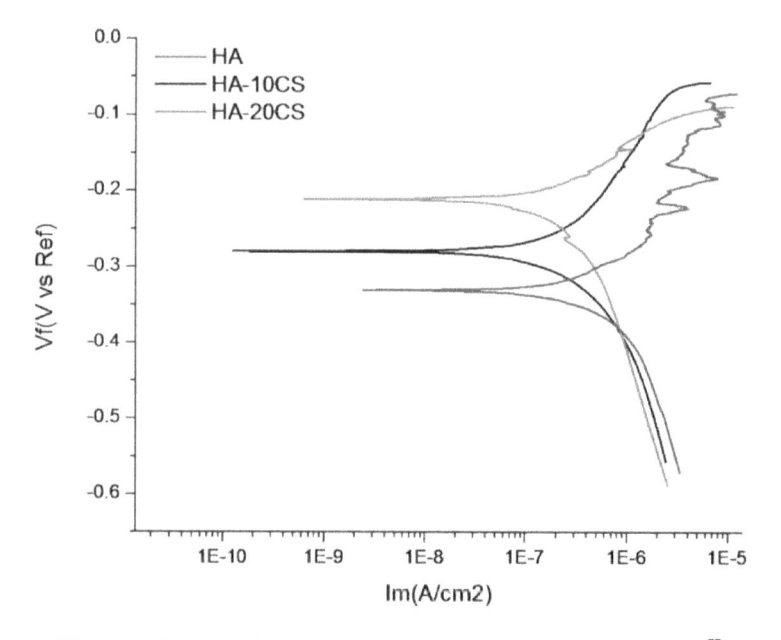

Fig. 22. Potentiodynamic polarization curves (Color online).[77]

properties of the PS coated samples were assessed by using electrochemical measures in an SBF solution. Figure 22 shows the potentiodynamic polarization curves of HAp–CS composite and pure HAp PS coatings. It can be seen that the HAp coating with CS reinforcement has greater corrosion resistance and it increases with increase in reinforcement (wt.%) of CS in HAp. The presence of a high portion of crystalline HAp phase in the HAp–CS coatings may explain their improved corrosion resistance as shown in Fig. 23. The pure HAp, HAp-10 CS, and HAp-20 CS PS coatings have crystallinities of 76.77%, 77.36% and 79.82%, respectively.

4.6. *Role of magnesium reinforcement on the HAp coatings*

One of the most critical factors affecting osseointe-gration is the presence of porosity on the implants surface. These pores are ideal for osteoblastic cell penetration and the development of connective

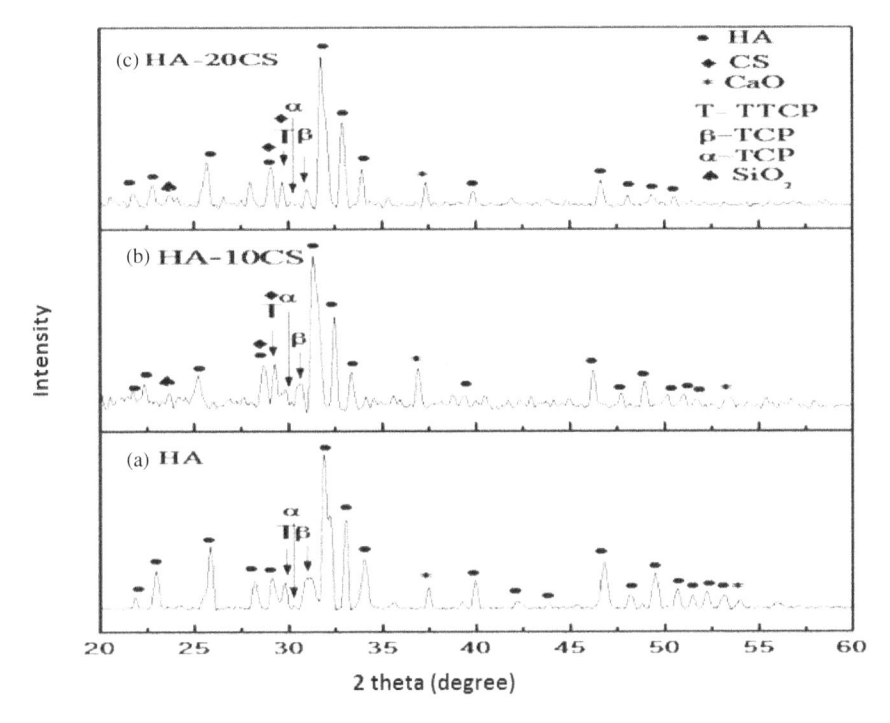

Fig. 23. XRD patterns of deposited coatings.[77]

tissues among the implant and the bone. The application of biodegradable materials on implant surfaces can aid in the creation of in vivo porosities. Magnesium (Mg) has the potential to be explored as a biodegradable phase among biodegradable materials.[78] Rezeai et al.[79] deposited a double layer coating in which middle layer consisting of HAp was applied by PS method to enhance corrosion resistance and thereafter top layer consisting of HAp/Mg was applied by cold spray (CS) technique. By generating porosities in the bodily fluid, the biodegradable Mg phase at the top can aid osseointegration. Figure 24 shows the potentiodynamic polarization curves of HAp/HAp–Mg double layer coatings. HAp-10 wt.% Mg shows highest corrosion resistance as middle dense HAp layer assists in enhancing the corrosion resistance. Porosities emerge from the dissolving of the Mg phase at the top of coated samples. These porosities are preferable areas for growth of bone. However, because

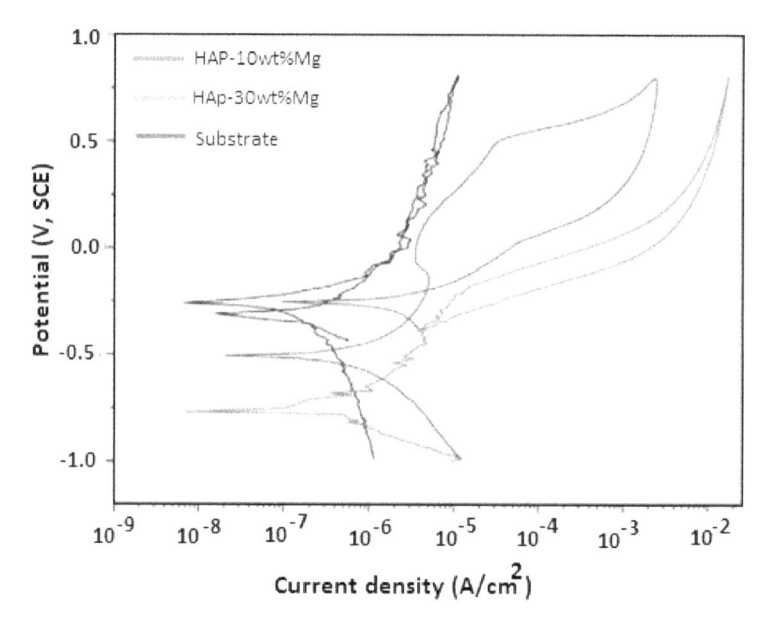

Fig. 24. Potentiodynamic polarization curves (Color online).[79]

of their porous nature, the Hap–Mg layers at top cannot improve the substrate corrosion resistance.

5. Role of HT on the Various Properties of HAp Coatings Deposited Using TS Processes

Knowledge of mechanical and microstructural properties is critical for improving HAp-based biomedical implant coatings. HT is an effective and promising approach to improve mechanical properties by reducing the amorphous phases from during thermal spraying of HAp coatings. Therefore, the effect of HT on various coatings properties has been discussed in this section.

5.1. Microhardness

Microhardness is a critical property for HAp coating that will be employed in the host body. Hardness depends upon phase ratio i.e. calcium/phosphorous ratio.[80] The hardness of the coatings decreases with an increase in amorphous phase in the coatings. The amorphous

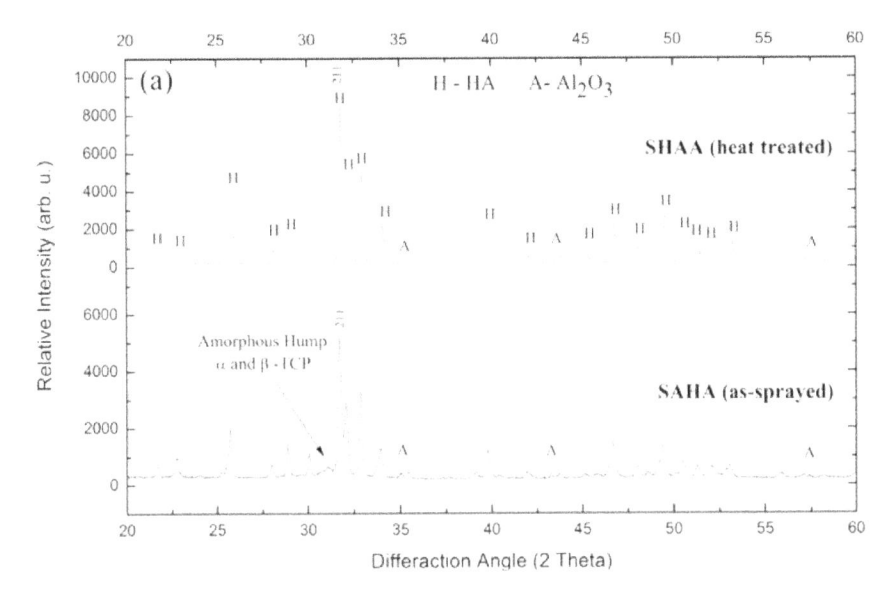

Fig. 25. XRD patterns of HT and as-sprayed coatings (Color online).[30]

phase in the as-sprayed coatings increases as a result of high temperature involved in thermal spraying. Therefore, HT may be adopted to reduce these amorphous phase present in the as-sprayed coatings. Singh et al.[30] performed HT at 700°C for 1 h and noticed that amorphous phase present in the as-sprayed coatings reduced completely. There occurs a transformation from the amorphous phase to crystalline phase in the as-sprayed HAp coatings during HT process and the same has been confirmed by the XRD patterns of the HT coatings (Fig. 25). Gross et al.[31] hypothesized the concept of amorphous phase to crystalline phase transformation during the HT process. The amorphous phase is mostly made up of dehydroxylated calcium phosphate, according to the findings. HT causes crystallization of hydroxyl-rich portions of the amorphous phase, which is then followed by hydroxyl ion diffusion. This resulted in the formation of HAp in the coatings which further leads to increasing the crystalline phase of the coatings.

Microhardness of some commonly used pure and reinforced (as-sprayed and HT coatings) are summarized in Table 7.

Table 7. Comparison of hardness values of commonly used HAp coatings before and after HT.

Feedstock material	HT conditions	Coating technique	Microhardness before HT (Hv)	Microhardness after HT (Hv)	Increase in crystallinity after HT (%)
HAp	HT temperature: 700°C	PS	486	507	
HAp + 10 wt.% (80 Al$_2$O$_3$ and 20	Time duration: 2 h[51]		598	618	
HAp	HT temperature: 800°C Time duration: 2 h[82]	LVOF	288 ± 10.5	367.2 ± 4.8	94.4
HAp	HT temperature: 700°C Time duration: 1 h[77]	APS	276 ± 91.2	316.3 ± 41.4	93.8
HAp + TiO$_2$	HT temperature: 750°C Time duration: 1 h[83]	PS	523 ± 10	874 ± 26	75.6
HAp	HT temperature: 800°C	PS	458	645	
HAp + 10 wt.% (80 Al$_2$O$_3$ and 20	Time duration: 2 h[84]		496	643	

5.2. Young's modulus

The YM, which is a measure of coating stiffness, is another important mechanical property worth considering. The significant modulus difference among the implant metal and the deposited coating or the bone suggests that strains may develop at these interfaces. These stiffness differences resulted in dissimilar deformations causing stress. As a result, deterioration of the bone matrix may occur at the coating–bone interface, which is driven fundamentally by a mismatch in modulus. Therefore, for better long-term performance of the coated implant, the modulus difference should be less. Li et al.[85] studied the influence of HT on YM of HVOF-sprayed HAp/ nano-zirconia composite coating. HT was carried out at temperature of 750°C for 30 min duration and the effect of HT is shown in Fig. 26. It can be noticed that before HT process, there is a significant difference among the YM within different regions in the

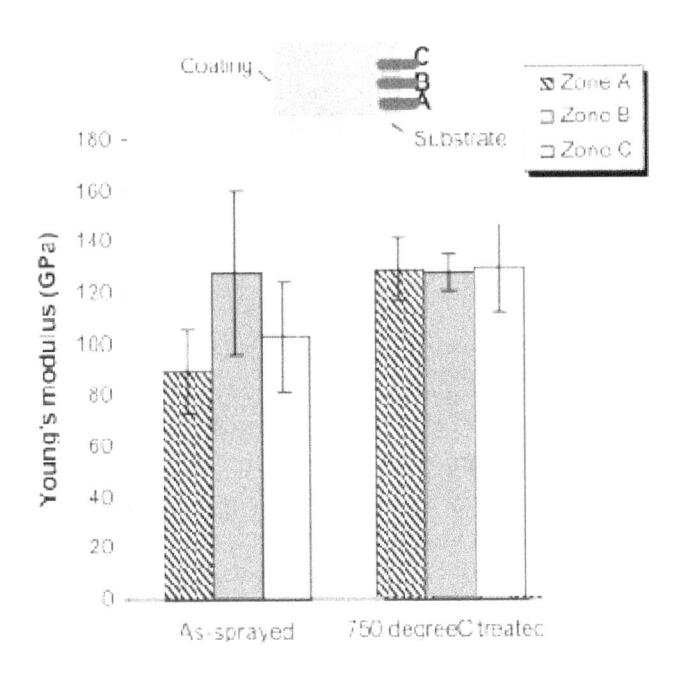

Fig. 26. YM's of the HAp/nano-zirconia composite coatings showing the influence of HT at different locations.[85]

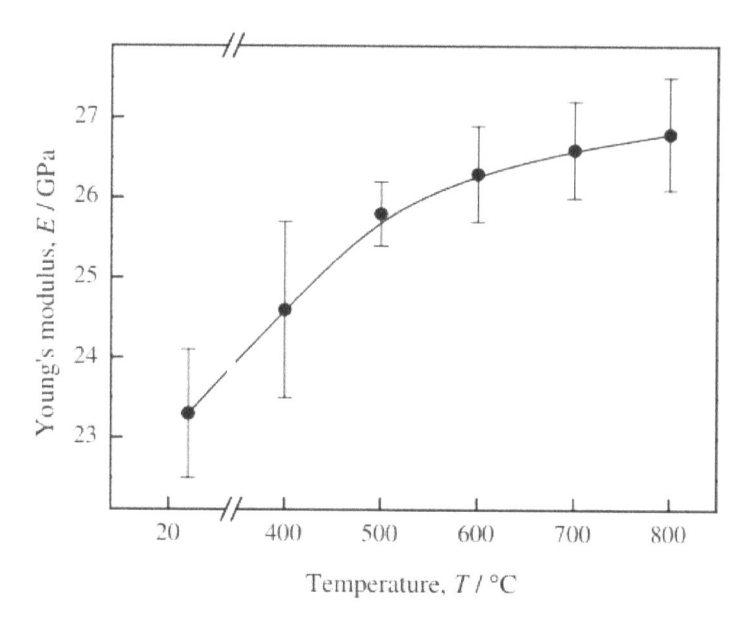

Fig. 27. Variation of YM's due to rise in HT temperature.[86]

as-sprayed coating. But HT improves the coating homogeneity and as a result, the uniform value of YM can be achieved. The maximum value of YM for Ti-6Al-4V substrate is 113 Gpa.

Therefore, before HT, significant differences may occur in YM within the as-sprayed coating. Hence YM is a function of HT and it increases with increase in HT temperature also. Yang et al.[86] also observed that the YM of the as-sprayed coatings increases with rise in HT temperature as shown in Fig. 27. The YM is a measure of interatomic binding forces and the HAp YM depends on crystallization. The YM and the crystallinity index of the HAp have a linear relationship, as shown in Fig. 28.

It is plausible to believe that the crystallinity of HAp coatings is a determining factor for the increase of YM, fracture toughness and the bonding strength of the HAp coatings.

Li et al.[87] also noticed that the change in YM's caused by crystallization (phase transformation) plays a vital role in enhancing the YM of the coating (Fig. 29). The significant difference can be noticed in the YM after HT. Another reason for improvement in

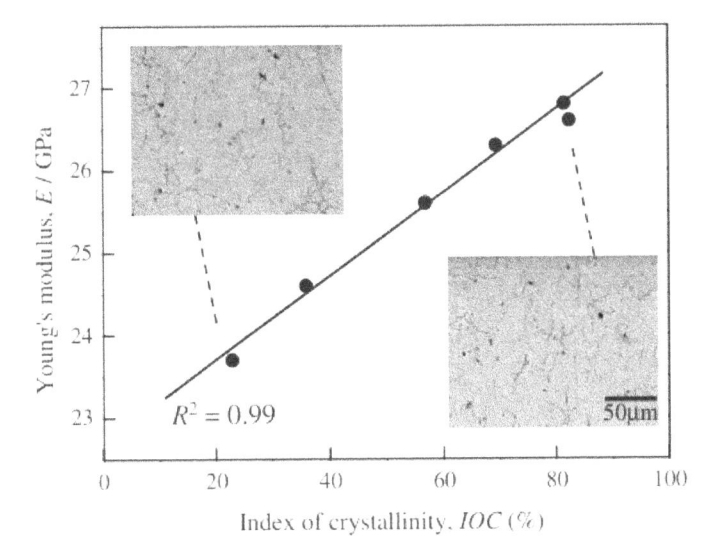

Fig. 28. A linear correlation among the YM and crystallinity of the HAp coatings.[86]

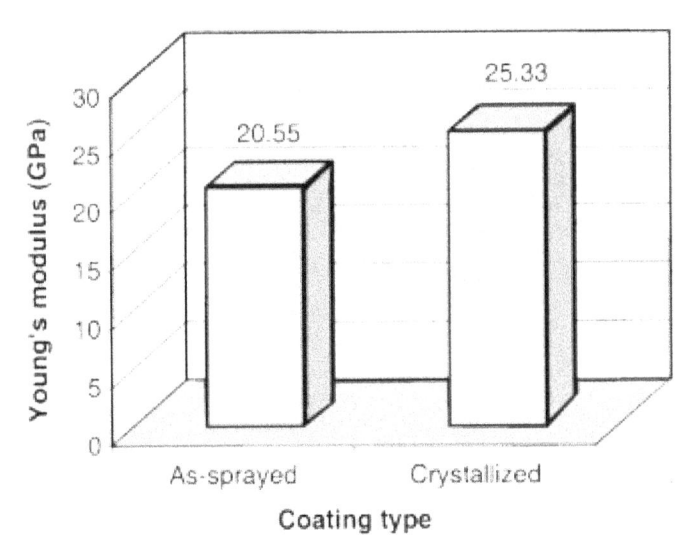

Fig. 29. YM of HAp coatings as a function of crystallization treatment.[87]

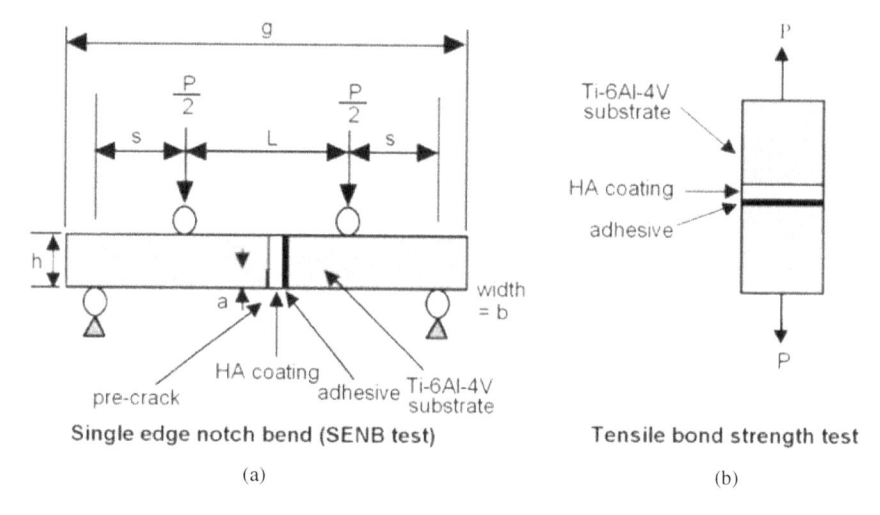

Fig. 30. Different set-ups used to measure adhesion strength. (a) SENB and (b) tensile bond strength test.[88]

YM may be diffusion among distinct splats caused by the HT at 750°C. In addition to this, the residual stresses generated during the deposition of coating can be released by annealing at 750°C which also contributes in improvement of YM.

5.3. *Adhesion strength*

Long-term stability of the HAp coated implants comes from better adhesion strength among the coating and substrate. Adhesion strength depends upon two aspects: chemical bonding and mechanical interlocking. Adhesion may be defined as an attractive force among various layers that holds the two surfaces together. This adhesive force provides resistance against the applied stresses that tries to separate the two surfaces. Various tests used to determine the coating's adhesion strength[88] is shown in Fig. 30. Adhesion strength is an important parameter in clinical applications like dentistry and artificial joints. Due to mismatch in co-efficient of thermal expansion at substrate and coating interface, there is a generation of residual stresses at the interface of coatings and substrate, which often leads to low adhesion strength and cracks in

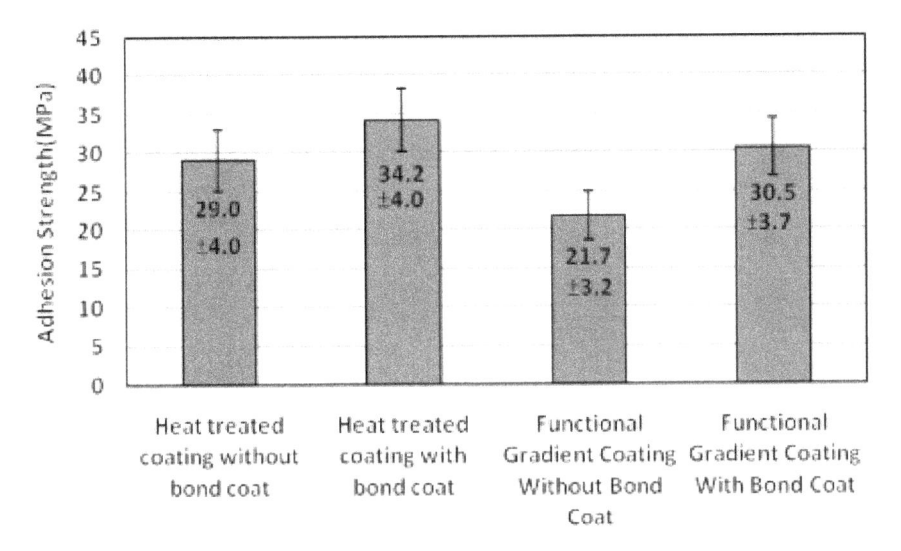

Fig. 31. Adhesion strength of HT and functional graded coatings (Color online).[77]

HAp coatings. Therefore, to increase the adhesion strength of the as-sprayed coatings, HT is an effective approach as HT promotes phase transformations and refines the microstructure which leads to improved adhesion and cohesion strength.[89] Singh et al.[77] conducted HT at 700°C for 1 h on the middle layer of PS HAp coatings.

The authors compared the adhesion strength of HT and functional graded coatings. As shown in Fig. 31, the HT coatings exhibit better adhesion strength in comparison to the functional gradient coatings. It is believed that HT leads to improvement in adhesive, cohesive and shear strength as well as fracture toughness of HAp coatings. Furthermore, HT coatings have less microcracks and low porosity which also result in improved adhesion strength of HT coating in comparison to functional graded coatings as shown in Fig. 32.

Li et al.[85] and Rocha et al.[83] also observed similar findings on improvement of adhesive strength of coatings after the HT process as shown in Fig. 33. The authors[83,85] reported that the annealing at 750°C significantly improves adhesion strength of coatings owning to crystallization of amorphous phase after HT.

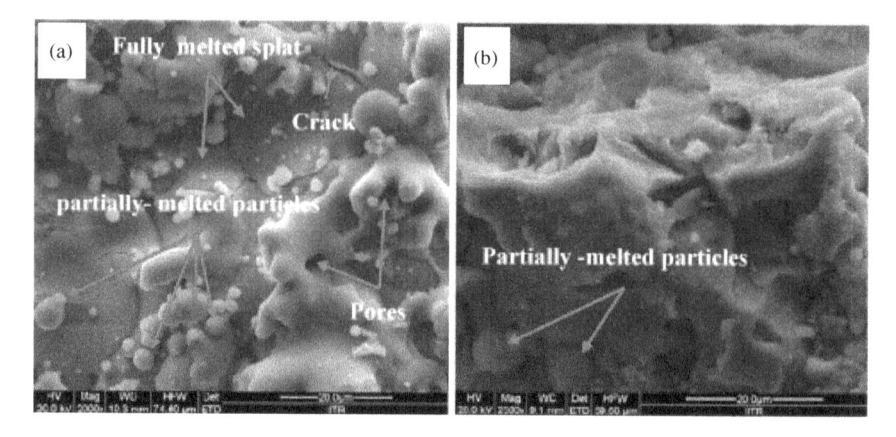

Fig. 32. SEM micrographs of (a) functional graded coating and (b) HT coating (Color online).[77]

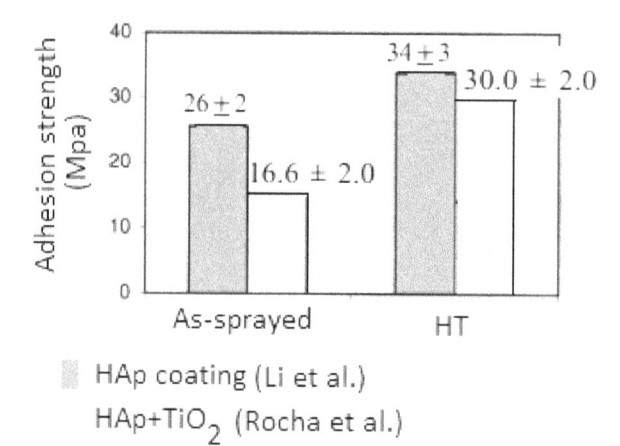

Fig. 33. Influence of HT on adhesive strength of HAp-based coatings.

6. Current Challenges in Coating of Metallic Implants by HAp Feedstock

When employed as bone-supporting implants in real applications, inappropriate coating on surface of the metallic implants caused by variations in process parameter during deposition process increases the risk of rapid collapse.[90,91] During the healing process of tissue, the poor layers of coating will exfoliate, causing significant damage.

Due to their weak strength, the pieces of coating will fall off from the surface and cause damage to the surrounding body parts.[92,93] The most difficult aspect of these coating implants is maintaining the HAp film's stability. The natural cell mechanisms occurring in the body in the presence of body environment factors, such as pH, osteoclast cells and water content, causes the complete degradation of HAp coatings in 4–5 years. Cell mediation or dissolution can initiate HAp degradation *in vivo*. The rate of degradation also relies upon the HAp chemical properties, Ca/P ratio, size of crystal, porosity and crystallinity.[94] After 4–5 years, dissolution begins on the surface and progresses down to the metallic substrate. Hence, a stable film of HAp will provide reliable implant in load bearing applications. Successful HAp coating with consistency on metallic implants is further hampered by operational issues during the coating process. The biomedical implant is also affected by porosity and cracks on the coated surface. Porosity is an important parameter as the porosity within coating layers will assure its bioactivity for implants. Pores and surface roughness can be increased to improve the adhesive strength even further. Porous implants coated with HAp have advantages in terms of increasing bioactivity. The YM of these implants is nearer to that of bone, addressing the effect of stress shielding. The migration and multiplication of osteoblast cells, as well as vascularization, are all possible with a porous coating for bone repair. *In vivo* experiments have revealed that the pores aid cell migration, tissue development, and waste transport.[95] Therefore, manufacturability of complicated shapes with appropriate porosity architecture is significant. Another challenge that should be given a thought is the HAp decomposition during thermal spraying that gives rise to secondary phases (OHA, TTCP and α-TCP/β-TCP), amorphous phase & calcium oxide due to high temperatures involved in these processes. To avoid decomposition and to achieve high-crystalline HAp coatings, precise control of TS conditions is essential and challenging task.

In addition to this, patients frequently require revision surgery after undergoing orthopedic surgery for the knee or hip joint, which is critical. Longer healing time, implants with no antibacterial

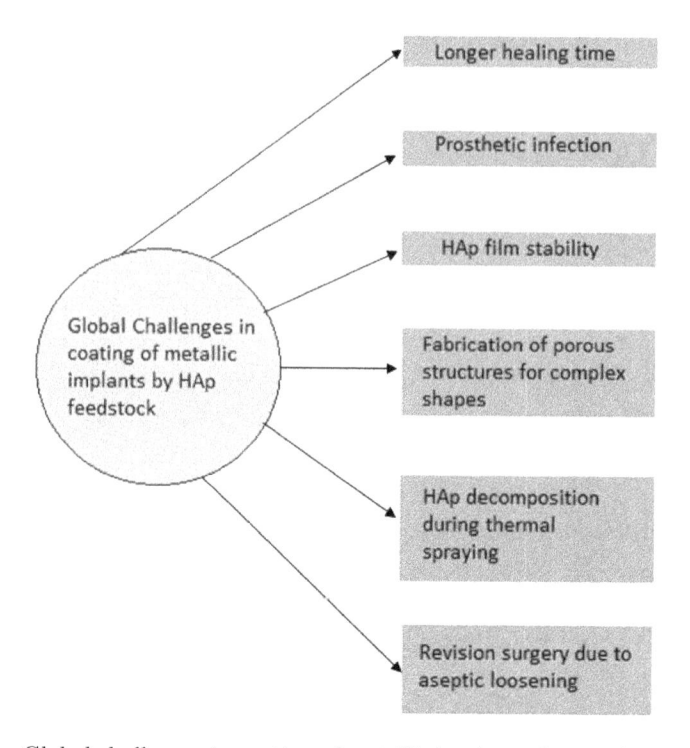

Fig. 34. Global challenges in coating of metallic implants during thermal spraying of HAp feedstock (Color online).

properties, such as prosthesis infection and chronic wound healing, which is also a concern. Figure 34 represents global challenges in coating of metallic implants by HAp feedstock.

7. Future Perspective of Thermal Spray Coatings in the Field of Biomedical Implants

The demand for precise functional and mechanical properties, as well as production flexibilities, has sparked interest in additive manufacturing (AM) in different industrial applications.[96–100] Presently, three-dimensional (3D) printing technology is mainly used to fabricate biomedical implants for repair of tissues, orthopedics and surgical tools (Fig. 35).[101] Hence AM holds promise for generating

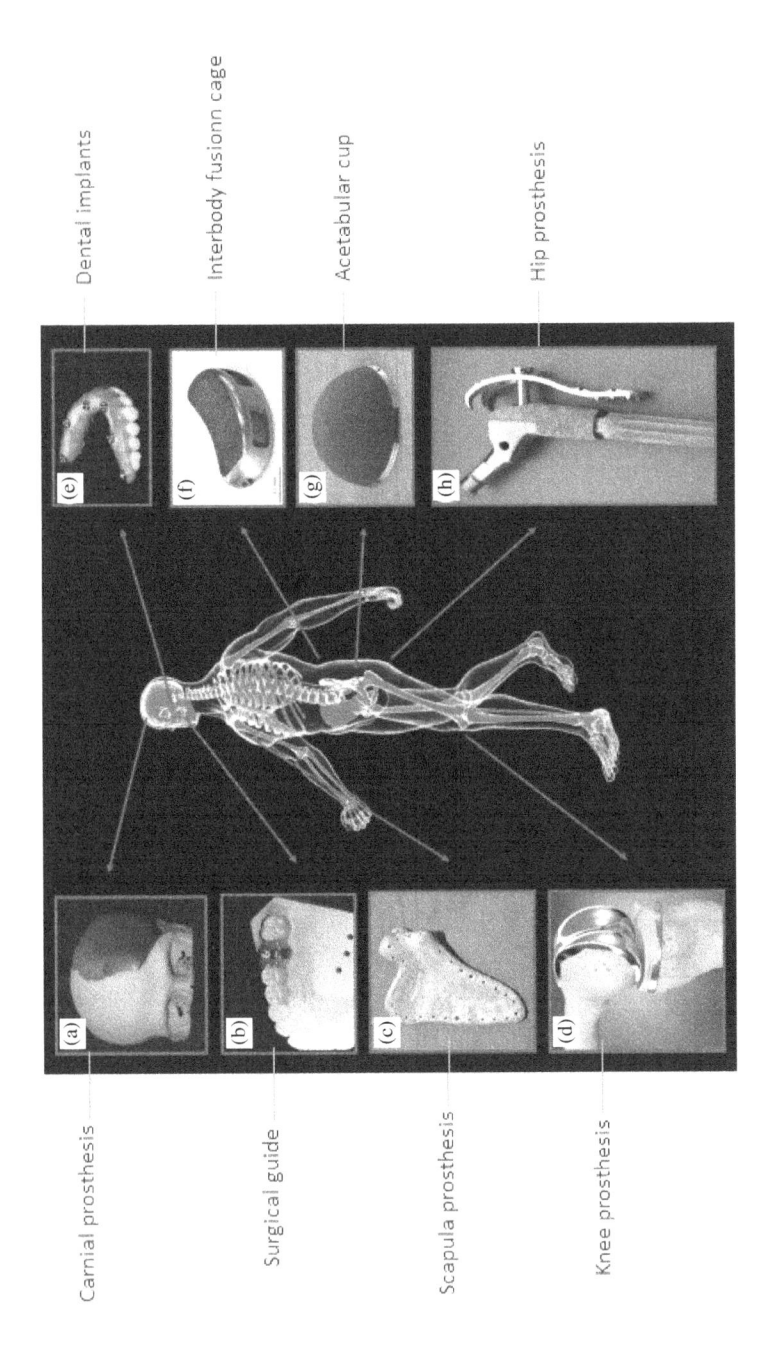

Fig. 35. Applications of 3D printing of the biomaterials for different parts of human body (Color online).[101]

complicated shapes. However, there are some limitations associated with AM mainly related to high-temperature metal processing, which frequently results in undesirable mechanical properties, like undesired phase transformations, part distortion, greater residual stresses and poor mechanical properties.[102-105] The technology of supersonic powder deposition (CS) is effectively utilized to overcome these short-comings of high temperature metal processing.[106] CSAM has enormous potential to change the size and distribution of Industry 4.0 with remarkable sustainable benefits and minimal impact on environment as it is a green technology. This differs from the conventional AM procedures in which the powders are laid down on a powder bed. The ability to 3D print organs is one of the futuristic applications of AM outside of traditional manufacturing. This could be one of the most transformative applications of 3D printing, having the potential to revolutionize the world as we know it. Millions of people in need of transplant donors would no longer have to wait. Therefore, AM allows for not only personalization but also, more importantly, cost savings and increased success.

Recently, Moridi et al.[107] used CSAM approach to generate solid state 3D printed Ti-6A1-4V implant having porous structure. When the compressive yield strength of solid-state 3D printed Ti-6A1-4V alloy was compared to that of other AM techniques, it was found that there is an increase in yield strength up to 42% in the solid-state 3D printed parts. The biocompatibility of 3D printed structure with MC3T3-E1 SC4 murine preosteoblast cells was compared. The result emphasizing the materials potential for the biomedical applications. Hence, employing a broad spectrum of metallic materials that are already being used in CS processing, one-step subcritical CS deposition can be used to build cellular structures in future. This may be easily adapted to manufacture 3D objects by combining a robot with a supersonic nozzle in CS, as businesses like impact Innovations,[108] Speed3D[109] and NRC Canada[110] have already done the same. As viable manufacturing approach for the fabrication of biomedical implant materials in near future, 3D solid state CS printing has been proposed.

8. Conclusion

The major goal of this review is to compile a large body of information on HAp coatings and coating techniques for biomedical implant development. Main points are summarized as follows:

- Corrosion and wear are the major problem that will degrade the performance of biomedical implants inside human bodies.
- Due to the ability to deposit homogeneous coating layer on surfaces of implants, the TS coatings are the most effective method and used widely on metallic implants. The high speed and temperature associated with TS technique allows the HAp-based ceramic particles to be deposited on the surfaces of implant with fewer defects.
- The reinforcement (wt.%) in suitable proportions in pure HAp resulted in an increase in mechanical and tribological properties like microhardness, fracture toughness, adhesive strength, corrosion resistance, etc.
- The use of an HT for the HAp coating and implant surface improves implant efficiency and makes the biomaterial more suitable for host body.
- By accelerating particles to supersonic impact velocities, CS may become one-step approach for production of porous Ti-6Al-4V biomedical structures.

Acknowledgment

The authors acknowledge the Science and Engineering Research Board (SERB) for financial support under sponsored project sanctioned to I. K. Gujral Punjab Technical University (No. SRG/2019/001182).

References

1. J. Park and R. S. Lakes, *Biomaterials: An Introduction* (Springer Science & Business Media, Berlin/Heidelberg, Germany, 2007).
2. S. Ramakrishna, J. Mayer, E. Wintermantel and K. W. Leong, *Compos. Sci. Technol.* **61** (2001) 1189.
3. D. L. Wise and D. J. Trantolo (eds.), *Biomaterials Engineering and Devices: Human Applications,* Vol. 2 (Humana Press, Totowa, 2000), p. 346.
4. M. Geetha, A. K. Singh, A. Asokamani and A. K. Gogia, *Prog. Mater. Sci.* **54** (2009) 397.
5. S. Kurtz, K. Ong, E. Lau, F. Mowat and M. Halpern, *J. Bone Joint Surg. Am.* **89** (2007) 780.
6. G. Prashar and H. Vasudev, *J. Thermal Spray Enq.* **2** (2020) 50.
7. Q. Chen and G. A. Thouas, *Mater. Sci. Eng. R* **87** (2015) 1.
8. J. J. Jacobs, A. K. Skipor, L. M. Patterson, N. J. Hallab, W. G. Paprosky, J. Black and J. O. Galante, *J. Bone Joint Surg. Am.* **80** (1998) 1447.
9. J. J. Jacobs, C. Silverton, N. J. Hallab, A. K. Skipor, L. Patterson, J. Black and J. O. Galante, *Clin. Orthopaedics Relat. Res.* **358** (1999) 173.
10. M. Brayda-Bruno, M. Fini, G. Pierini, G. Giavaresi, M. Rocca and R. Giardino, *Int. J. Artif. Orqans* **24** (2001) 41.
11. G. Manivasagam, D. Dhinasekaran and A. Rajamanickam, *Recent Pat. Corros. Sci.*
12. Y. H. Zhu, K. Y. Chiu and W. M. Tang, *J. Orthopaedic Surg.* **9** (2001) 91.
13. H. C. Amstutz, P. Campbell, N. Kossovsky and I. C. Clarke, *Clin. Orthopaedics Relat. Res.* **276** (1992) 7.
14. M. H. Huo and S. M. Cook, *J. Bone Joint Surg.* **83** (2001) 1598.
15. M. Al-Amin, A. M. Abdul-Rani, M. Danish, S. Rubaiee, A. B. Mahfouz, H. M. Thompson, S. Ali, D. R. Unune and M. H. Sulaiman, *Materials* **14** (2021) 3597.
16. G. Bolelli, N. Stiegler, D. Bellucci, V. Cannillo, R. Gadow, A. Killinger, L. Lusvarghi and A. Sola, *Surf. Coat. Technol.* **210** (2012) 28.
17. P. Krieg, A. Killinger, R. Gadow, S. Burtscher and A. Bernstein, *Bioact. Mater.* **2** (2017) 162.

18. W. Zhang, G. Wang, Y. Liu, X. Zhao, D. Zou, C. Zhu, Y. Jin, Q. Huang, J. Sun, X. Liu and X. Jiang, *Biomaterials* **34** (2013) 3184.
19. G. Bolelli, D. Bellucci, V. Cannillo, R. Gadow, A. Killinger, L. Lusvarghi, P. Müller and A. Sola, *Surf. Coat. Technol.* **280** (2015) 232.
20. G. Bolelli, D. Bellucci, V. Cannillo, L. Lusvarghi, A. Sola, N. Stiegler, P. Müller, A. Killinger, R. Gadow, L. Altomare and L. De Nardo, *Mater. Sci. Eng. C* **34** (2014) 287.
21. A. Fomin, M. Fomina, V. Koshuro, I. Rodionov, A. Zakharevich and A. Skaptsov, *Ceram. Int.* **43** (2017) 11189.
22. L. Sun, C. C. Berndt, K. A. Gross and A. Kucuk, *J. Biomed. Mater. Res.* **58** (2001) 570.
23. G. Zhao, L. Xia, B. Zhong, G. Wen, L. Song and X. Wang, *Surf. Coat. Technol.* **251** (2014) 38.
24. R. S. Pillai, M. Frasnelli and V. M. Sglavo, *Ceram. Int.* **44** (2018) 1328.
25. E. Garcia, P. Miranzo and M. A. Sainz, *Ceram. Int.* **44** (2018) 12896.
26. Y. Liang, Y. Xie, H. Ji, L. Huang and X. Zheng, *Appl. Surf. Sci.* **256** (2010) 4677.
27. W. Xue, X. Liu, X. Zheng and C. Ding, *Surf. Coat. Technol.* **185** (2004) 340.
28. X. Liu, S. Tao and C. Ding, *Biomaterials* **23** (2002) 963.
29. T. P. S. Sarao, H. Singh and H. Singh, *J. Thermal Spray Technol.* **27** (2018) 1388.
30. A. Singh, G. Singh and V. Chawla, *J. Mech. Behav. Biomed. Mater.* **85** (2018) 20.
31. M. B. Nasab, M. R. Hassan and B. B. Sahari, *Trends Biomater. Artif. Organs* **24** (2010) 69.
32. D. Arcos and M. Vallet-Regí, *J. Mater. Chem. B* **8** (2020) 1781.
33. C. A. Beevers and D. D. McIntyre, *Miner. Mag. J. Miner. Soc.* **27** (1946) 254.
34. DIN EN 657, *Thermal Spraying: Terminology, Classification* (Beuth, Berlin, 1994).
35. I. C. Roata, C. Croitoru, A. Pascu and E. M. Stanciu, *AIMS Mater. Sci.* **6** (2019) 335.
36. E. Pfender, *Thin Solid Films* **238** (1994) 228.
37. J. G. Thakare, C. Pandey, M. M. Mahapatra and R. S. Mulik, *Met. Mater. Int.* **27** (2021) 1947.

38. J. E. Tercero, S. Namin, D. Lahiri, K. Balani, N. Tsoukias and A. Agarwal, *Mater. Sci. Eng. C* **29** (2009) 2195.
39. K. Balani, Y. Chen, S. P. Harimkar, N. B. Dahotre and A. Agarwal, *Acta Biomater.* **3** (2007) 944.
40. G. Prashar and H. Vasudev, Application of thermal spraying techniques used for the surface protection of boiler tubes in power plants: Thermal spraying to combat hot corrosion, in *Advanced Surface Coating Techniques for Modem Industrial Applications* (IGI Global, Hershey, Pennsylvania, 2021), pp. 112–134.
41. I. B. Grafts and B. Substitutes, Three-dimensionally engineered hydroxyapatite ceramics with interconnected pores as a bone substitute and tissue engineering scaffold, in *Biomaterials in Orthopedics* (CRC Press, Boca Raton, Florida, 2004), p. 287.
42. J. Schrooten and J. A. Helsen, *Biomaterials* **21** (2000) 1461.
43. Y. C. Yang and B. Y. Chou, *Mater. Chem. Phys.* **104** (2007) 312.
44. A. Vend, S. Arostegui, G. Favaro, F. Zivic, M. Mrdak, S. Mitrović and V. Popovic, *Tribol. Int.* **44** (2011) 1281.
45. A. Latifi, M. Imani, M. T. Khorasani and M. D. Joupari, *Appl. Surf. Sci.* **320** (2014) 471.
46. P. Hameed, V. Gopal, S. Bjorklund, A. Ganvir, D. Sen, N. Markocsan and G. Manivasagam, *Colloids Surf. B, Biointerfaces* **173** (2019) 806.
47. J. Rauch, G. Bolelli, A. Killinger, R. Gadow, V. Cannillo and L. Lusvarghi, *Surf. Coat. Technol.* **203** (2009) 2131.
48. M. Monsalve, E. Lopez, H. Ageorges and F. Vargas, *Surf. Coat. Technol.* **268** (2015) 142.
49. L. Sun, C. C. Berndt and K. A. Gross, *J. Biomater. Sci., Polym. Edn.* **13** (2002) 977.
50. M. F. Morks, *J. Mech. Behav. Biomed. Mater.* 1 (2008) 105.
51. G. Singh, S. Singh and S. Prakash, *Surf. Coat. Technol.* **205** (2011) 4814.
52. S. Yugeswaran, C. P. Yoganand, A. Kobayashi, K. M. Paraskevopoulos and B. Subramanian, *J. Mech. Behav. Biomed. Mater.* **9** (2012) 22.
53. P. C. Rath, L. Besra, B. P. Singh and S. Bhattacharjee, *Ceram. Int.* **38** (2012) 3209.
54. Y. C. Yang, C. C. Chen, J. B. Wang, Y. C. Wang and F. H. Lin, *Ceram. Int.* **43** (2017) S829.

55. H. L. Yao, H. T. Wang, X. B. Bai, G. C. Ji and Q. Y. Chen, *Surf. Coat. Technol.* **342** (2018) 94.
56. S. Singh, K. K. Pandey, O. A. Rahman, S. Haidar, D. Lahiri and A. K. Keshri, *Mater. Res. Exp.* **7** (2020) 015415.
57. L. Fu, K. A. Khor and J. P. Lim, *Surf. Coat. Technol.* **127** (2000) 66.
58. K. Balani, R. Anderson, T. Laha, M. Andara, J. Tercero, E. Crumpler and A. Agarwal, *Biomater.* **28** (2007) 618.
59. Y. W. Gu, K. A. Khor, D. Pan and P. Cheang, *Biomaterials* **25** (2004) 3177.
60. B. Y. Chou and E. Chang, *Surf. Coat. Technol.* **153** (2002) 84.
61. Y. H. Meng, C. Y. Tang and C. P. Tsui, *J. Mater. Sci., Mater. Med.* **19** (2002) 75.
62. C. Kaya, *Ceram. Int.* **34** (2008) 1843.
63. Y. Chen, T. H. Zhang, C. H. Gan and G. Yu, *Carbon* **45** (2007) 998.
64. E. M. Carlisle, *Science* **167** (1970) 279.
65. T. Gao, H. T. Aro, H. Ylänen and E. Vuorio, *Biomaterials* **22** (2001) 1475.
66. N. Hijón, M. V. Cabanas, J. Pena and M. Vallet-Regí, *Acta Biomater.* **2** (2006) 567.
67. L. L. Hench and J. M. Polak, *Science* **295** (2002) 1014.
68. H. Li, K. A. Khor and P. Cheang, *Biomaterials* **23** (2002) 85.
69. S. Mohajernia, S. Pour-Ali, S. Hejazi, M. Saremi and A. R. Kiani-Rashid, *Ceram. Int.* **44** (2018) 8297.
70. B. Singh, G. Singh and B. S. Sidhu, *J. Thermal Spray Technol.* **27** (2018) 1401.
71. B. Singh, G. Singh and B. S. Sidhu, *J. Mater. Res.* **34** (2019) 1678.
72. H. Chen, E. Zhang and K. Yang, *Mater. Sci. Eng. C* **34** (2014) 201.
73. P. Tian, H. Hu, H. Wang, X. Liu and C. Ding, *Mater. Lett.* **117** (2014) 98.
74. M. Y. Zakaria, A. B. Sulong, N. Muhamad, M. R. Raza and M. I. Ramli, *Mater. Sci. Eng. C* **97** (2019) 884.
75. H. H. Beheri, K. R. Mohamed and G. T. El-Bassyouni, *Mater. Des.* **44** (2013) 461.
76. K. Lin, M. Zhang, W. Zhai, H. Qu and J. Chang, *J. Am. Ceram. Soc.* 94 (2011) 99.
77. J. Singh, S. S. Chatha and H. Singh, *Ceram. Int.* **47** (2021) 782.

78. Y. Chen, Z. Xu, C. Smith and J. Sankar, *Acta Biomater.* **10** (2014) 4561.
79. A. Rezaei, R. B. Golenji, F. Alipour, M. M. Hadavi and I. Mobasherpour, *Ceram. Int.* **46** (2020) 25374.
80. C. E. Mancini, C. C. Berndt, L. Sun and A. Kucuk, *J. Mater. Sci.* **36** (2001) 3891.
81. K. A. Gross, V. Gross and C. C. Berndt, *J. Am. Ceram. Soc.* **81** (1998) 106.
82. S. Tiwari and S. B. Mishra, *Surf. Coat. Technol.* 405 (2021) 126500.
83. R. C. Rocha, A. G. D. S. Galdino, S. N. D. Silva and M. L. P. Machado, *Mater. Res.* **21** (2018) e20171144.
84. G. Singh, S. Singh and S. Prakash, *J. Min. Mater. Charact. Eng.* **10** (2011) 173.
85. H. Li, K. A. Khor, R. Kumar and P. Cheang, *Surf. Coat. Technol.* 182 (2004) 227.
86. C. W. Yang and T. S. Lui, *Mater. Trans.* **48** (2007) 211.
87. H. Li, K. A. Khor and P. Cheang, *Biomaterials* **23** (2002) 2105.
88. Y. C. Tsui, C. Doyle and T. W. Clyne, *Biomaterials* **19** (1998) 2015.
89. G. L. Zhao, G. W. Wen and W. Kun, *Trans. Nonferr. Met. Soc. China* **19** (2009) s463.
90. S. Bauer, P. Schmuki, K. Von Der Mark and J. Park, *Prog. Mater. Sci.* **58** (2013) 261.
91. Y. Tao, G. Ke, Y. Xie, Y. Chen, S. Shi and H. Guo, *Appl. Surf. Sci.* **357** (2015) 8.
92. Y. Huang, M. Hao, X. Nian, H. Qiao, X. Zhang, X. Zhang, G. Song, J. Guo, X. Pang and H. Zhang, *Ceram. Int.* **42** (2016) 11876.
93. Z. Zhao, L. Du, Y. Tao, Q. Li and L. Luo, *Ultra-sonicssonochemistry* **33** (2016) 10.
94. G. Hannink and J. C. Arts, *Injury* **42** (2011) S22.
95. D. Bellucci, E. Veronesi, V. Strusi, T. Petrachi, A. Murgia, I. Mastrolia, M. Dominici and V. Cannillo, *Materials* **12** (2019) 3633.
96. C. B. Williams, J. K. Cochran and D. W. Rosen, *Int. J. Adv. Manuf. Technol.* **53** (2011) 231.
97. S. I. Roohani-Esfahani, P. Newman and H. Zreiqat, *Sci. Rep.* **6** (2016) 1.
98. W. J. Sames, F. A. List, S. Pannala, R. R. Dehoff and S. S. Babu, *Int. Mater. Rev.* **61** (2016) 315.

99. Y. M. Wang, T. Voisin, J. T. McKeown, J. Ye, N. P. Calta, Z. Li, Z. Zeng, Y. Zhang, W. Chen, T. T. Roehling and R. T. Ott, *Nat. Mater.* **17** (2018) 63.
100. J. H. Martin, B. D. Yahata, J. M. Hundley, J. A. Mayer, T. A. Schaedler and T. M. Pollock, *Nature* **549** (2017) 365.
101. J. Ni, H. Ling, S. Zhang, Z. Wang, Z. Peng, C. Benyshek, R. Zan, A. K. Miri, Z. Li, X. Zhang and J. Lee, *Mater. Today Bio.* **3** (2019) 100024.
102. D. C. Hofmann, S. Roberts, R. Otis, J. Kolodziejska, R. P. Dillon, J. O. Suh, A. A. Shapiro, Z. K. Liu and J. P. Borgonia, *Sci. Rep.* **4** (2014) 1.
103. W. E. Frazier, *J. Mater. Eng. Perform.* 23 (2014) 1917.
104. B. Vayre, F. Vignat and F. Villeneuve, *Mech. Ind.* **13** (2012) 89.
105. A. Moridi, *Powder Consolidation Using Cold Spray: Process Modeling and Emerging Applications* (Springer, Salmon Tower Building New York City, 2016).
106. A. Moridi, S. M. Hassani-Gangaraj, M. Guagliano and M. Dao, *Surf. Eng.* **30** (2014) 369.
107. A. Moridi, E. J. Stewart, A. Wakai, H. Assadi, F. Gartner, M. Guagliano, T. Klassen and M. Dao, *Appl. Mater. Today* **21** (2020) 100865.
108. https://impact-innovations.com/en/applications/rocket-nozzle/.
109. https://www.spee3d.com/.
110. https://3dprintingindustry.com/news/nrc-canada-to-advance-adoption-of-cold-spray-additive-manufacturing-158862/.

Deep cryogenic treated high carbon steel blades: Tribological, morphological, and economic analysis*

Chander Jakhar[†], Anil Saroha[†], Parvesh Antil[‡,††], Vishal Ahlawat[§],
Asha Rani[¶], Dharam Buddhi[∥] and Vinay Kumar[**]

[†]*Department of Farm Machinery and Power Engineering,
COAE&T, CCS HAU, Haryana, India*
[‡]*Department of Basic Engineering,
COAE&T, CCS HAU, Haryana, India*
[§]*Department of Mechanical Engineering,
UIET, Kurukshetra University, Haryana, India*
[¶]*Department of Physics, Maharishi Dayanand University,
Haryana, India*
[∥]*Division of Research and Innovation,
Uttaranchal University, Uttarakhand, India*
[**]*Department of Mathematics and Statistics,
CCS HAU, Haryana, India*
[††]*parveshantil@hau.ac.in*

Straw combines are intended to process the remaining harvested straw. When cut at high temperatures and in abrasive conditions, the cutting blade of straw combines undergoes substantial surface deterioration. This deterioration shortens the blade's lifespan and increases the cutting cost of the machine. In recent decades, cryogenic treatments have played a significant role in enhancing material properties. In this paper, cryogenic treatment is utilized to boost the wear resistance of straw

---

[*]To cite this article, please refer to its earlier version published in Surface Review and Letters, Vol. 30, No. 1 (2023) 2241002 (13 pages) DOI: 10.1142/S0218625X22410025
[††]Corresponding author.

combine blades in the current investigation. The performance of cryogenic treatment was tested in the laboratory using the pin-on-disc wear tester with sample type, load, sliding velocity, and time serving as process factors and wear loss as the response parameter. The smoothness of the cryogenically-treated sample's surface is certified through morphological examination. Specific wear rate and field emission scanning electron microscope (FE-SEM) indicated that cryogenic treatment enhances the grain structure and intermolecular interaction of the specimen, resulting in an increase in wear resistance. As opposed to the untreated specimen, the wear on the treated surface is uniform over the entire surface, as demonstrated by FE-SEM analysis. The grain structure and intermo-iecuiar bonding of the specimen were improved as a result of the cryogenic treatment. The cryogenic treatment increased the cost of the cutter bar and chopping cylinder blades by 9.38% and 13.61%, respectively, compared to untreated blades, but the increased cost was fuiiy offset by the longer blade life.

Keywords: Cryogenic; energy dispersive spectroscopy; FE-SEM; high carbon steel; straw combine; Taguchi; wear.

1. Introduction

Automated combine harvesters are used in developed countries around the world to harvest grain and straw for threshing. Cutting at high temperatures and in an abrasive environment, these cutting materials are subject to severe surface degradation, which eventually limits the blade's life and raises the cost of the machine. The deteriorating tool life at high temperatures and abrasive attacks is a matter of concern. The replacement of blades is time-consuming as well as an economic strain on marginal farmers. The aim of the research is to improve the deteriorating tool life through surface treatment. Cryogenic (shaiiow and deep) treatment has become a prominent surface treatment approach for increasing the surface qualities of materiais to overcome these limits of high carbon steel blades or other cutting alloy blades. A number of scientists have studied the effects of cryogenic treatment on the performance of materiais. Akincioglu *et al.*[1] observed that deep cryogenic treatment (DCT) enhanced the life of the cutting tool by 27% and reduced the main cutting wear by 11%. Using tungsten carbide inserts that had undergone shallow cryogenic treatment and DCT, Gill et al.[2] found that the cutting life of the inserts increased by 27% and 37%, respectively. It has also been found that the hardness of cryogenically-reated steel used for rotavator blades improved by 260.73%, and the wear rate was reduced by 60% compared to material that had not been cryogenically treated.[3] There is a significant amount of weight loss from blade wear in the middle of a blade, according to a study by Marian et al.[4] In the case of cutting tools, cryogenic treatment improved wear resistance while decreasing cutting forces and surface roughness.

By homogenizing the carbide distribution within the carbide tools, the cryogenic treatment boosts their toughness and hardness properties. There are many variables that need to be taken into consideration when implementing cryogenic treatment on cutting tools in order to get the most benefit out of the process. These variables include the holding temperature, holding time, identifying what heat treatment will be applied, the cooiing rate and so on. Using DCT, Arslan et al.[5] and Vidyarthi et al.[6] explored the improvement of steel characteristics. Tool steels, carburized steels, cast iron, and other materials have shown enhanced performance. Although the mechanics behind this therapy are still uncertain, it is impossible to predict the consequences of this treatment on a specific alloy. The DCT has evolved as a more effective means of treating materials than traditional approaches. Blocky retained austenite (RA) is eliminated, and the stability of filmy RA is increased without any change in nanobainitic microstructure.[7] Because of this, DCT gives better surface residual stress as well as wear resistance compared to traditional heat treatments. Improved wear resistance is achieved by precipitating sufficient carbides in large volume during cryogenic treatment.[8] Austenite levels are primarily maintained by austenitization time and appiied temperature, according to Cardoso et al.[9] When Kang et al.[10] performed shallow cryogenic treatment and DCT on $Fe_{75}Ni_{25}$ alloy, they determined that the rise in martensite volume fraction likely explains the enhanced hardness and wear resistance. Researchers found that the wear track cracks and fractures in dry cryogenic, dry sliding wear were higher than those seen in the fluid condition.[11] Mode-I (the proposed coolant strategy) considerably reduced T and V wear to a maximum of 61% and 29% reduction compared to cryogenic cooling with a Mode-II approach, according to Siviah et al.[12] Using cryogenic treatment on Cu-Be$_2$ alloy, Ahmed et al.[13] observed that the cryogenic-treated sample had a lower wear rate and coefficient of friction (COF) than the untreated samples. For example, a mixture of abrasive and adhesive wear, as well as wear caused by oxygen, is created by the cryogenic treatment process. As a result, the fracture is now classified as a "quasi-cleavage fracture"[14] Most scientists assumed that at ambient temperature, a small proportion of austenite was still present after the heat treatment. Retained austenite, a soft phase in steels, has the potential to shorten product life. DCT now transforms austenite into martensite, increasing cutting tool life, although secondary fine carbides precipitation boosted cutting tool life in high carbon steel. Consequently, it was observed that cryogenic treatments improve wear resistance and increase hardness by precipitating fine secondary carbides and changing residual austenite into martensite. As a result, each treatment material necessitates its own set of tests. In tribology,[15-19] the wear, friction, and lubrication are all included. Cryogenic treatment is widely used for improving surface characteristics. The current research aims to provide a solution for better tool life for straw combine blades. Tool wear and breakage are common occurrences during the Indian summer harvest season. The harvesting procedure has been hampered by the high temperature and abrasive particles in the agricultural area. Cryogenic treatment of straw combination blades' effect on wear behavior is being investigated by researchers in light of this problem. Recent findings focus on the application of cryogenic treatment to high carbon steel blades in an effort to improve their wear resistance. Field emission scanning electron microscopes (FE-SEMs) have been used to detect microstructural changes. Cryogenic treatment's higher cost is shown to be compensated by the longer tool life, according to the economic analysis.

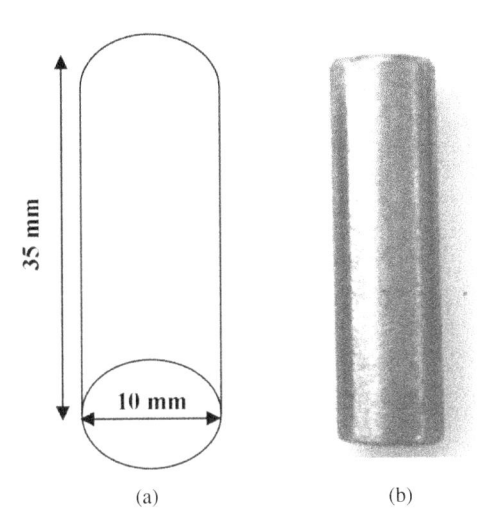

Fig. 1. Wear test specimen, (a) Schematic representation, (b) Actual.

2. Materials and Methods

Specimens of cylindrical shape made of high carbon steel were tested in this investigation (Fig. 1). High carbon steel blades that are commonly utilized in straw combines were used in the experiment. The specimens were made with a high carbon steel rod. Figure 1 shows the measurements for the specimens used in the wear tests, which were 35 mm in length and 10 mm in diameter.

3. Experimental Planning and Design

3.1. *Cryogenic treatment*

To begin the process of cryogenic treatment, the temperature of the component must be gradually lowered. A period of time (freezing) is followed by gradual warming back to room temperature. According to Table 1, DCT was performed. On the basis of past findings,[1] the DCT parameters were chosen for this study. Acetone was used to clean the test specimens before DCT. For the cryogenic treatment, a computer-controlled insulated chamber with liquid nitrogen as the working fluid was used.

Table 1. Deep cryogenic treatment process parameters.

Sr. No.	Parameters	Value
1	Soaking temperature (°C)	−190
2	Cooling temperature (°C min^{-1})	1
3	Soaking period (h)	16
4	Heating rate (°C min^{-1})	1

Table 2. Parameters and their levels for wear analysis.

Sr. No.	Parameters	Levels
1	Sample (type)	Untreated, treated
2	Load (N)	15, 20, 25, 30
3	Time (s)	300, 600, 900, 1200
4	Sliding velocity (ms^{-1})	0.5, 1.0, 1.5, 2.0

3.2. Taguchi's methodology

Using Taguchi's methods,[20-24] an experimental strategy was devised. Abrasive wear testing was carried out on test specimens of high carbon steel with a diameter of 10 mm and a length of 35 mm. Cryo-treated and untreated specimens were the two types of samples used for the wear analysis. With the use of a pin-on-disc wear tester, the wear analysis was carried out by loading a pin (specimen) against a flat rotating disc (the counter surface). This equipment was used to test the wear and friction qualities of materials in pure sliding situations. Before commencing the testing machine, the sample was inserted into the holder with a set screw and then pressed vertically. Table 2 shows the wear analysis parameters that were selected. Table 3 shows the pin-on-disk wear tester's specifications. Each specimen's weight before and after the experiment was measured using a weighing balance with 0.0001 g accuracy. Table 4 shows the designed experimental plan for the wear analysis.

Table 3. Specifications of the pin-on-disc wear testing machine.

Sr. No.	Particulars	Specifications
1	Pin on disc	Ducom TR 20 LE
2	Disc material	EN 31 steel disc hardened to 64 HRC
3	Disc size (mm)	165 diameter
4	Track diameter (mm)	0–120
5	Speed (rpm)	0–2000
6	Speed (ms^{-1})	0–12,560
7	Normal load (kg)	0–20
8	Drive	1.1 KW DC motor with constant torque
9	Power supply	230V, 15A single phase 50Hz AC
10	Software	Winducom 2010

Table 4. Experimental plan using L_{32} orthogonal array.

Experiment No.	Sample (type)	Load (N)	Time (s)	Sliding velocity (m/s)	Wear (mg)
1	Untreated	15	300	0.5	0.011
2	Untreated	15	600	1.0	0.018
3	Untreated	15	900	1.5	0.024
4	Untreated	15	1200	2.0	0.028
5	Untreated	20	300	0.5	0.014
6	Untreated	20	600	1.0	0.019
7	Untreated	20	900	1.5	0.025
8	Untreated	20	1200	2.0	0.031
9	Untreated	25	300	1.0	0.030
10	Untreated	25	600	0.5	0.032
11	Untreated	25	900	2.0	0.038
12	Untreated	25	1200	1.5	0.040
13	Untreated	30	300	1.0	0.039
14	Untreated	30	600	0.5	0.042
15	Untreated	30	900	2.0	0.045

(*Continued*)

Table 4. (Continued)

Experiment No.	Sample (type)	Load (N)	Time (s)	Sliding velocity (m/s)	Wear (mg)
16	Untreated	30	1200	1.5	0.048
17	Treated	15	300	2.0	0.014
18	Treated	15	600	1.5	0.017
19	Treated	15	900	1.0	0.019
20	Treated	15	1200	0.5	0.020
21	Treated	20	300	2.0	0.018
22	Treated	20	600	1.5	0.020
23	Treated	20	900	1.0	0.023
24	Treated	20	1200	0.5	0.026
25	Treated	25	300	1.5	0.024
26	Treated	25	600	2.0	0.027
27	Treated	25	900	0.5	0.029
28	Treated	25	1200	1.0	0.032
29	Treated	30	300	1.5	0.034
30	Treated	30	600	2.0	0.035
31	Treated	30	900	0.5	0.037
32	Treated	30	1200	1.0	0.039

4. Results and Discussion

4.1. Analysis of variance

The analysis of variance (ANOVA) was utilized to determine which parameter had a significant impact on response.[25,26] Table 5 shows the ANOVA as well as the percentage contribution of process parameters to wear loss. The ANOVA revealed that all procedure parameters had a significant impact on wear loss for treated specimens. According to the available data, load is the most important factor, followed by time, sample type, and sliding velocity. The model summary revealed that the model has a 99.26% R^2 value, indicating that it is significant.

Table 5. Analysis of variance for wear analysis.

Source	Code	DF	Adj SS	Adj MS	F-value	P-value	% Contribution
Sample	A	1	153.12	153.125	30.30	0.000	5.20
Load	B	3	2180.13	726.708	143.80	0.000	74.00
Time	C	3	456.38	152.125	30.10	0.000	15.50
Sliding velocity	D	3	50.13	16.708	3.31	0.040	1.70
Error		21	106.12	5.054			3.60
Total		31	2945.88				100

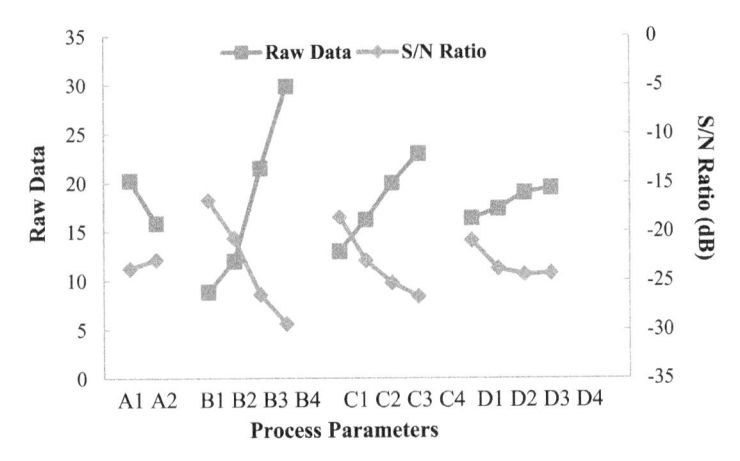

Fig. 2. S/N ratio and data means plots for specimen (s) (Color online).

The signal-to-noise (S/N) ratio and data mean plots for wear analysis are presented in Fig. 2. The graphs show variation of individual parameters with respect to increase in the levels on wear behavior. Three sorts of S/N ratio criteria are used by researchers based on response requirements: higher, nominal, and lower are all considered preferable. In this test, the lower the better S/N ratio criterion was selected and calculated as follows:

$$\eta_{ij} = -10 \log \left\{ \frac{1}{n} \sum_{i=1}^{n} y_{ij}^2 \right\}$$

Here, y_{ij} represents the ith experiment at the jth test, and n is the overall number of tests.[18]

Cryogenic-treated samples wear out more slowly than untreated samples in a dry sliding test, according to the data on wear loss based on sample type. The rise in toughness and wear resistance is to blame for this development. Improved wear resistance can be achieved with DCT, which converts austenite to martensite.[27] In addition to the formation of crystal defects like twins, the cryogenic treatment also developed a "new martensite" phase that mostly consisted of thin plate martensite and lenticular martensite. The creation of thermally-activated martensite, as reported by Villa et al.,[28] can be understood as the transformation of thin plate martensite into lenticular martensite via thermal nucleation during cooling. Compared to a CHT sample, the microstructure of a material treated with liquid nitrogen becomes much finer. Furthermore, the microstructure is improved when the cryogenic treatment temperature drops. The martensite transition at low temperatures leads to plastic deformation, which can lead to crystal shattering.[29] Load effect on wear shows a rise in wear loss or a drop in specimen resistance to wear with an increase in applied load, according to the available trend. While there was a noticeable rise in wear loss from 15 N to 20 N, there was a far faster increase in wear loss when the applied load increased from 20 N to 25 N. The surface appears to degrade faster due to molecular debonding at higher loads.[30,31] The wear behavior of the specimen is greatly influenced by the amount of time spent sliding. The amount of wear that occurs is directly related to the amount of time spent sliding. The surface wear behaves remarkably identically as time progresses. Long sliding times raise the surface temperature, which is why this happens. A similar tendency is shown when plotting the effect of sliding velocity variation on wear loss, albeit at a much slower rate. In the range of 0.5–1.0ms^{-1} sliding velocity, wear loss increased at a slower rate. There was a dramatic increase in wear loss when the sliding velocity was increased from 1.5 ms^{-1} to 2.0 ms^{-1}. This is because the abrasive layer is more likely to be in contact with the interior grains due to the higher sliding velocity of this material. The results that were available at the time showed that the wear increased initially with time but decreased with time at the following level. Figure 2. depicts the S/N ratio data, which reveal that $A_2B_1C_1D_1$ is the most wearresistant parametric combination.

4.1.1. Specific wear rate and coefficient of friction

Wear is the progressive loss of materials from contacting interfaces. The study of mechanism of contacting interfaces under motion is referred to tribology. The specific wear rate (SWR) is the amount of wear loss per unit distance per unit load. The SWR can be calculated as wear loss divided by the product of sliding distance, density of material and normal load.[32]

$$SWR = M/\rho LS$$

where,
  M is the total mass loss, g
  L is the normal load, N
  ρ is the density of material used in this work, g cm^{-3} (7.85)
  S is the sliding distance, m
  SWR, mm^3 Nm^{-1}

The SWR was calculated and presented in Table 6 for specimens. Table 6 represents the value of SWR of specimens at variable load, sliding velocity, time, and sliding distance.

The COF is calculated by dividing the friction force by normal load.

At varying sliding speeds and sliding times, the COF corresponding to four different load circumstances is depicted in Fig. 3. The untreated samples had a higher COF at low speeds and short sliding times than the treated samples. To improve surface hardness, samples were cryogenically treated. This resulted in an increase in the coefficient of friction compared to untreated samples. As the surface hardness increases, brittle fractures appear on

Table 6. Specific wear rate.

Experiment No.	Sample (type)	Sliding distance (m)	Volume loss (m^3)	Wear rate (m^3/m)	Specific wear rate (m^3 N^{-1} m^{-1})
1	Untreated	150	0.001401274	0.00000934	6.23E–19
2	Untreated	600	0.002292994	0.00000382	2.55E–19
3	Untreated	1350	0.003057325	0.00000226	1.51E–19
4	Untreated	2400	0.003566879	0.00000149	9.91E–20
5	Untreated	150	0.001783439	0.00001189	5.94E–19
6	Untreated	600	0.002420382	0.00000403	2.02E–19
7	Untreated	1350	0.003184713	0.00000236	1.18E–19
8	Untreated	2400	0.003949045	0.00000165	8.23E–20
9	Untreated	300	0.003821656	0.00001274	1.02E–18
10	Untreated	300	0.004076433	0.00001359	2.72E–19
11	Untreated	1800	0.004840764	0.00000269	1.43E–19
12	Untreated	1800	0.005095541	0.00000283	8.49E–20
13	Untreated	300	0.004968153	0.00001656	1.10E–18
14	Untreated	300	0.005350318	0.00001783	2.97E–19
15	Untreated	1800	0.005732484	0.00000318	1.42E–19
16	Untreated	1800	0.006114650	0.00000340	8.49E–20
17	Treated	600	0.001783439	0.00000297	7.93E–19
18	Treated	900	0.002165605	0.00000241	2.41E–19
19	Treated	900	0.002420382	0.00000269	1.20E–19
20	Treated	600	0.002547771	0.00000425	7.08E–20
21	Treated	600	0.002292994	0.00000382	7.64E–19
22	Treated	900	0.002547771	0.00000283	2.12E–19
23	Treated	900	0.002929936	0.00000326	1.09E–19
24	Treated	600	0.003312102	0.00000552	6.90E–20
25	Treated	450	0.003057325	0.00000679	8.15E–19
26	Treated	1200	0.003439490	0.00000287	2.29E–19
27	Treated	450	0.003694268	0.00000821	1.09E–19
28	Treated	1200	0.004076433	0.00000340	6.79E–20

Table 6. (*Continued*)

Experiment No.	Sample (type)	Sliding distance (m)	Volume loss (m^3)	Wear rate (m^3/m)	Specific wear rate (m^3 N^{-1} m^{-1})
29	Treated	450	0.004331210	0.00000962	9.62E–19
30	Treated	1200	0.004458599	0.00000372	2.48E–19
31	Treated	450	0.004713376	0.00001047	1.16E–19
32	Treated	1200	0.004968153	0.00000414	6.90E–20

the surface. A larger COF value can be achieved by cryogenically treating the surface, which creates tougher asperities with more wear resistance. Due to their softer asperity and low hardness, untreated samples have a lower COF value than treated ones.

Figure 4 and Table 6 depict the mean value, standard deviation, and individual specific wear rate data for each test condition. The majority of low-sliding-speed tests demonstrate a higher specific wear rate compared to high-sliding-speed and load tests. Cryogenically-treated samples had a lower SWR at high loads than untreated samples, as shown by the SWR data. Compared to untreated samples (5.27×10^{-18} m^3/Nm), treated samples have a reduced mean SWR value (4.99×10^{-18} m^3/Nm). As a result, cryogenic-treated materials outperform untreated samples in terms of wear resistance when subjected to greater weights and sliding speeds.

4.2. *Morphological analysis*

SEM images of specimen surface wear behavior were generated using a scanning electron microscope and proved to be a valuable tool for analyzing the surface behavior in prior studies.[33–35] They were both scanned and examined at a magnification of 100× and 500×. SEM images of untreated and treated specimens are depicted in Figs. 5(a) and 5(b). The metallurgical analysis of the untreated

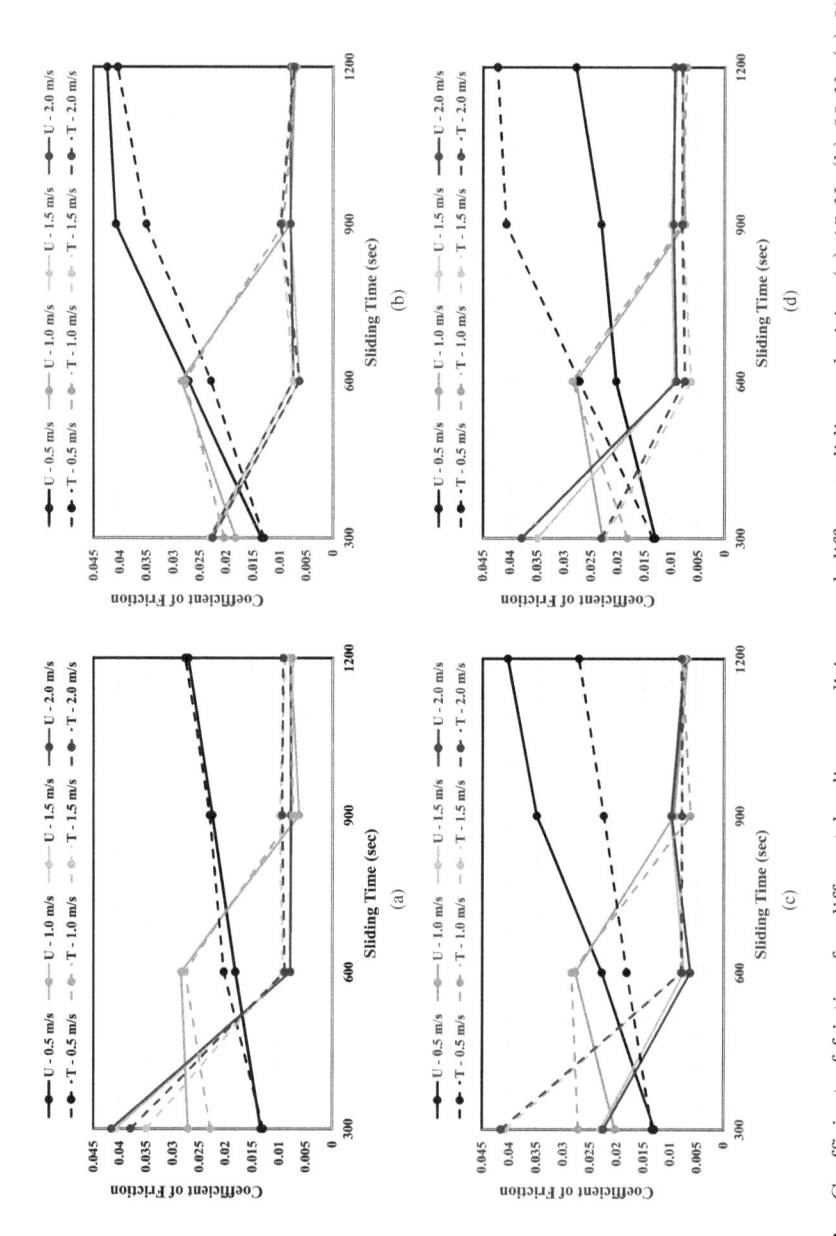

Fig. 3. Coefficient of friction for different loading conditions and different sliding velocities (a) 15 N, (b) 20 N, (c) 25 N, and (d) 30 N (Color online).

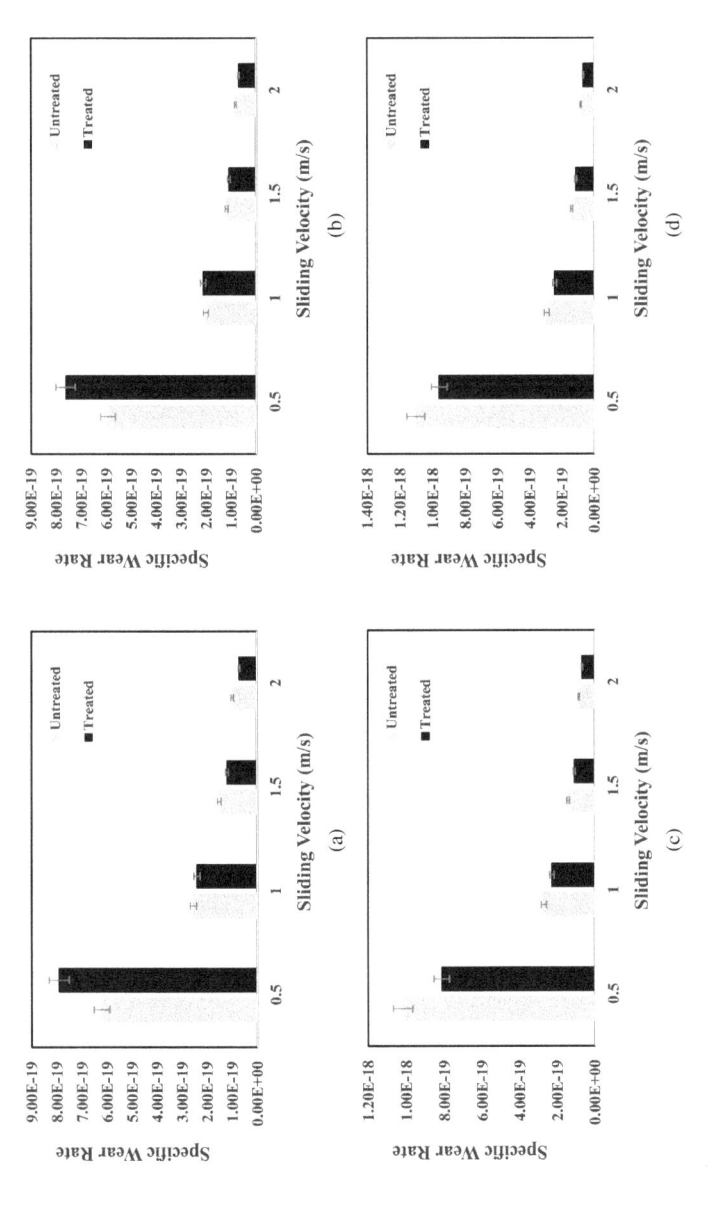

Fig. 4. Specific wear rate (mean value and standard deviation) for different loading conditions and different sliding velocities (a) 15 N, (b) 20 N, (c) 25 N, and (d) 30 N.

and treated specimens was carried out using SEM. Because the treated sample has better hardness qualities, it is obvious from the SEM that the untreated surface has deeper grooves than the treated sample. Because of their hardness, the more durable materials showed lower wear rates and had a smoother surface finish. A closer look at the surface in Figs. 5(a) and 5(b) reveals that the surface has a nonuniform texture and roughness, with deep pits and adhesion marks evident all over. The untreated specimen's surface material is bluntly removed. This homogeneity is shown in the case of deep cryogenic treatment of the surface Fig. 5(c). It is clear that the surface has been deteriorated, yet it's clearer than the untreated surface shown in Fig. 5(b). Images captured with a SEM reveal adhesion marks and oxides on high carbon steel that was subjected

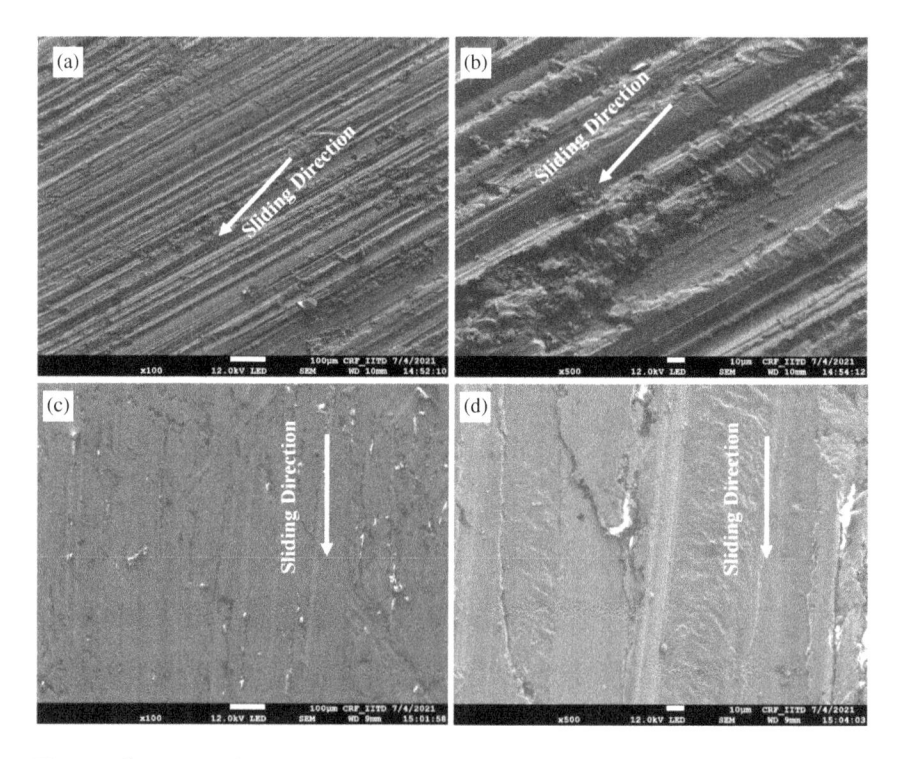

Fig. 5. Scanning electron microscope at 30 N load of (a) untreated specimen at 100× resolution, (b) untreated specimen at 500× resolution, (c) treated specimen at 100× resolution, and (d) treated specimen at 500× resolution.

to a wear test under a normal load of 30 N (Figs. 5(a)–5(c)) (appearing bright). With the increased magnification, the adhesion marks become more pronounced, and the signs of plastic flow and cracking become more apparent (Figs. 5(b)–5(d)). The highly work-hardened layer, especially at depth, is prone to cracking. Delamination occurs in the form of metallic sheets when cracks expand and become interconnected.[36,37] Further supporting the idea of "delamination" occurring during adhesive wear is the presence of sheet-like wear debris. Additional evidence for the occurrence of an oxidative wear regime comes from the presence of oxides (a bright look) at the worn surfaces and in the wear debris. However, when subjected to heavy loads, the high carbon steel's worn surface shows signs of abrasion and nodular oxides. According to existing literature,[38] the worn look is consistent with that of a material surface that has been subjected to abrasive wear by a micro-cutting technique. This is consistent with the fact that abrasive wear produces nodular oxide particles as a by-product of its micro-cutting activity. As a result, when subjected to a greater load, high carbon steel exhibits a shift from adhesive wear to abrasive wear. FE-SEM was used to perform energy dispersive spectroscopy (EDS). The EDS was used to determine the composition of the components and newly generated elements following cryogenic treatment. Figures 6 and 7 show the EDS analysis of untreated and treated specimens, respectively.

Before and after the cryogenic treatment, the specimens' elemental composition was analyzed with an energy dispersive spectroscope. Compared to the untreated samples, the treated specimens revealed no variation in composition. As a result of the cryogenic treatment, the treated specimens' intermolecular bonding and grain structure both improved due to the elemental composition. Grain structure and intermolecular forces in the treated specimens improved as a result of this process.

5. Economic Analysis

The economic analysis was done to analyze the predictable improvement in performance of straw combine blade during field

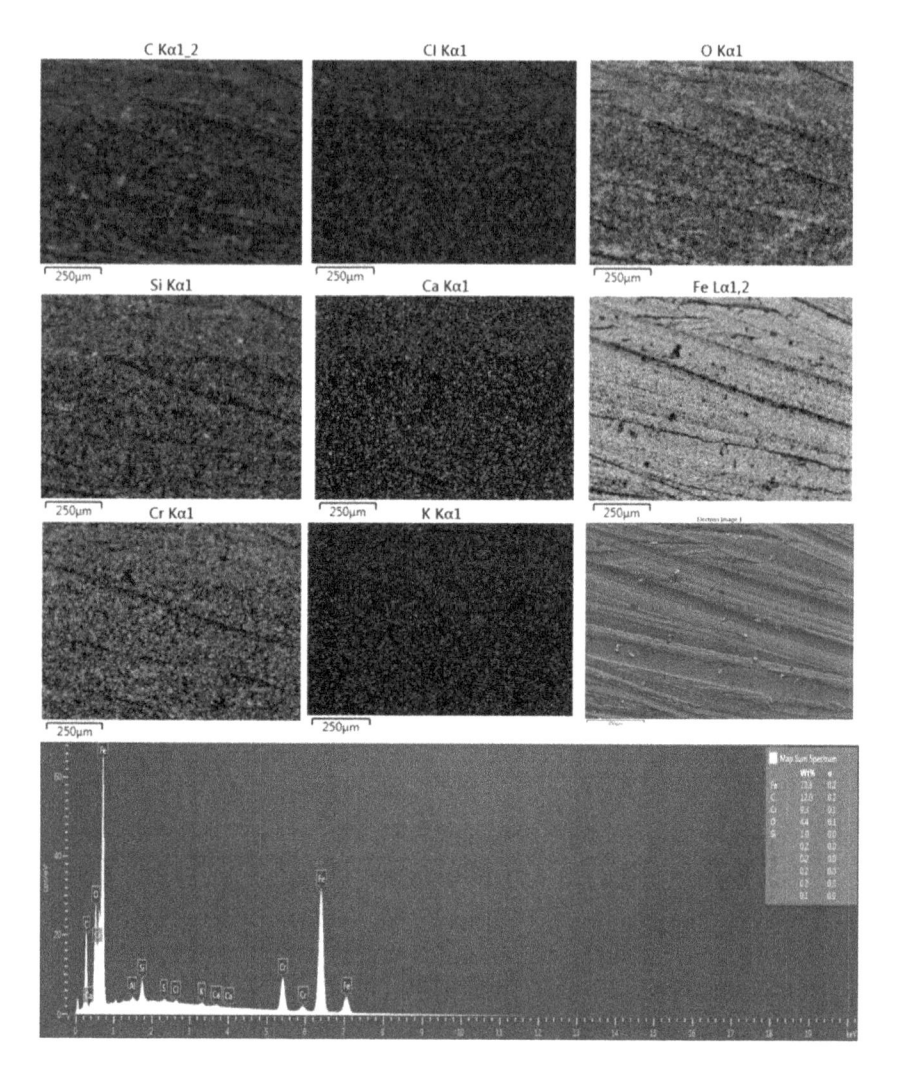

Fig. 6. Energy dispersive spectroscopy analysis of untreated specimen (Color online).

operation against the added cost calculated in Indian Rupees (Rs.) due to cryogenic treatment. The initial cost of cutter bar blade (A type) and chopping cylinder blade (M type) was Rs. 24 and Rs. 18, respectively. The cost of cryogenic treatment was calculated with respect to the batch. The capacity of machine in terms

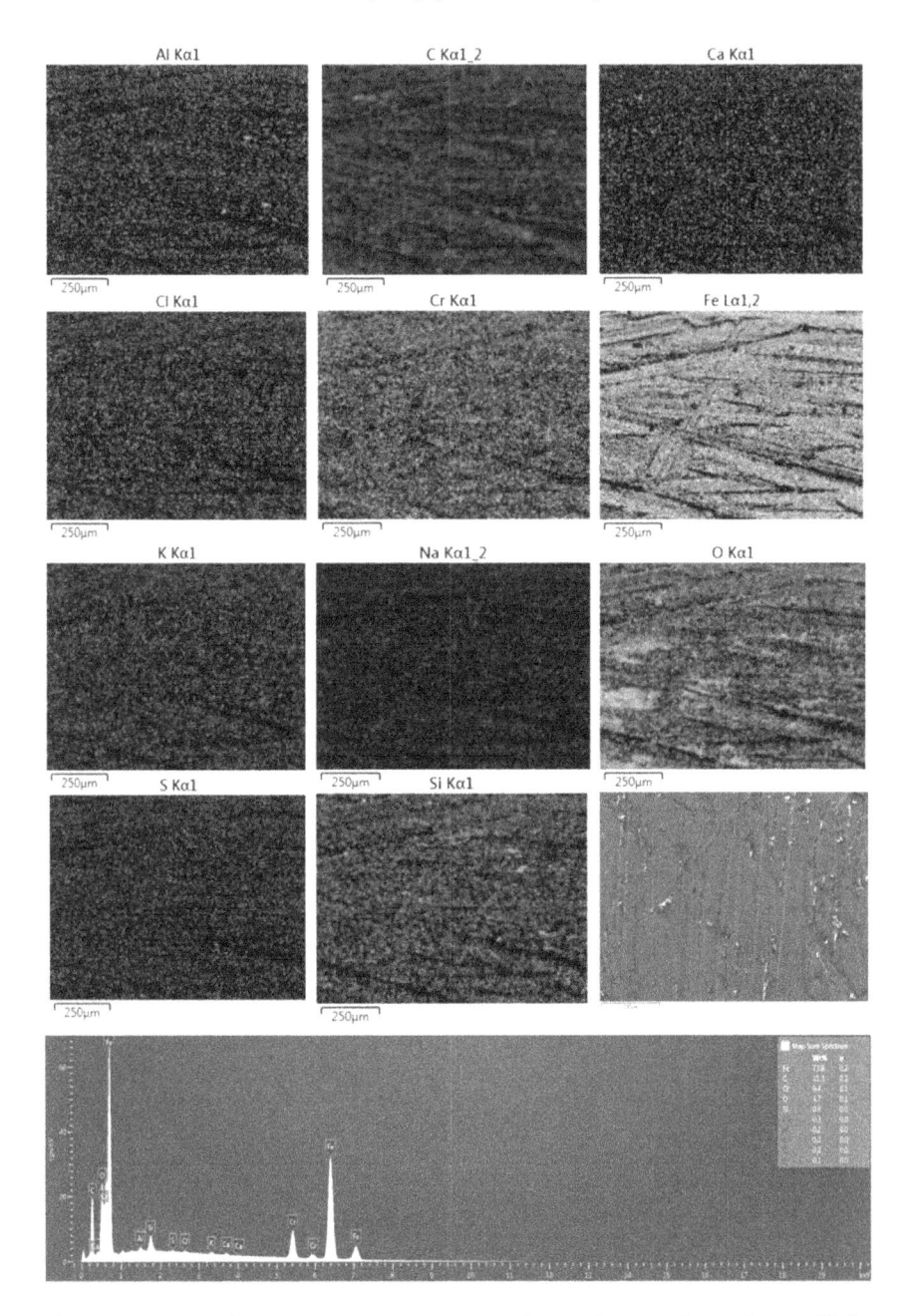

Deep cryogenic treated high carbon steel blades 205

Fig. 7. Energy dispersive spectroscopy analysis of treated specimen (Color online).

Table 7. Cost analysis of cryogenic treatment per blade of cutter bar blade (A type).

S. No.	Particulars	Values
1	Cost of blade (Rs.)	24
2	Number of blade in cutter bar	30
3	Weight of one blade (g)	65
4	Number of blades to be cryogenic treated in one lot	4000
5	Additional cost for proposed cryogenic treatment on one lot (Rs.)	9000
6	Additional cost per blade (Rs.)	2.25
7	Percentage increase in cost per blade due to cryogenic treatment (%)	9.38
8	Percentage improvement in wear rate resistance of blade after cryogenic treatment as compared to untreated in field condition (%)	75.04
9	Life of blade before treatment (h)	400
10	Life of blade after treatment (h)	700
11	Percentage increase in the life of blade due to cryogenic treatment (%)	75

of weight per batch was 260 kg. The cost of cryogenic coating for one batch was Rs. 9000. The average weight of one cutter bar blade (A type) was 65 g; therefore, 4000 number of blades can be treated in one time. The average weight of the chopping cylinder blade (M type) was 71 g; therefore, 3600 blades can be treated per batch. The cost of the cryogenic treatment was Rs 2.25 and Rs. 2.50 for cutter bar blade (A type) and chopping cylinder blade (M type), respectively. The percentage increase in cost of cutter bar and chopping cylinder blades due to cryogenic treatment was 9.38% and 13.61%, respectively. The increased cost of the blades was completely compensated due to the increase in the life of the blades by 75% and 70% for cutter bar and chopping cylinder blades, respectively. The DCT improves the mechanical properties and reduces cost by 50% because of less wear during working.[39] In addition, the improvement on tool life by 45% was reported after

Table 8. Cost analysis of cryogenic treatment per blade of chopping cylinder blade (M type).

S. No.	Particulars	Values
1	Cost of blade (Rs.)	18
2	Number of blade in chopping cylinder blade	304
3	Weight of one blade (g)	71
4	Number of blades to be cryogenic treated in one lot	3600
5	Additional cost for proposed cryogenic treatment on one lot (Rs.)	9000
6	Additional cost per blade (Rs.)	2.50
7	Percentage increase in cost per blade due to cryogenic treatment (%)	13.61
8	Percentage improvement in wear rate resistance of blade after cryogenic treatment as compared to untreated in field condition (%)	71.15
9	Life of blade before treatment (h)	150
10	Life of blade after treatment (h)	256
11	Percentage increase in the life of blade due to cryogenic treatment (%)	70

performing DCT which makes the process cost-effective.[40] The economic analysis clearly justified the additional cost of cryogenic treatment (Tables 7 and 8).

6. Conclusions

Based on the laboratory and field results, the following conclusions were drawn:

- At 74%, load was most influential in determining wear loss, with 15.50% of the variance coming from time, 5.20% from sample type, and 1.70% from sliding velocity.
- SWR and FE-SEM demonstrated that cryogenic treatment increases the specimen's grain structure and intermolecular interaction, resulting in an enhancement in the wear resistance.

- In contrast to the untreated specimen, the wear on the treated surface is uniform over the entire surface, as demonstrated by FE-SEM analysis. Grain structure and intermolecular bonding of the specimen were improved as a result of the cryogenic treatment.
- Using the EDS, the constituent components were identified, along with the corresponding percentage change between the untreated and treated specimens. The cryogenically-treated surface had a greater hardness than the untreated specimens because of this change in the percentage of elements in the specimens.
- Because of the cryogenic treatment, the price of cutter bars and chopping cylinder blades increased by 9.38 and 13.61%, respectively.
- In comparison to untreated cutter bar blades (A type), the wear resistance of those treated with cryogenics increased by 75.04%. For the M type chopping cylinder blades, the cryogenic treatment resulted in a 71.15% increase in wear resistance.
- There has been a 75% increase in cutter bar blade life (A type) and 70% increase in chopping cylinder blade life (M type).

References

1. S. Akincioglu, H. Gokkaya and I. Uygur, *Int. J. Adv. Manuf. Technol.* **78** (2015) 1609.
2. S. S. Gill, H. Singh, R. Singh and J. Singh, *Mater. Manuf. Process.* **26** (2011) 1430.
3. T. P. Singh, A. K. Singla, J. Singh, K. Singh, M. K. Gupta, H. Q. Song, Z. Liu and C. I. Pruncu, *Materials.* **13** (2020) 436.
4. E. Marin, A. Rondinella, W. Zhu, B. J. McEntire, B. S. Bal and G. Pezzotti, *J. Mech. Behav. Biomed. Mater.* **65** (2017) 616.
5. F. K. Arslan, I. Altinsoy, A. Hatman, M. Ipek, S. Zeytin and C. Bindal, *Vacuum.* **86** (2011) 370.
6. M. K. Vidyarthi, A. K. Ghose and I. Chakrabarty, *Cryogenics.* **58** (2013) 85.
7. F. Shichao, H. Hai and Z. Xingguo, *Steel Re. Int.* **92** (2021) 2000554.
8. D. Senthilkumar, *Adv. Mater. Process. Technol.* https://doi.org/10.1080/2374068X.2021.1878696 (2021).

9. P. H. S. Cardoso, C. L. Israel, M. B. Da Silva, G. A. Klein and L. Soccol, *Wear.* **456** (2020) 203382.
10. Y. Kang, J. Yang, J. Ma, Q. Bi, L. Fu and W. Liu, *J. Eng. Tribol.* **226** (2011) 71.
11. S. K. Josyula and S. K. R. Narala, *J. Eng. Tribol.* **230** (2015) 919.
12. P. Sivaiah and D. Chakradhar, *Silicon.* **11** (2019) 25.
13. M. P. Ahmed and H. S. Jailani, *Silicon.* **11** (2019) 105.
14. Z. Haidong, Y. Xianguo, H. Qiang and C. Zhi, *Adv. Mater. Sci. Eng.* **2021** (2021) 1.
15. S. Goswami, R. Ghosh, H. Hirani and N. Manda, *Ceram. Int.* **48** (2022) 11879.
16. K. Ghosh, S. Mazumder, H. Hirani, P. Roy and N. Manda, *J. Tribol.* **143** (2021) 061401.
17. G. Paul, S. Shit, H. Hirani, T. Kuila and N. C. Murmu, *Tribol. Int.* **131** (2019) 605.
18. H. Singh, C. Prakash and S. Singh, *J. of Bion. Eng.* **17** (2020) 1029.
19. S. Singh, C. Prakash, H. Wang, X. F. Yu and S. Ramakrishna, *Eur. Polym. J.* **118** (2019) 561.
20. S. S. Kharb, P. Antil, S. Singh, S. K. Antil, P. Sihag and A. Kumar, *Silicon.* **13** (2021) 1113.
21. P. Antil, *Silicon.* **11** (2019) 1791.
22. P. Antil, S. Singh, S. Singh, C. Prakash and C. I. Pruncu, *Meas. Control.* **52** (2019) 1167.
23. P. Antil, S. K. Antil, C. Prakash, G. Królczyk and C. Pruncu, 2020. *Meas. Control.* **53** (2020) 1902.
24. S. Singh, S. C. Prakash, P. Antil, R. Singh, G. Królczyk and C. I. Pruncu, *Materials* **12** (2019) 1907.
25. M. Karim, J. B. Tariq, S. M. Morshed, S. H. Shawon, A. Hasan, C. Prakash, S. Singh, R. Kumar, Y. Nirsanametla and C. I. Pruncu, *Sustainability.* **13** (2021) 7321.
26. A. Babbar, C. Prakash, S. Singh, M. K. Gupta, M. Mia and C. I. Pruncu, *J. Mater. Res. Technol.* **9** (2020) 7961.
27. G. F. Dieison, T. P. Cleber, S. R. Tonilson, R. M. Afonso and D. Tier, *J. Mater. Res. Technol.* **9** (2020) 12354.
28. M. Villa, K. Pantleon and M. A. J. Somers, *Acta Mater.* **65** (2014) 383.
29. T. V. Kumar, R. Thirumurugan and B. Viswanath, *Mater. Manuf. Process.* **32** (2017) 1789.

30. T. P. Singh, A. K. Singla, J. Singh, K. Singh, M. K. Gupta, H. Ji, Q. Song, Z. Liu and C. I. Pruncu, *Materials.* **13** (2020) 436.
31. S. Jagtar, S. L. Pal and K. Ankur, *Frict. Wear Res.* **1** (2013) 22.
32. H. Hirani, *Fundamental of Engineering Tribology with Applications.* Cambridge University Press, Cambridge (2016).
33. D. N. Nguyen, T. P. Dao, C. Prakash, S. Singh, A. Pramanik, G. Krolczyk and C. I. Pruncu, *J. Mater. Res. Technol.* **9** (2020) 5112.
34. C. Prakash, S. Singh, M. K. Gupta, M. Mia, G. Królczyk and N. Khanna, *Materials.* **11** (2018) 1602.
35. C. Prakash, S. Singh, A. Basak, G. Królczyk, A. Pramanik, L. Lamberti and C. I. Pruncu, *J. Mater. Res. Technol.* **9** (2020) 242.
36. K. L. Sahoo, C. S. S. Krishnan and A. K. Chakrabarti, *Wear* **239** (2000) 211.
37. S. Kumar, A. Bhattacharyyaa, D. K. Mondal, K. Biswas and J. Maity, *Wear* **270** (2011) 413.
38. K. H. Z. Gahr, Book on metallurgical aspects of wear, German Society of Metallurgy, Germany (1981), ISBN 9783883550466, pp. 73–103.
39. A. Molinari, M. Pellizzari, M. S. Gialanella, G. Straffelini and K. H. Stiansny, *J. Mater. Process. Technol.* **118** (2001) 350.
40. M. D. Lal, S. Renganarayanan and A. Kakanidhi, *Cryogenics.* **41** (2000) 149.

Preparation and biological evaluation of PLD-based forsterite-hydroxyapatite nanocomposite coating on stainless steel 316L*

P. Shakti Prakash[†,§], Suryappa Jayappa Pawar[‡,¶] and Ravi Prakash Tewari[‡,||]

[†]*Department of Biomedical Engineering,*
School of Engineering & Technology, Mody University,
Lakshmangarh, Sikar, Rajasthan 332311, India
[‡]*Department of Applied Mechanics,*
Motilal Nehru National Institute of Technology Allahabad,
Prayagraj, Uttar Pradesh 211004, India
[§]*pspsbme@gmail.com*
[¶]*sjpawar@mnnit.ac.in*
[||]*rptewari@mnnit.ac.in*

The present work deals with the fabrication of forsterite–hydroxyapatite (FS–HA) hybrid coatings on stainless-steel 316L using the pulsed laser deposition (PLD) technique. The stainless steel (SS 316L) as a metallic implant is widely used in hard tissue applications. The XRD studies have confirmed the crystalline behavior of synthesized FS powder with an average crystallite size of 54 nm. The synthesized FS powder was mixed in different compositions (10, 20, 30 wt.%) into HA for preparing PLD targets (pellets). The XRD of the prepared pellets by UTM has confirmed both phases of FS and HA. The Scanning Electron Microscopy (SEM) of the coated samples depicted the successful deposition of composite powders on the substrates (SS 316 L). The Ellipsometer was used to investigate the thickness of

---

*To cite this article, please refer to its earlier version published in Surface Review and Letters, Vol. 30, No. 1 (2023) 2141002 (12 pages) DOI: 10.1142/S0218625X2141002X
[§]Corresponding author.

different substrates and it was found as 243, 251, 255, and 257 nm for CP1, CP2, CP3, and CP4, respectively. The bioactivity of the coated substrates with different compositions (pure HA, 10%, 20%, 30%, and pure FS) was investigated by immersing the samples in simulated body fluid (SBF) for 14 days. The same samples were then characterized by SEM which confirms the apatite layer formation that reflects the bioactivity. The addition of FS powder into HA will stimulate the apatite formation which enhances the bioactivity. The Raman Spectroscopy of coated samples reveals the successful deposition of different compositions of FS–HA nanocomposite. The peaks of Raman spectroscopy were corresponding to the XRD results of the pellets (different compositions of FS–HA). The antimicrobial activity of different compositions of FS–HA against Escherichia coli (*E. coli*) bacteria also showed a significant zone of inhibition. The bioactivity and antimicrobial behavior of FS–HA along with successful deposition by PLD have shown better potential applications for biomedical implant coating.

Keywords: SS 316L; forsterite; implant; pulsed laser deposition; corrosion; bioactivity.

1. Introduction

The development of biomaterials started many years before (600 BC). According to Sushruta Samhita, the first nose reconstruction was done with the help of a patch of living flesh from the region of the cheek. Later in the 18–19th century, the use of various metal devices made of iron, silver, gold, and platinum to fix bone fractures can be found.[1] The metals-based biomedical implants have been widely used for more than 10 decades. The Lane first introduced steel plates and screws for fixation of bone fracture.[2] The metal implants faced corrosion problems in the early development until Sherman in 1920s first developed and introduced metal alloys to overcome the corrosion, which attracted the interest of clinicians.[3] Thereafter, vast development in the field of biomaterials and their clinical use can be found in the literature.[4-10]

The foremost and predominant requirement for the selection of the biomaterial is its acceptance by the human body.[11] Implants should not cause any unfavorable effects like inflammation, infection, pain, and cytotoxicity.[12] A biomaterial ought to stay intact for a long time and should not fail until the death of the person.[13] Its success highly depends on various factors like material properties (chemical, mechanical, and tribological), the implant's biocompatibility, the health condition of the beneficiary, and the expertise of the surgeon.[11]

Presently, the austenitic stainless steels (mostly SS 316L) are widely used for making implants as well as internal fixation devices such as plates and screws because of having exceptionally good mechanical properties and being inexpensive as compared to other metals and alloys.[14,15] However, stainless steels have limited bio-functionalities such as bone conductivity, hemocompatibility, osseointegration, bioactivity, etc. Using metallic implants in the biological system have several other limitations too such as its biologically inert behavior which do not promote the formation of an apatite layer on its surface[16] and it releases metallic ions which may combine with biomolecules and cause adverse biological reaction.[17,18] Hence, the surface modification becomes necessary using bioactive materials which can accelerate bone healing and bonding of the coatings with tissues.[19-21] Nowadays, several bioceramics, bioactive glasses, and polymers as coating materials are being used to enhance the biofunctionalities of implants.[22]

There are various physical and chemical techniques available to produce thin film coatings for biomedical implants. The dip coating, plasma spray, pulsed laser deposition (PLD), and magnetron sputtering are few examples. From these techniques, the PLD consider a significant technique to produce a thin film of high homogeneity and crystallinity[23,24] also many materials that are normally difficult to deposit by other methods especially multi-element oxides, nitrides, metallic multilayers can be successfully deposited by it.[25-29] There are several other advantages of PLD such as (i) easy to implement, (ii) growth in any environment, (iii) stoichiometry of complicated materials can be maintained, (iv) variable growth rate and (v) greater control of growth.[29-31]

Currently, the hydroxyapatite (HA) is considered as one of the desired material for coating over implants[32-34] due to its basic resemblance and chemical structure to bone and teeth,[35,36] and other applications of HA as it is used in the form of powders in bone cement,[37,38] and as composites.[39] However, some problems like poor fatigue resistance, poor strength, and inherent brittleness associated with bulk HA ceramic restrict its use for load-bearing applications.[12,37] Moreover, poor adhesion of HA as coatings on metal implants,[40] and high dissolution of pure HA in a physiological environment restrict its use for biomedical implant coating.[41] HA is used primarily in the form of a porous coating on metal implants in dental and orthopedic applications, which combines both the bioactivity of HA as well as the excellent mechanical properties of metals, simultaneously.[42–44] On the other hand, the hybrid coatings have proven much as they combine properties of different constituent materials and to overcome the limitation of HA.[45] Therefore, to improve the biological and mechanical properties of HA[46–48] researchers have been adding various materials such as bioactive glass, bioceramics, polymers, etc. to it.[49]

Forsterite (FS) is one of the materials with superior mechanical and biological properties, showing higher values of fracture toughness as compared to HA and bioactive glass.[50–62] The addition of FS nanopowder could improve the properties, both the dissolution rate and the stimulation for apatite formation for the composite.[50,63,64] FS exhibits good bioactivity when used for coating. In addition, if amalgamated with HA, the resulting products are expected to have better bioactivity compared to HA if used alone. The aim of this work is the fabrication of FS–HA nanocomposite coating using PLD technique and evaluation of biological properties of prepared composites coating associated with the composition of FS used in HA in the composite.

2. Materials and Methods

2.1. *FS powder synthesis*

The FS powder was synthesized using the sol–gel method as described by Prakash *et al.*[12] In this method, the reagent-grade precursors namely magnesium nitrate hexahydrate (MNH), nitric acid (from MERCK Ltd, USA), colloidal silica, sucrose, and polyvinyl alcohol (PVA) (from Sigma-Aldrich, USA) were used. To prepare FS powder, initially, three different solutions were prepared namely solutions A, B, and C. Solution A was magnesium nitrate which dissolved in deionized water and added with colloidal silica, solution B was a sucrose solution and solution C was PVA solution. Thereafter, all three solutions were mixed and kept on stirring for 4h at 80°C temperature. The pH of the solution was maintained to one by using the HNO_3 solvent. Afterward, the solution was kept for the gelation and drying process at 200°C in a hot air oven which resulted in complete dehydration and changed into a brown mass. The obtained mass was crushed and grounded into a fine powder using a mortar and pestle. The powders were again heat treated at 1000°C to get FS powder. Finally, an average crystallite size of 54 nm of FS powder was obtained.

2.2. Preparation of PLD target

In order to prepare PLD targets, the HA powder with Ca/P molar ratio 1.61–1.71 standard[65] was procured from Clarion Pharma, India, and the FS powder which was prepared from the above process was used. These powders were rigorously mixed with 1 wt.% of PVA solution (used as a binder). Powders with several compositions used for preparing the target are listed in Table 1. The palettes with different compositions were prepared by compressing the powder up to 35 kN using a steel mold (diameter, 25 mm approx.)

Table 1. Different compositions of powder used to make palettes.

Material / Palettes	P1	P2	P3	P4	P5
FS powder (in %)	—	10	20	30	100
HA powder (in %)	100	90	80	70	—

Notes: P1, P2, P3, P4, and P5 are Palettes (Targets).

in the UTM machine. Subsequently, the palettes were heat-treated at 1000°C for 3h for densification and removal of PVA contents.

2.3. Surface preparation of the substrate

The substrate material, stainless steel 316L was taken for the present study. Before the coating process, the substrates were prepared according to an ASTM D2651 standard for adhesive bonding. Substrates of $10 \times 20 \times 2$ mm^3 dimensions were cut from a SS 316L sheet for the study. These pieces were hand rubbed through a series of emery paper up to 2000 grits followed by a mirror-polished on polisher machine and cleaned with deionized water. Afterward, acid etching was done using the etchant's constituents (hydrochloric acid, orthophosphoric acid, and hydrofluoric acid) according to the ASTM standards. For this process, the substrates were immersed for 2 min at approximately 93°C in the above etchant solution heated by a boiling water bath. Then, the substrates were dried in an oven at 60°C for 1 h and cleaned with acetone.

2.4. The coating on substrates (SS 316L)

The construction and laser-material interaction of the PLD technique have been described elsewhere.[34,36,66,67] The coating chamber and laser beam emitter are the two major parts of the PLD equipment (Instruments-Model: Quanta System Q1 DNA was used in this study). In this coating process, the Nd: YAG type laser was operated on the third harmonic at a wavelength of 355 nm. The coating chamber is made up of stainless steel which has a rotating target holder, substrate holder, and substrate heater equipment which are depicted in Fig. 1(a).

The palettes of different compositions were placed in the target holder. For each deposition, the distance between the target holder and substrate was maintained at approximately ~4.5 cm. The deposition was carried out in the vacuum chamber of stainless steel at maintained pressure of 10^{-5} mbar. The target holder was continuously rotated for uniform laser ablation. The HA–FS films were

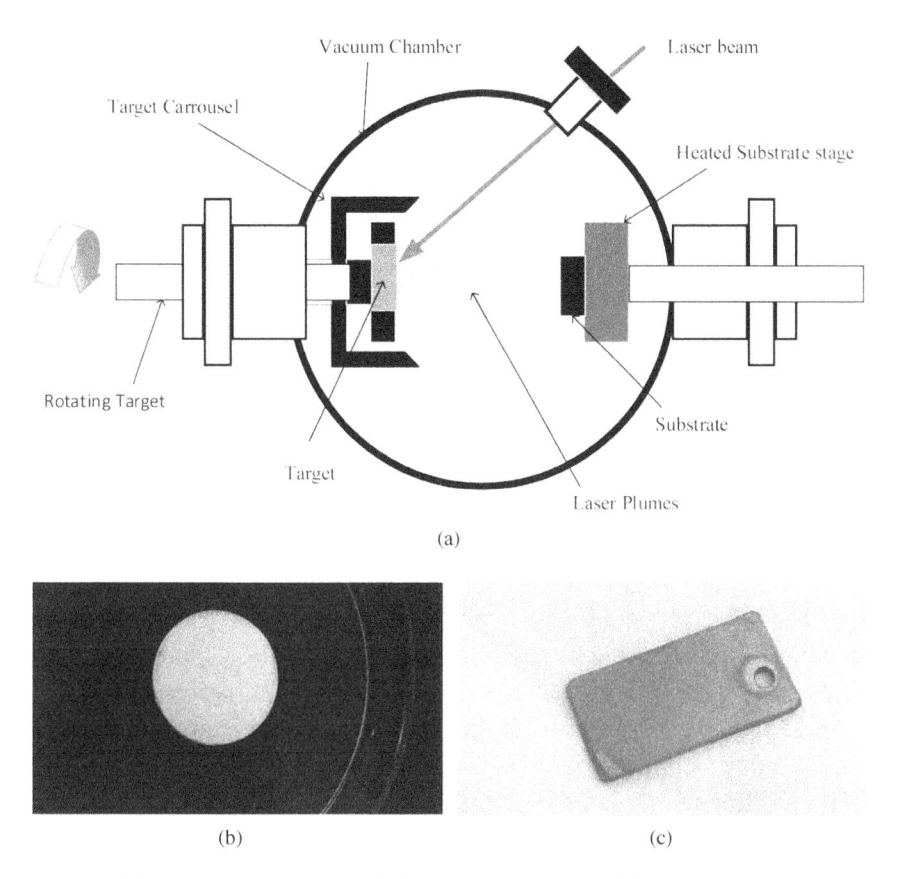

Fig. 1. (a) Schematic of PLD, (b) PLD target, and (c) the coated specimen showing uncoated regions due to clamps (Color online).

grown on a heated substrate with a laser fluence of 5.5J/cm^2. The optimized parameters for deposition such as substrate temperature, deposition time, laser energy level, pulse width, and pulse frequency were set at 650°C, 1 h, 225 mJ, 5 ns, and 10 Hz, respectively. Further, the coated substrates were annealed at 600°C temperature for 2 h.

2.5. Antimicrobial assay

The nutrient Luria broth (LB) agar was dissolved in deionized water (DI water) and sterilized by autoclaving at 121–124°C, 15 psi for 20 min. The media was then poured (25 mL) into the sterile Petri plates. Subsequently, 80 μL culture *Escherichia* coli (*E. coli*, gram-negative bacteria) was inoculated in the plates, after solidifying the LB. The respective wells were punched by sterile micro tips, and 20 μL of the individual samples were added. The samples were kept for incubation at 37°C for 24 h. The powder concentrations (as mentioned in Table 1) were taken as 1000 μg/mL. The DI was a negative control, and the ciprofloxacin antibiotic was taken as a positive control. The zone of inhibition was measured (mm) at overnight incubation.

2.6. *Bioactivity test*

An *in vitro* method was used to evaluate the bioactivity of prepared samples. It started with exposure of the material to an acellular solution termed "simulated body fluid" (SBF) also called "Hank's solution" and it was first developed by Kokubo.[68] This solution comprises a quantity and type of ions in a similar concentration found in human blood plasma. The SBF was prepared according to the standard compositions and dissolved in DI water and kept at 37°C.[69] Afterward, the SBF solution was poured into small containers and then coated samples were kept in this solution. It was followed by placing these containers in a water bath kept at 37°C (the same temperature of the human body) for the periods of 14 days. After the exposure with the SBF, the samples were dipped in acetone solution to stop all reactions. Thus, the surface of these samples was evaluated by Scanning Electron Microscope (SEM).

3. Results and Discussion

3.1. *Surface morphology of coatings*

The FS and HA coatings have been applied on the substrate using the PLD technique of different compositions as mentioned above in Table 1. The SEM images for the four different compositions of FS and HA have been shown in Fig. 2. Figures 2(a)–2(d) showed the coated surface of substrate for CP1 (100% HA), CP2 (FS10%HA90%),

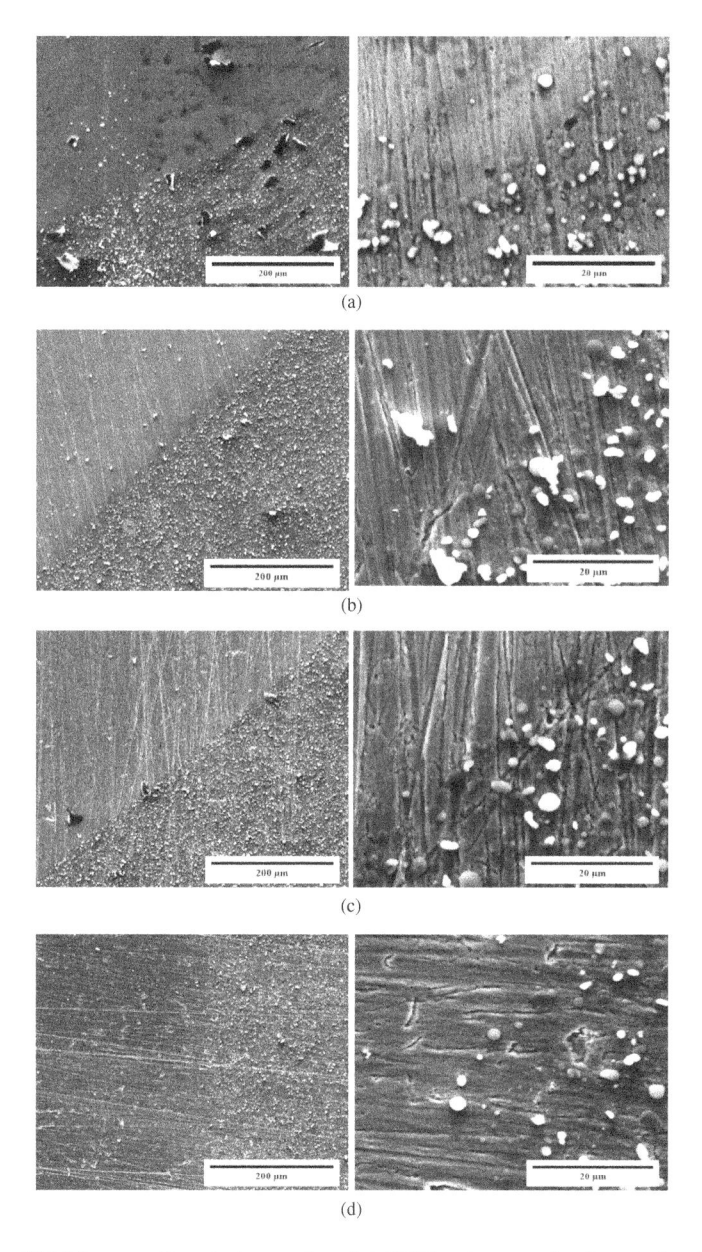

Fig. 2. SEM image of coated samples. (a) SEM micrograph of coated substrate CP1; scale at 500× and 5000×. (b) SEM micrograph of coated substrate CP2; scale at 500× and 5000×. (c) SEM micrograph of coated substrate CP3; scale at 500× and 5000×. (d) SEM micrograph of coated substrate CP4; scale at 500× and 5000×.

CP3 (FS20%HA80%), and CP4 (FS30%HA70%), respectively. In all four conditions, the two corner ends are covered inside the clamps for holding the specimen during coating as shown in Fig. 1(c). Hence, the coating was not seen on those two corners ends. The difference between the coated and uncoated regions can be seen in all the images in Fig. 2. In addition, uniform coatings have also been observed on the surface of the substrates. The micrographs showed that the deposition has taken place in a columnar fashion in all four cases. A similar type of surface morphology was manifested by various researchers.[70–72] It is evident from Fig. 2, the deposition of particles in the form of droplets over the surface of all substrates obtained might be referred to as the laser fluence level used in the PLD technique. This result is an example of an improved stage of classical absorption that increases the plasma temperature around the focus of the laser,[73] and which corresponds to an increase in surface roughness of coating. High surface roughness is desirable for biomedical applications and that can be achieved by a large number of deposited particles in this process. Because higher surface roughness leads to an extension of the surface and increases proliferation and cell growth,[74] it results in effective bioadhesion.[74–77]

3.2. *Bioactivity testing*

In vitro bioactivity of coated substrates of FS-HA composition was studied by immersing it into the SBF solution (Henk's solution). The compositions and the procedure for the preparation of the SBF solution were adopted from the available literature.[68,69] FS[50,53,60,62,78] and HA[79–82] powders are known to have excellent bioactivity properties. Figure 3 shows the surface microstructures of coated substrates i.e. CP1, CP2, CP3, CP4, and CP5 after 14 days of soaking in the SBF solution. A large number of clusters of granular precipitates of apatite formation were noticed around the areas of the coated surface in all the specimens. It can be clearly visible in Fig. 3(e) that pure FS specimen having a higher density of granular precipitation of apatite formation results in better bioactivity in comparison to CP1 i.e. HA-coated specimen. It is reflected from the micrographs

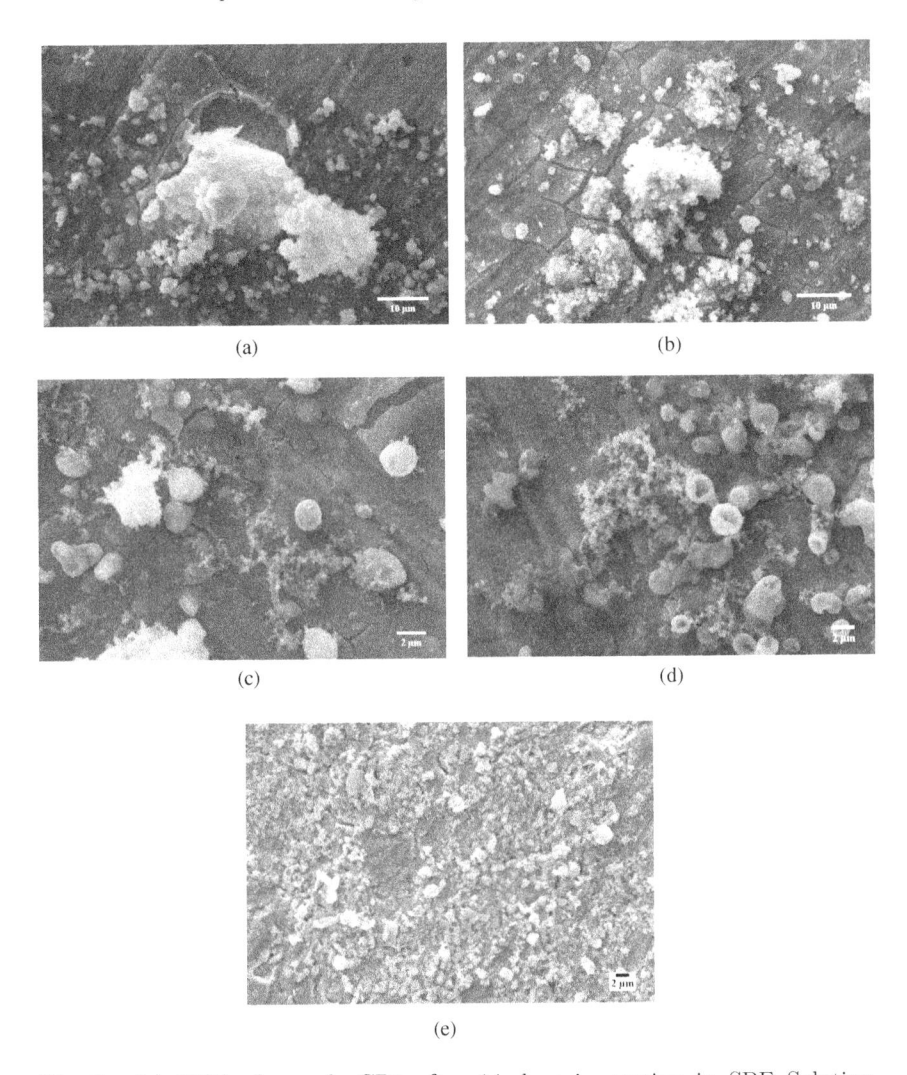

Fig. 3. (a) SEM of sample CP1 after 14 days immersion in SBF Solution. (b) SEM of sample CP2 after 14 days immersion in SBF Solution. (c) SEM of sample CP3 after 14 days immersion in SBF Solution. (d) SEM of sample CP4 after 14 days immersion in SBF Solution. (e) SEM of sample CP5 after 14 days immersion in SBF Solution.

(CP2, CP3, and CP4) that an increasing amount of FS powder in HA enhances bioactivity. The probable reason for enhanced bioactivity/formation of apatite on the surface is attributed to the ion exchange between FS–HA hybrid and SBF solution.[83]

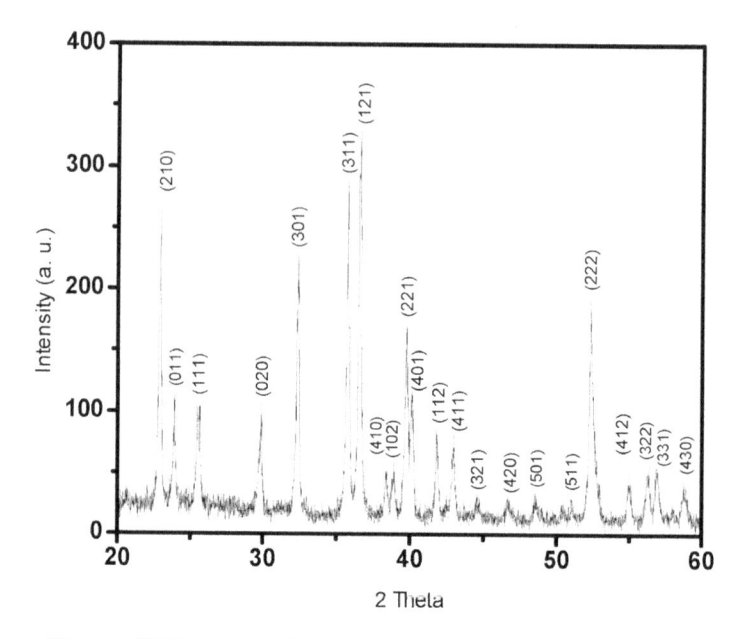

Fig. 4. XRD pattern for the synthesized nano FS at 1000°C.

3.3. *XRD analysis*

Figure 4 illustrated the X-ray diffraction (XRD) patterns of synthesized FS nanopowder annealed at 1000°C for 8h. XRD confirmed the crystal phases of pure FS and their peaks were exactly matched with the standard JCPDS no. 163295. Using the Scherrer equation, the average crystallite size of FS was measured to be 54 nm. XRD analysis of targets prepared using different compositions is illustrated in Table 1. The targets were sintered at 1000°C. Before sintering, each different composition was rigorously hand grounded for 3h to get uniform distribution/mixing.

Figure 5 shows the stacked XRD of PLD targets (P1–P5) to show the presence and maintained crystallinity of both materials. The five different prominent XRD peaks and respective planes for pure HA (P1) and FS (P5) were found in prepared targets which are described in Tables 2 and 3, respectively. XRD pattern of P1 and P5 exactly matched with the standard data i.e. JCPDS card no. 87727 and 163295, respectively. The identified peaks marked in

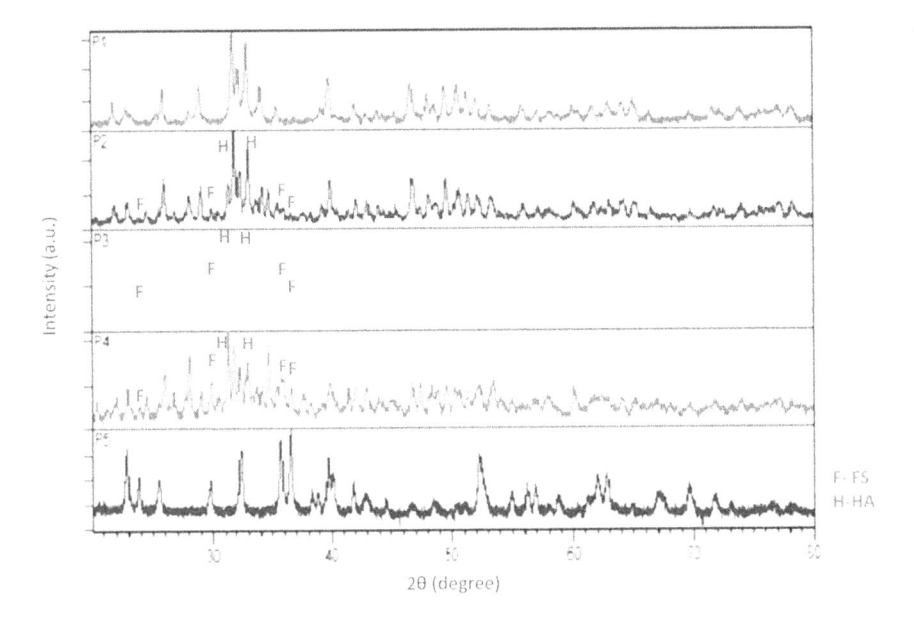

Fig. 5. XRD peaks for palettes: P1 (100% HA), P2 (90% HA, 10% FS), P3 (80% HA, 20% FS), P4 (70% HA, 30% FS), and P5 (100% FS) (Color online).

Table 2. Prominent XRD peak in HA (P1).

Peak position (2θ)	(h k l)	Peak intensity (%)
31.8	1 2 1	100
32.2	1 1 2	59.6
32.9	0 3 0	60.5
39.8	1 3 0	34.3
46.7	2 2 2	31.9

Table 3. Prominent XRD peak in FS (P5).

Peak position (2θ)	(h k l)	Peak intensity (%)
23.8	2 1 0	62.9
29.8	3 0 1	59.4
35.8	3 1 1	77.3
36.5	1 2 1	100
52.2	2 2 2	64.4

Fig. 5 of PLD target P2, P3, P4 by alphabets H and F which represents HA and FS part's in the mixture of PLD targets shows the crystallinity and planes at similar 2θ position as described in Tables 2 and 3. Similar planes will appear since both FS and HA were mixed for preparing the targets and the same can be observed in Fig. 5.

3.4. *Raman spectroscopic characterization and analysis*

In order to study confirmative coatings at the surface of substrates, the Raman spectra were collected from coated substrates using a high-resolution Raman spectrometer (Renishaw, UK) equipped with an Nd-YAG induced laser beam of wavelengths (532 nm). The solid-state diode laser pump provides the power of 3 mW/mm^2, out of which only 1% was used to excite samples. The equipped 50× short distance objective of Leica DM 2500M microscope focused the incident laser beam on the sample. Raman signal was collected in back scattering geometry and 2400 grooves/mm grating was used for dispersion element.

The Raman spectra of different deposition on stainless steel 316L substrate were shown in Fig. 6. A broad spectrum was found between 600 and 800 cm^{-1} which corresponds to the bare substrate (uncoated stainless steel), while pure HA (CP1) peaks were at 226, 291, 426, 495, 1050, and 1073 cm^{-1}. It can be deduced from the graph shown in Fig. 6 that there is a confirmative deposition on the substrate. It has been observed that no peaks were found in the case of pure FS coating (CP5). The reason behind is that the synthesized FS was Raman inactive and showed a fluorescence kind of feature, instead of scattering. So, when the system is fluorescing and whenever it mix in Raman active material, it will affect the Raman activity of the original one, normally suppressing the peak intensity, the same we can observe in the case of CP2, CP3, and CP4. It can be depicted that coating has both HA and FS present at the surface of the substrate.

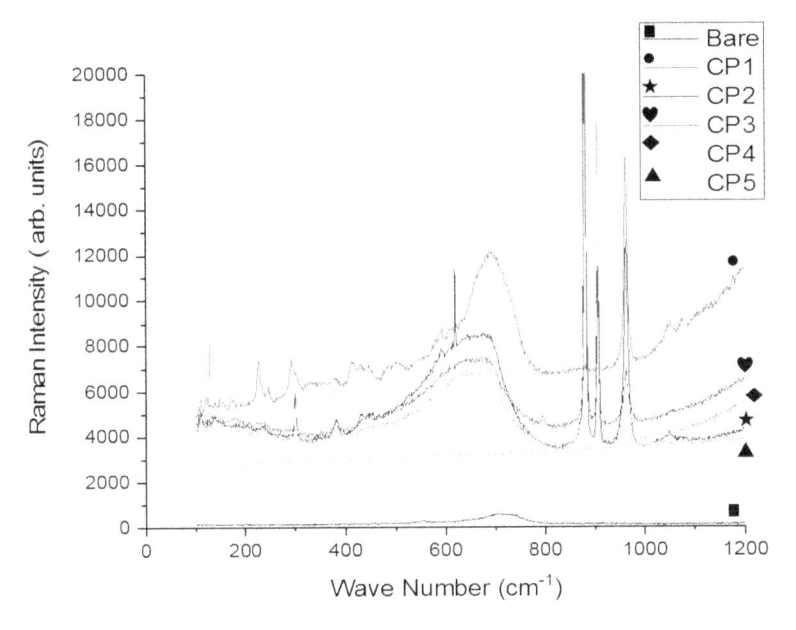

Fig. 6. Raman Spectra of different coated substrates: CP1 (100% HA), CP2 (90% HA, 10% FS), CP3 (80% HA, 20% FS), CP4 (70% HA, 30% FS), CP5 (100% FS), and Bare (uncoated SS 316L) (Color online).

3.5. *Ellipsometry*

The thickness of coated samples was evaluated by Ellipsometer (from J. A. WoollamCo.Inc.Model: VB-400, VASE). It is a non-invasive, non-contact, and non-destructive optical technique that uses a model-based approach to analyze the thin film. The thickness of deposition on different substrates was found to be 243, 251, 255, and 257 nm for CP1, CP2, CP3, and CP4, respectively.

3.6. *Antibacterial testing*

The antibacterial test is generally used to detect the possible bactericidal activity of any material against the common pathogens. In the present work, the agar well diffusion method was used for the

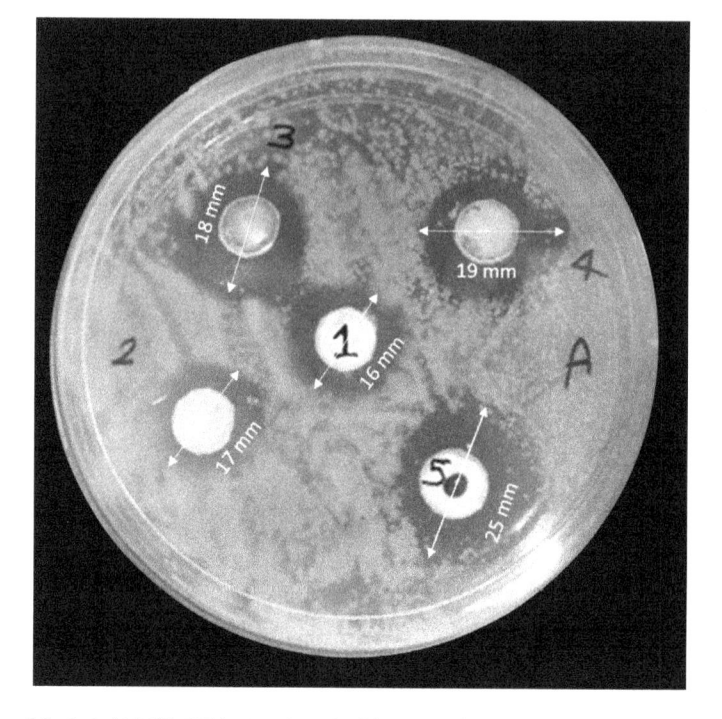

Fig. 7. Mark 1 (100% HA), Mark 2 (90% HA, 10% FS), Mark 3 (80% HA, 20% FS), Mark 4 (70% HA, 30% FS), and Mark 5 (100% FS) (Color online).

study using LB Agar broth plates. The zone of inhibition was evaluated around the materials. The zone of inhibition is the area where the growth of bacteria (*E. coli*) inhibits due to interaction with the given material. It has been found that FS has good antibacterial property.

It is evident from Fig. 7 that the zone of inhibition was increasing as the FS powder ratio increased in the prepared FS–HA hybrid.

4. Conclusion

The composite coatings have been successfully achieved by the PLD technique using targets of FS–HA on SS 316L substrates. The crystalline structure of FS nanopowder with an average crystallite size of 54 nm, synthesized by the sol–gel route was confirmed by the XRD pattern. The SEM results showed that the uniform coating was obtained over the entire surface of substrates in the form of a droplet structure. The obtained morphology revealed that the surface roughness (by PLD technique) can be achieved which might be desirable for cell growth or proliferation. The Raman spectroscopy of coated samples also confirmed the successful deposition of FS–HA on SS 316L substrates and the thicknesses were found to be 243, 251, 255, and 257 nm measured by the Ellipsometer. Among the coated samples, the addition of FS percentage (10, 20, and 30 wt.%) into HA promoted the better apatite formation as compared to HA, consequently obtained better bioactivity. Moreover, the same behavior can also be seen in the case of antibacterial activity, where the percentage increases of FS into HA showed the greater zone of inhibition, which confirms a significant antimicrobial behavior of the FS. Hence, the results reflected that the addition of FS into HA could enhance the bioactivity as well as antibacterial property. It can be concluded from the above results that FS can be a prominent coating material for a biomedical implant due to its excellent bioactivity as well as antibacterial property. Furthermore, the PLD technique obtained the promising results in line with the requirements for the advancement of implant materials. The results were found remarkably well and signify the potential application of FS–HA hybrid coating over medical implants.

Conflict of Interest

The author(s) declare no potential conflict of interest with respect to the research, authorship, and/or publication of this paper.

Acknowledgments

This work is supported by Centre for Interdisciplinary Research (CIR) and Technical Education Quality Improvement Programme, Phase-II (TEQIP-II) and Phase-III (TEQIP-III), Motilal Nehru

National Institute of Technology Allahabad, Prayagraj, UP, India. Special thanks go to Dr. Animesh Ojha and Dr. Naresh Kumar from the Department of Physics, MNNIT Allahabad, Prayagraj, India for their valuable contribution to the successful conduction of the experiments and characterizations.

References

1. S. V. Bhat, *Biomaterials* (Springer, 2002), pp. 1–11.
2. W. A. Lane, *Br. Med. J.* **1** (1895) 861.
3. W. O. Sherman, *Surg. Gynecol. Obstet.* **14** (1912) 629.
4. S. V. Bhat, *Biomaterials* (Kluwer Academic Publishers, Boston, 2002).
5. N. A. Peppas and R. Langer, *Science* **263** (1994) 1715.
6. M. Niinomi, M. Nakai and J. Hieda, *Acta Biomater.* **8** (2012) 3888.
7. J. B. Brunski, D. A. Puleo and A. Nanci, *Int. J. Oral. Maxillofac. Implants* **15** (2000) 15.
8. Y. Chen, Z. Xu, C. Smith and J. Sankar, *Acta Biomater.* **10** (2014) 4561.
9. M. Geetha, A. K. Singh, R. Asokamani and A. K. Gogia, *Prog. Mater. Sci.* **54** (2009) 397.
10. Q. Chen and G. A. Thouas, *Mater. Sci. Eng. R Rep.* **87** (2015) 1.
11. H. Hermawan, D. Ramdan and J. R. P. Djuansjah, *Biomedical Engineering — From Theory to Applications*, ed. R. Fazel (InTech, 2011).
12. P. S. Prakash, S. J. Pawar and R. P. Tewari, *Proc. Inst. Mech. Eng. L* **233** (2019) 1227.
13. G. Manivasagam, D. Dhinasekaran and A. Rajamanickam, *Recent Pat. Mater. Sci.* **2** (2010) 40.
14. J. A. Disegi and L. Eschbach, *Injury* **31** (2000) D2.
15. J. Walczak, F. Shahgaldi and F. Heatley, *Biomaterials* **19** (1998) 229.
16. M. A. McGee, D. W. Howie, K. Costi, D. R. Haynes, C. I. Wildenauer, M. J. Pearcy and J. D. McLean, *Wear* **241** (2000) 158.
17. D. L. Mitchell, S. A. Synnott and J. A. VanDercreek, *Int. J. Oral Maxillofac. Implants* **5** (1990) 83.
18. K. Bessho, K. Fujimura and T. Iizuka, *J. Biomed. Mater. Res.* **29** (1995) 901.

19. P. A. Ramires, A. Wennerberg, C. B. Johansson, F. Cosentino, S. Tundo and E. Milella, *J. Mater. Sci.: Mater. Med.* **14** (2003) 539.
20. T. Li, J. Lee, T. Kobayashi and H. Aoki, *J. Mater. Sci.: Mater. Med.* **7** (1996) 355.
21. C. S. Chai and B. Ben-Nissan, *J. Mater. Sci.: Mater. Med.* **10** (1999) 465.
22. J. Lahann, D. Klee, H. Thelen, H. Bienert, D. Vorwerk and H. Ho, *J. Mater. Sci.: Mater. Med.* **10** (1999) 443.
23. G. Popescu-Pelin, F. Sima, L. E. Sima, C. N. Mihai-lescu, C. Luculescu, I. Iordache, M. Socol, G. Socol and I. N. Mihailescu, *Appl. Surf. Sci.* **418** (2017) 580.
24. J. L. Arias, M. B. Mayor, J. Pou, Y. Leng, B. León and M. P. Amor, *Biomaterials* **24** (2003) 3403.
25. A. A. Menazea and M. K. Ahmed, *J. Mol. Struct.* **1217** (2020) 128401.
26. A. A. Menazea and N. S. Awwad, *Radiat. Phys. Chem.* **177** (2020) 109112.
27. A. A. Menazea, A. M. Abdelghany, N. A. Hakeem, W. H. Osman and F. H. Abd El-kader, *J. Electron. Mater.* **49** (2020) 826.
28. A. A. Menazea, A. M. Abdelghany, N. A. Hakeem, W. H. Osman and F. H. A. El-kader, *Silicon* **12** (2020) 13.
29. A. A. Menazea and A. M. Mostafa, *J. Environ. Chem. Eng.* **8** (2020) 104104.
30. E. Mohseni, E. Zalnezhad and A. R. Bushroa, *Int. J. Adhes. Adhes.* **48** (2014) 238.
31. Y. Liu, B. Rath, M. Tingart and J. Eschweiler, *J. Biomed. Mater. Res. A* **108** (2020) 470.
32. P. Habibovic, F. Barrere, C. A. Van Blitterswijk, K. de Groot and P. Layrolle, *J. Am. Ceram. Soc.* **85** (2002) 517.
33. M. K. Ahmed, R. Ramadan, M. Afifi and A. A. Menazea, *J. Mater. Res. Technol.* **9** (2020) 8854.
34. A. M. Fathi, M. K. Ahmed, M. Afifi, A. A. Menazea and V. Uskoković, *ACS Biomater. Sci. Eng.* **7** (2021) 360.
35. F. H. Albee, *Ann. Surg.* **71** (1920) 32.
36. L. L. Hench, *J. Am. Ceram. Soc.* **74** (1991) 1487.
37. M. H. Fathi and A. Hanifi, *Mater. Lett.* **61** (2007) 3978.
38. Y. Sung, *J. Cryst. Growth* **262** (2004) 467.
39. Y.-M. Sung and D.-H. Kim, *J. Cryst. Growth* **254** (2003) 411.

40. X. Zheng, M. Huang and C. Ding, *Biomaterials* **21** (2000) 841.
41. S. Zhang, Z. Xianting, W. Yongsheng, C. Kui and W. Wenjian, *Surf. Coat. Technol.* **200** (2006) 6350.
42. H. Kusakabe, T. Sakamaki, K. Nihei, Y. Oyama, S. Yanagimoto, M. Ichimiya, J. Kimura and Y. Toyama, *Biomaterials* **25** (2004) 2957.
43. K. A. Gross, C. S. Chai, G. S. K. Kannangara, B. Ben-Nissan and L. Hanley, *J. Mater. Sci.: Mater. Med.* **9** (1998) 839.
44. J. L. Ong and D. C. Chan, *Crit. Rev. Biomed. Eng.* **28** (2000) 667.
45. M. H. Fathi, M. Salehi, A. Saatchi, V. Mortazavi and S. B. Moosavi, *Dent. Mater.* **19** (2003) 188.
46. B. M. Hidalgo-Robatto, M. López-Álvarez, A. S. Azevedo, J. Dorado, J. Serra, N. F. Azevedo and P. González, *Surf. Coat. Technol.* **333** (2018) 168.
47. S. J. Ding, Y. M. Su, C.-P. Ju and J. C. Lin, *Biomaterials* **22** (2001) 833.
48. S. Singh, C. Prakash and H. Singh, *Surf. Coat. Technol.* **398** (2020) 126072.
49. A. A. Menazea, S. A. Abdelbadie and M. K. Ahmed, *Appl. Surf. Sci.* **508** (2020) 145299.
50. M. Kharaziha and M. H. Fathi, *Ceram. Int.* **35** (2009) 2449.
51. M. H. Fathi and M. Kharaziha, *Mater. Lett.* **63** (2009) 1455.
52. M. H. Fathi and M. Kharaziha, *J. Alloys Compd.* **472** (2009) 540.
53. M. A. Naghiu, M. Gorea, E. Mutch, F. Kristaly and M. Tomoaia-Cotisel, *J. Mater. Sci. Technol.* **29** (2013) 628.
54. M. Stewart, J. F. Welter and V. M. Goldberg, *J. Biomed. Mater. Res.* **69** (2004) 1.
55. R. Emadi, F. Tavangarian, S. I. R. Esfahani, A. Sheikhhosseini and M. Kharaziha, *J. Am. Ceram. Soc.* **93** (2010) 5.
56. C. Wu, Z. Chen, Q. Wu, D. Yi, T. Friis, X. Zheng, J. Chang, X. Jiang and Y. Xiao, *Biomaterials* **71** (2015) 35.
57. S. Ligot, T. Godfroid, D. Music, E. Bousser, J. M. Schneider and R. Snyders, *Acta Mater.* **60** (2012) 3435.
58. Y. Xie, W. Zhai, L. Chen, J. Chang, X. Zheng and C. Ding, *Acta Biomater.* **5** (2009) 2331.
59. V. P. Orlovskii, V. S. Komlev and S. M. Barinov, *Inorg. Mater.* **38** (2002) 12.
60. S. Ni, L. Chou and J. Chang, *Ceram. Int.* **33** (2007) 83.
61. H. Ghomi, M. Jaberzadeh and M. H. Fathi, *J. Alloys Compd.* **509** (2011) L63.

62. M. Kharaziha and M. H. Fathi, *J. Mech. Behav. Biomed. Mater.* **3** (2010) 530.
63. M. M. Sebdani and M. H. Fathi, *J. Alloys Compd.* **509** (2011) 2273.
64. M. Mazrooei Sebdani and M. Fathi, *Int. J. Mod. Phys. Conf. Ser.* **5** (2012) 510.
65. H. Khandelwal, G. Singh, K. Agrawal, S. Prakash and R. D. Agarwal, *Appl. Surf. Sci.* **265** (2013) 30.
66. P. Kuppusami and V. S. Raghunathan, *Status of Pulsed Laser Deposition: Challenges and Opportunities* (Taylor & Francis, 2006).
67. H. Zeng, W. R. Lacefield and S. Mirov, *J. Biomed. Mater. Res. A* **50** (2000) 248.
68. T. Kokubo and H. Takadama, *Biomaterials* 27 (2006) 2907.
69. K. J. Bundy, *Crit. Rev. Biomed. Eng.* **22** (1994) 139.
70. R. K. Singh, F. Qian, V. Nagabushnam, R. Damodaran and B. M. Moudgil, *Biomaterials* **15** (1994) 522.
71. P. Baeri, L. Torrisi, N. Marino and G. Foti, *Appl. Surf. Sci.* **54** (1992) 210.
72. C. M. Cotell, D. B. Chrisey, K. S. Grabowski, J. A. Sprague and C. R. Gossett, *J. Appl. Biomater.* **3** (1992) 87.
73. R. Kassing, P. Petkov, W. Kulisch and C. Popov, *Functional Properties of Nanostructured Materials* (Springer Science & Business Media, 2007).
74. G. P. Dinda, J. Shin and J. Mazumder, *Acta Biomater.* **5** (2009) 1821.
75. O. Blind, L. H. Klein, B. Dailey and L. Jordan, *Dent. Mater.* **21** (2005) 1017.
76. D. M. Brunette, P. Tengvall, M. Textor and P. Thomsen, *Titanium in Medicine*, 1st edn. (Springer-Verlag, Berlin Heidelberg, 2001).
77. L. Torrisi, *Thin Solid Films* **237** (1994) 12.
78. F. Tavangarian and R. Emadi, *Ceram. Int.* **37** (2011) 2275.
79. C.-S. Chien, T.-Y. Liao, T.-F. Hong, T.-Y. Kuo, C.-H. Chang, M.-L. Yeh and T.-M. Lee, *J. Med. Biol. Eng.* **34** (2014) 109.
80. A. V. Popkov, E. N. Gorbach, N. A. Kononovich, D. A. Popkov, S. I. Tverdokhlebov and E. V. Shesterikov, *Strategies Trauma Limb Reconstr.* **12** (2017) 107.
81. S. L. Aktuğ, S. Durdu, E. Yalçm, K. Çavuşoğlu and M. Usta, *Mater. Sci. Eng. C* **71** (2017) 1020.
82. S. C. Eum, J. H. Kim and J. K. Lee, *J. Nanosci. Nanotechnol.* **17** (2017) 3909.
83. Y. W. Gu, K. A. Khor and P. Cheang, *Biomaterials* **25** (2004) 4127.

Comparison and performance of α, $\alpha+\beta$ and β titanium alloys for biomedical applications[*]

Pralhad Pesode[†] and Shivprakash Barve

*School of Mechanical Engineering,
Dr. Vishwanath Karad MIT-World Peace University,
Pune 411038, Maharashtra, India*
[†]*pralhadapesode@gmail.com*

As implant materials, titanium and its alloys have been extensively utilized because of their exceptional mechanical properties and biocompatibility. Despite this, corporations and researchers alike have kept up their aggressive pursuit of better alloys since there are still issues that require immediate attention. One of these causes a problem with stress shielding as a noticeable variation in the elastic modulus of the implant material. Ti alloys release harmful ions after extended usage. The poor bioactivity of the Ti alloy surface slows the healing process. In order to address these problems, additional research has concentrated on developing Ti alloys for the 21st century that contain a more suitable phase and change the surface of the alloy from inherently bioinert to bioactive. This study assesses the knowledge presently existing on the biological, chemical, mechanical, and electrochemical characteristics of important β-Ti alloys created in recent years with the objective to provide scientific justification for using β-titanium-based alloys as a substitute for cpTi. Dental implants might be made using β-Ti alloys as an alternative. The enhanced alloy qualities, which include a lower modulus of elasticity, improved strength, suitable biocompatibility, and good abrasion and excellent resistance to corrosion, offer the

---

[*]To cite this article, please refer to its earlier version published in Surface Review and Letters, Vol. 30, No. 12 (2023) 2330012 (26 pages) DOI: 10.1142/S0218625X23300125
[†]Corresponding author.

essential proof. Additionally, structural, chemical, and thermomechanical modifications to β-Ti alloys allow for the production of materials that may be tailored to the needs of unique instances for clinical practises. By researching the paper, the performance and attributes of β-titanium alloy are compared to those of other forms of titanium alloy, such as α titanium alloys. To support their usage as cpTi substitutes, *in vivo* studies are required to assess new β-titanium alloys.

Keywords: Ti alloys; biomaterials; β-titanium alloys; biocompatibility; osseointegrations.

1. Introduction

Titanium (Ti) is employed in technological area including biomedical, automotive, aviation, and other industries because of its high strength and lower density. Menaccanite was found to be the initial Ti mineral and albeit fabrication started in 1910. Titanium's properties may be improved even more by alloying substances. Around 1960, Ti and its alloys have been widely used frequently as metallic biomaterials in a range of biomedical treatments, comprising craniofacial, orthopaedic, dental, prosthetic, and joint replacement surgery.[1,2] The first article on internal fastening plates was published in 1895.[3] Since then, surgical stainless steel (316L), cobalt–chromium (Co–Cr), and titanium alloys have witnessed a significant increase in their usage for angioplasty and the treatment of bone fractures.[4] Outstanding mechanical properties, long-term stability, and good biocompatibility are all the characteristics of these materials. Biomedical implants, particularly bone implants, have been created utilizing Ti alloys because of their exceptional physicochemical stability, high biocompatibility, superior corrosion resistance, mechanical stability, and good osseointegration.[4] Some titanium alloys, nevertheless, are incompatible with the human body.

Dental implants are often fabricated from commercially pure titanium (cpTi).[6] Nevertheless, only areas with higher degrees of abrasion, tensile strength, and fatigue are allowed to apply it.[6-8] Pure titanium is a rather soft metal,[9] making it vulnerable to fatigue, particularly when utilized for small-diameter surgical implants, that must comply to stringent mechanical stability criteria to avoid overloading and implant breakage.[10] Additional challenges limiting the usage of this material as a dental implant material include the higher Young's modulus of cpTi and the challenge of improving the material's mechanical qualities without sacrificing biocompatibility.[11]

Ti can be alloyed with a wide range of different elements to create biomaterial with unique properties and dental implants that are almost perfect. Ti–6Al–4V alloy offers a wide variety of applications because of its exceptional mechanical performance.[8] Cell survival was negatively impacted and implant biocompatibility was dramatically reduced by the release of Al and V from this alloy.[5] Al can indeed have significant neurotoxic consequences, as evidenced by research linking it to bone brittleness,[12] Alzheimer's disease,[13] and possible local inflammatory triggers. Due to this research, the use of Ti–6Al–4V has been discouraged and alloys free of toxic materials that are inert in the oral cavity have also been developed. To increase their clinical usefulness, experimental alloys need to have suitable mechanical characteristics, be durable and stable in corrosive conditions, and suitable for use *in vivo*.[14] Moreover, they must be biocompatible. Titanium alloys have shown to be very attractive for the biomedical industry due to their remarkable strength and great biocompatibility[15,82] as well as features such as increased strength, improved corrosion resistance,[16] and modulus of elasticity close to actual human bone.[17,18]

The extraordinary properties of β-Ti alloys make it an ideal alternative to cpTi for the fabrication of dental implants as well as other uses.[19,83] In spite of being true that a number of materials are being created for usage in the biomedical industry, various studies on whether it would be feasible to use these innovative materials in place of cpTi have come up empty. In order to promote their use, a few scientists have also looked into experimental alloys in living things. We provide a review of many significant titanium alloys for issues affecting dentistry in this study. To promote the application of β-titanium alloys in place of cpTi and related alloys for therapeutic uses, the biological, mechanical, chemical and electrochemical properties of the main β-Ti developed recently have been investigated in detail.

Metastable β-Ti alloys are exceptionally versatile substances with important applications in the medical industries. Due to their light weight and great strength, that may be attained by precipitation hardening, these alloys are ideal structural materials for use in aeroplanes. They may be created using non-toxic elements. Because it possesses shape memory capabilities and is extremely elastic, β-microstructure is valuable in biological and different technological uses.[20,84] The choice of alloy constituents as well as desired qualities in β titanium alloys is governed by design requirements that are particular to the use at hand. For example, the nontoxicity of β-Ti alloy is a crucial need for its application in biomedicine. Nevertheless, the density of β-Ti alloys used in aerospace shouldn't increase as a result of the alloying additions. Due to their wide availability and variety, β-Ti alloys are the focus of this study. These alloys are of a biomedical grade. Two groups may be used to categorize the bulk of β-Ti alloys used in biomedicine. The group-1 includes structural alloys for orthopedic usage to withstand loading, whereas the group-2 includes functional alloys utilized in cardiovascular and orthodontic purposes. These two kinds of applications' design objectives are very different from one another. Functional alloys require superior elasticity or shape memory properties, while orthopaedic alloys, for instance, need high strength and low modulus.[20,84,89]

1.1. *Need for β Ti alloys*

Only by creating new materials will it be possible to prevent the intrinsic toxicity of current biomedical alloys, which is caused by metal ion discharge and stress shielding effects.[21,22] The answers to the aforementioned issues will determine how metastable β-titanium alloys are designed for orthopaedic usage. Recent β-titanium alloy formulations include the harmless elements Fe, Mo, Ta, Nb, Zr, and

Sn.[23,88] Additionally, Ti–6Al–4V, Co–Cr–Mo alloys, and 316L stainless steel have a lower modulus than known metallic biomaterials.[24]

Three aspects of alloy manufacturing are the subject of research: alloy composition design, thermo-mechanical processing, and performance evaluation. The aim of alloy composition development is low elastic modulus. β-Ti alloys undergo thermomechanical processing to increase their strength while preserving a lower modulus of elasticity. It is necessary to assess the performance parameters for biocompatibility, corrosion, fatigue, and wear. In the investigation which has monitored advancements in area of metallic biomaterials, β-Ti alloys have only been briefly described.[23,25,85] This paper's major objective is to provide a thorough topical overview of β-Ti alloys for orthopaedic applications, which is still lacking despite substantial progress in the field. The performance of these alloys is thoroughly examined in this paper along with the techniques for creating alloy compositions and thermo-mechanical processing techniques.

2. Classification of Titanium Alloys

It is possible to create solid solutions by mixing materials with the same-sized atoms with the transition metal titanium. It exhibits what is referred to as the "α-structure," a HCP geometry up to 882.5°C. Above this temperature, solid β-titanium crystallizes into a body-centered cubic crystal and adopts a joint structure before melting at 1688°C.[27] Titanium alloys come in a variety of forms, including pure, impure, and mixed forms.[28] Either α-stabilizers, such as oxygen and aluminium, or β-stabilizers, including vanadium, iron, nickel, and cobalt, are used to alloy titanium.[26] Zirconium is one example of a neutral metallic substance that has no impact of any phase. Processing has the potential to influence the microstructure, which impacts the mechanical characteristics (ductility, strength, fatigue resistance, and so on). Table 1 lists the many stabilizing substances that have been added to titanium alloys to enhance their characteristics and stabilize certain structural flaws.

Table 1. The impacts and various features of a stabilizer for titanium alloys.[5]

Types of stabilizers	Effects	Influence on titanium characteristics	Constituent elements
α-stabilizers	Transition temperature increase	Hardening	O, C, Al, N
β-stabilizers	Transition temperature decrease	Grain Refiners	Cr, Mo, Co V, Fe, Ni, Nb
Neutral	No observable influence transition temperature	Hardening	Zr and Sn

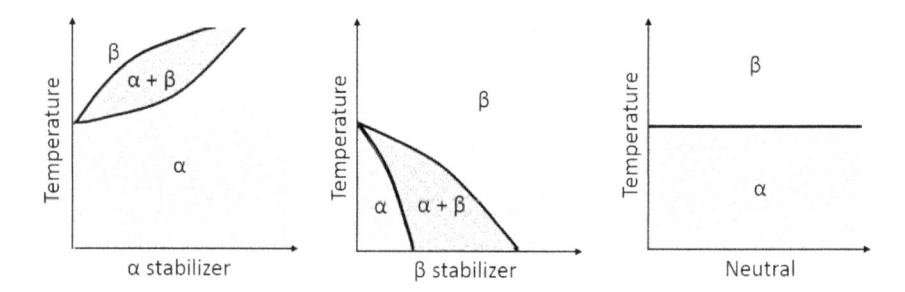

Fig. 1. Titanium phase diagram.[5]

Characterizing Ti alloys' microstructure, concentration, kind of alloying components, and crystalline phases which are existing at room temperature, is important.[29,30] The following three types of materials can be used to make titanium alloys: Examples of α-stabilizers include N, Al, O, and C, which work to stabilize the phase by raising the transition temperature; Fe, Mo, Ni, V, Cr, Nb, and Co, which work for stability of the β-phase by reducing the transition temperature; and Zr and Sn are neutral elements. The data are summarized in Table 1 for your convenience. To comprehend their operation, Fig. 1 shows the effects of stabilizing elements on the phase diagram of titanium. It is simple to understand how changing the atom count affects α and β structures and temperatures at which they change.

Ti can be more divided into "α", "near α", "α + β", "near β", and "β" phases based on the proportionate proportions of each phase.[31] Near α-alloys have 1–2% of β-stabilizers and 5–10% of

β phases; in contrast, ($\alpha + \beta$) Ti alloys, which generally have microstructures with 10–0% of β phases, have larger levels of β-stabilizers. The volume percentage, shape, size, and distribution of the phase precipitates inside the matrix have been found to have a substantial impact on the structure, which in turn has an enormous effect on the substance's physical and chemical characteristics.[32,33]

Consider that many scientists are aware of the elements that have an impact on titanium's structure and how this connection works. They mixed components into pure titanium to produce implants that performed more effectively than implants made with cpTi. For instance, it was believed that the combined action to create Ti–6Al–4V would build a biphasic microstructure ($\alpha + \beta$) as a consequence of the stabilization actions of α and β, V and Al. High strength, ductility, and decreased cycle fatigue are features of alloys with $\alpha+\beta$ microstructure.[19] Ti64 or grade-5 is a typical Ti alloy used for implant devices wherever superior strength is necessary.[34] When alloys containing Al are employed, the titanium matrix hardens fast in solid solution.[35] Additionally, Bi and Zr have been used to create solid solution hardening effects in a manner similar to that of Al.[36,37] When titanium is cast entirely with Zr, alloys may be made in a range of ratios that generally increase titanium's mechanical strength along with its propensity for corrosion and wear.[31] On the other hand, alloys consist of grain refiners known as β-stabilizers.[14] The development of alloys that combine superior mechanical properties with excellent biocompatibility has drawn the greatest attention to Nb, Mo, and Ta.[32,37] These characteristics make them one of the best alloys for manufacturing implants.[29,32] The subsequent topics are the most appropriate for discussing the connection between chemical elements, microstructure, and alloy characteristics.

2.1. α-type and near-α Ti alloys

cpTi and titanium alloy in a variety of grade make up α-titanium alloys, that have the α phase only. The four grades of cpTi that are available generally from 0.18–0.40 wt.% of O and 0.20–0.50 wt.%

of Fe.[38] Near-α titanium, in contrast to α-type titanium, primarily includes α-microstructure and a small amount of β-microstructure (which is lower than 5 Vol.%), which results from the addition of incredibly minute quantities of β-stabilizers (1–2 wt.%).[39] Both α-titanium and near-α titanium have traits including exceptional corrosion performance, and strong resistance to creep. They are therefore suitable for uses that call for high temperatures. Since the HCP structure is significantly more solid, their strength is undoubtedly lower at room temperature and frequently cannot be increased by heat treatment.

2.2. β-type Ti alloys

In comparison to (α + β)-type Ti alloys, β-type Ti alloys include more β-stabilizers (such as Mo, Ta, and Zr) and less α-stabilizers with no intermetallic phase development.[40] Due to the lack of a micro-galvanic interaction between various phases, β-titanium is predicted to have higher resistance to corrosion in human body than to (α + β) titanium. They also have comparable strength and improved biocompatibility.[41] As a result, β-type Ti alloys have the potential to overcome obstacles in their usage as biomedical materials.

2.3. (α + β)-type Ti alloys

In comparison to near-α titanium, (α + β) titanium includes a larger amount of "β-stabilizers." Accordingly, (α + β) titanium has a greater amount of the β-phase (in the range of 5 and 30 Vol. %).[42] (α + β) titanium has good manufacturability, higher strength at room temperature strength, and reasonable strength at higher temperature.[43] As compared to α-phase, properties of (α + β)-phase can be improved by heat treatment techniques,[44] permitting for the optimization of its mechanical characteristics. Among the most well-known (α + β)-Ti alloys utilized in biomedical field is Ti–6Al–4V (Ti64), which accounts for 50% of all Ti production in addition to cpTi.[39]

2.4. ω-type Ti alloys

A metastable phase known as the omega structure forms in Ti alloys that include components that have a tendency to help stabilize the BCC structures at higher temperatures. As was already established, the omega phase is present in any Ti alloy where the fast cooling of the higher temperature BCC phase may maintain it in a metastable condition at lower temperatures.[45] Recent research has demonstrated that the presence of the ω-phase, particularly beginning the process of aging, has a major impact on the nucleation of precipitates. The term "ω-assisted nucleation" is typically used to describe this process.[46,47] After solid-solution treatment, the ω-structure is often a typical metastable brittle in nature that increases the β titanium alloy's strength.[48] The phases produced considering both of these circumstances are known as athermal ω (ω_{ath}) and isothermal ω (ω_{iso}), respectively. Such precipitated phase is frequently produced through the cooling phase that follows solid-solution treatment or while in the low-temperature phase of aging.

In general, the existence of ω_{iso} decreases the driving power for nucleation and encourages old precipitates to form heterogeneous crystals.[49] According to Zheng et al.[50] coherent strain and compositional variation close to the ω/β contact are responsible for ω-assisted nucleation. According to Xiao et al.[51] the length of low-temperature pre-aging was favorably connected with the level of α-phase refining caused by nucleation with assistance. In a recent study by Barriobero-Vila et al.[52] it was shown that the skipped ω_{iso} transition causes intragranular α plates to gradually coarsen as the aging heating rate increases. All of these findings suggest that the α precipitation behavior is somewhat influenced by the features of the ω phase. So, it is possible to use ω-assisted nucleation to improve the mechanical characteristics of β titanium alloys.

2.5. Alloying elements in titanium

When it comes to enhancing the mechanical characteristics of advanced alloys like titanium alloys and nickel-based superalloys, refractory elements like Mo, Nb, Ta, and W are crucial.[53] Typically, substances whose melting points are greater than those of Ti, are also classified as refractory elements. W and Mo are β stabilizers and have a potent solid solution strengthening property in Ti alloys.[53] To produce Ti alloys with exceptional impact properties, it is essential to have a thorough understanding of the relationships among alloying composition, dislocation movement, deformation twinning, and resistance to fracture. How the β-stabilizers — Mo and Nb influence the impact toughness and fracture resistance of titanium alloys is discussed in this section. In detail, the study of impact specimens is insufficient to evaluate the deformation processes and the impacts of Mo and Nb elements on impact characteristics since deformation under impact loading is invisible. Overall, practical and theoretical research on the toughening effects of the stabilizers Mo and Nb was conducted in an effort to shed light on the mechanisms behind toughening and to inform the development of novel titanium alloys with superior impact characteristics.[54]

The beta phase of metastable Ti alloys is produced and heated to a temperature where they keep about 100% of the bcc b-phase while quenched to room temperature. This is accomplished by combining enough stabilizing elements for the beta phase to prevent the development of the hexagonal closely packed (HCP) a martensite phase. The bulk of these stabilizing elements are transition metals and are either isomorphous or eutectoid based on how alloying influences Ti. Some binary phase diagrams don't display a miscibility gap whenever combined with certain isomorphous elements, such as niobium (Nb) and tantalum (Ta).[55] On the opposite hand, some binary phase diagrams display a monotectoid reaction when the element is combined with different isomorphous elements such as Mo, V, and W. Eutectoid elements are referred to as the second category of beta phase stabilizing elements. They include Si, Fe, Mn, Cu, Ni, Co, and Cr.[55] The binary alloy contains the beta phase stabilizing element V, but the amount is much lower as compared to the critical amount of 15 wt.% needed to stabilize the BCC β-phase. Theoretically, a stable beta phase Ti alloy may be created by adding enough stabilizing elements to decrease the amount of Tb below ambient temperature. Contrary to metastable beta phase Ti alloys, the BCC β-phase in this situation is unchanging and is not divided into phases as it ages. As a result, it cannot be hardened. Mo amount larger than 10 wt.% martensitic transformations are totally repressed, leaving only the metastable BCC β-phase.[56] It was observed that increasing the amount of Mo results in change of martensitic phases' morphology from enormous (lamellar colonies) to acicular. The binary Ti–V alloy had a similar outcome.[55,57]

3. Properties Titanium Alloys

3.1. *Mechanical properties*

A new alloy's major design concerns are its mechanical and biological compatibility qualities.[15] To sustain the stresses and stress cycles they experience while in operation, dental implants must have outstanding mechanical qualities.[58,59] It is possible to improve the mechanical properties of titanium and its alloys using techniques like solid solution strengthening by substitutional and interstitial atoms, precipitation, grain modification, dispersion strengthening, and work hardening including lamellar and dispersed phases.[60] Greater strength and a lower elastic modulus are predicted characteristics of biomaterials.[61] These qualities seem to be incompatible for solid materials, especially metals and alloys.[18] Mo, Nb, Ta, and Zr are frequently added to titanium alloys to boost their strength and elastic modulus.[36] Finding a balance between strength and elastic modulus is essential for implants to perform better than those made with cpTi. In order to predict how biomaterials would behave when attached to bone, it is crucial to understand their features.[62] It is crucial to use caution when taking conclusions from the findings in Table 2 since it summarizes a number of research whose methods were not standardized.

Table 2. The mechanical properties of human bones with several Ti alloys.

Structure	Materials	Yield strength (MPa)	Ultimate tensile strength (UTS) (MPa)	Elastic modulus (E) (GPa)	References
Human Bone	Cortical bone	30–70	194–195	5–23	63
	Cancellous bone	—	0.9–8.80	0.01–1.57	63–65
α-structure	cpTi (Grade 1)	170	240	115	66
	cpTi (Grade 2)	275	344	105	67
	cpTi (Grade 3)	380	450	115	68
	cpTi (Grade 4)	480	550	105	2 and 68
(α+β)-structure	Ti–3Al–2.5V	585	690	100	25
	Ti–6Al–7Nb	921	1024	105	25,66
	Ti–5Al–2.5Fe	914	1033	110	66
	Ti–6Al–4V (annealed)	825–869	895–930	110–114	25 and 64
β-structure	Ti–12Mo–6Zr–2Fe	1000–1060	1060–1100	74–85	45 and 67
	Ti(10–80) Nb	760–930	900–1030	65–93	25
	Ti–15Mo–5Zr3–Al 870	870–968	882–975	75	66
	Ti–16Nb–10Hf	730–740	740–850	81	2 and 25
	Ti–15Mo–2.8Nb–3Al	771	812	82	66 and 67

Ti–13Nb–13Zr	900	1030	79	66
Ti–15Mo	544	874	78	2 and 64
Ti–24Nb–0.5O	665	810	54	2
Ti–24Nb–0.5N	665	665	43	2
Ti–29Nb–13Ta–4.6Zr	368	593	65	67
Ti–23Nb–0.7Ta–2Zr	280	400	55	2
Ti–36Nb–2Ta–3Zr0.3O	670–1150	835–1180	32	2
Ti–23Nb–0.7Ta–2Zr1.2O	830	880	60	2
Ti–35Nb–5Ta–7Zr	530	590	55	66

3.2. Ductility and strength

Given that an implant may be subjected to loads such as bending, torsion, compression, and tension,[32] the alloy's strength should be adequate to resist these forces. In order to preserve the integrity of the implant, avoid plastic deformation after implantation, and ensure stability between the implant and prosthetic elements, the material utilized for replacing hard tissue must have a variety of characteristics, such as tensile strength and fracture toughness.[11] Contrarily, because implants have complex geometries, ductility facilitates a variety of manufacturing processes.[61] The alloy's microstructure and size of grain can be altered to boost its ductility and strength.

The introduction of V and Al changes the solid solution alloy, causing particle precipitation and a change in the structure from α to the $(\alpha + \beta)$.[11,69] As a result, the tensile strength of Ti64 or grade-5 is significantly higher as compared to cpTi. Comparable outcomes were attained by combining Al with TiNb and Si with TiNbZrTa, which improved the grain and reinforced the solid solution.[70,71] When Si was introduced, grain growth was stifled because silicide intermetallic particles stabilized the grain boundaries.[70] The yield strength and tensile strength of Ti grains modified by introducing Nb are increased by 1.5–1.6 times in comparison to cpTi.[67] The UTS and ductility of the alloy are shown to increase with the addition of Ta to Ti–Nb–Zr, which includes a material that is elastic and fully plastic.[72] Higher-pressure torsion was effective in enhancing the structure and tensile strength of an alloy having equivalent content by enhancing the dislocations density.[16] The inclusion of alloying components may not necessarily improve the alloy's mechanical qualities.[73] As Sn content increases to the β-phase, it is likely to decrease an alloy's ductility and tensile strength in the instance of TiNbSn.[15,61] When compared to Ti–6Al–4V, this alloy made of Ti, Mo, Al, Ta, Zr, Cr, Sn, and Nb had lower strength, which was initially seen in a model for forecasting. The resistance levels of the remaining two kinds of alloys, which included less β-stabilizer, were higher. To produce a microstructure with lower amounts of β stabilizers, a thermomechanical alloy process must be solidified.

The interstitial content of materials is closely connected with the strength and ductility of Ti alloys.[11,74] Interstitial solutes reduce the hardness of cpTi while increasing its strength and decreasing its elongation values.[11] Intermetallic phases created by replacement of components such as Cu, Fe, and Ni in alloys cause identical behaviors. One kind of study that looked at two alloys with different amounts of CuNi discovered that lowering the concentration of Cu and Ni improved the elongation of alloy and increased the tensile strength to 1050 MPa.[74] The production method also has an influence on properties of an alloy. Ti alloys can be strengthened at the same time can have good elongation by cold rolling annealing, and small-scale integrating particles of ceramic into the titanium matrix.[16,32,70] In order to ensure the item's long-term durability, the material's biomechanical compatible with bone must be confirmed.[19] According to Table 2, the tensile strengths of Ti alloys produced using various methods are in range of 360–3267 MPa. It was discovered that ($\alpha+\beta$) titanium alloys, in particular, had good yield strength and UTS. It was observed that presence of Nb, Sn, or Bi has no appreciable impact on this property. The processing method has an impact on alloy strength as well, with SPS producing the highest values.

3.3. Hardness and elastic modulus

A material's hardness is determined by its ability to withstand repeated indentation deformation.[37,75] To achieve the best machinability while creating implants, it should be modest, but it also has to be rigid enough to shield the bone from stressors.[11] A key component of the biomechanical contact between the bone and the implant is elastic modulus. By lowering the elastic modulus, stress is better distributed at the implant-bone contact and bone atrophy is reduced.[11] The Young's modulus must be as close to the bone as is possible since "stress shielding" has been associated to bone resorption and osteoporosis around implant sites.[8,76] Young's modulus should not be too low since the metallic implant would move in tiny ways, which might lead to the failure of prosthesis and free falling of an implant.[67] Despite the fact that the presence of a α-microstructure is linked to higher hardness numbers, the modulus of elasticity might not desire the existence of a α phase. They raise the specimen's hardness because α-stabilizers work as a substitutional solute that lowers the material's atomic mobility[31] or because they encourage the precipitation of the phase when alloys are aged at low temperatures.[19] With increasing β content, the elastic modulus should first increase before rapidly decreasing at higher concentrations.[15]

Low concentrations of β-stabilizers often result in α+β alloys with higher elastic moduli since the α phase is present. Low hardness may result from an alloy's structure changing as a result of a rise in β-stabilizers,[37,61] which reduces the modulus of elasticity. As compared to cpTi, the hardness values of Ti increased when Zr and Bi were incorporated[36,31,77]; however, the hardness slightly decreased as Zr concentration raised.[21,58] When substantial quantities of Zr between 5% and 10% were incorporated, the findings first started to go downhill; nevertheless, at concentrations greater than 15%, this characteristic was boosted above that of cpTi.[31] The difference was caused by the greater atomic radius of Zr over Ti, which researchers attribute to a shift in the spacing among the alloy's constituent atoms. As a result, the force that regulates the atomic level elastic modulus is changed. By incorporating Nb into a TiZr alloy, a biphasic α+β microstructure and rise in hardness were created, which were later improved by aging heat treatments.[35] The elastics modulus is often increased by this procedure to amounts greater than 100 GPa.[70]

The dispersion of Nb throughout the matrix of Ti and the addition of Sn in Ti30Nb alloy give an alloy with a higher degree of β-microstructure, with the exception of lowering the elastic modulus.[78] One of the lowest elastic moduli of titanium that has been observed in a biomaterial (36 MPa) was found in a Ti–33Nb–4Sn alloy created by cold-rolling and then annealing techniques.[18]

The development of a β structure with low β-stabilizer content increased the likelihood of this. Although the elastic moduli of cpTi and related alloys are less than those of SS and Co–Cr alloy,[8,16] they are nevertheless slightly higher than those of human bone (10–30 GPa).[11,57] Less favorable results than β-alloys have been found for elastic modulus, cpTi, α and (α + β) titanium alloys.[62,80] It is possible that this is the case because the β phase's body centered cubic structure and low atomic lattice density varies from the α phase's hexagonal closely packed (HCP) structure.[16] This assures a greater plastic deformability for the alloy.[76,79] The atom spacing of these various topologies causes differences in the atomic-bonding force, which therefore causes variations in elastic modulus.[31] The elastic modulus of an alloy may fluctuate significantly during thermomechanical processing.[15,16] For instance, Young's modulus values for samples quenched in water were lower,[15,32] this is accurate according to Table 2. Faster cooling hinders the development of α-structure, that reduces elastic modulus and prevents the α phase from converting to the β phase.[32]

According to the information available, the elasticity moduli of titanium alloys vary between 35 Gpa to 175 GPa. The researchers notably point out Ti36Nb4Sn and Ti23.72Nb4.83Zr1.74Ta(Si) alloys, which had low modulus of elasticity value and those that are closer to that of bone. The addition of Fe, Zr, Mo, and Al causes the cpTi's hardness rating, which is already among the lowest, to rise sharply. The three hardest metals are TiFe(Ta), TiAlV, and TiZr.

3.4. Fatigue behaviors

Because cyclic loading simulates a far wider range of real-world situations than pure monotonic loading, it is essential to use it to evaluate the fatigue characteristics of dentistry material.[59,81] By hastening the appearance of surface flaws and their expansion to a dangerous extent, which results in implant fracture, the surroundings where implants are placed may also have an effect on their resistance to fatigue.[82–84] Notwithstanding this, there isn't plenty of studies that examine the behavior of Ti alloys in this sort of scenario. Ti–24Nb–Zr–7.6Sn withstood more fatigue cycles in the 0.9% NaCl solution before failure compared to air due to the medium's cooling action, this might stop materials from melting during fatigue testing if the temperature rises.[85] In addition, the oxide coat formed by the NaCl solution can make the alloy more resistant to corrosion. During strain-controlled fatigue testing, Ti–24Nb4Zr7.6Sn demonstrated much higher fatigue strength as compared to Ti–6Al–4V ELI61.[61] Eagle's approach was put to the test utilizing Ti alloys for a corrosion fatigue test. At 10^8 cycles, Ti–Al–Nb–Ta had a fatigue strength of around 700 MPa, whereas TiZrNbTaPdON and TiSnNbTaPdO exhibited a fatigue value of roughly 600 MPa. The Ti–Mo–Zr–Al alloy performed worse at 10^7 cycles than prior alloys.[86] Dental implants and Ti–15Zr discs performed much better under fatigue compared to Ti-Grade IV. The maximum limits for the β Ti alloy's and cpTi's fatigue resistance were 560 and 435 MPa, respectively.[87]

In comparison to Ti–6Al–4V and cpTi and, Ti–7.5Mo and Ti–13Nb–13Zr have lower fatigue strengths. The highest fatigue strength, however, was seen in Ti–7.5Mo.[88] Implant fatigue behavior can be improved by integrating material parameters, surface characteristics, and design optimization.[87] How well Ti alloys handle fatigue in this environment depends significantly on different heat treatments used, mechanical treatment and the application of hard thinner coating.[82] Cold rolling boosts fatigue resistance by significantly shrinking microstructures.[85] Alloy fatigue fracture may be primarily caused by surface porosity created by the method of casting.[88] In a manner similar to this, it has been shown that surface treatments, such as SLA, that enhance roughness have a detrimental effect on a material's fatigue performance and raise the risk of initiating fractures.[87] It is frequently recommended to carry out more research on the fatigue characteristics of titanium alloys under cyclic loads in human-relevant fluids like synthetic saliva and SBF. What's more, dental implant materials are more susceptible than any other material to collective impacts corrosive surrounding and fluctuating mechanical loading,[81] thus it's crucial that they meet these requirements. To avoid production faults which can act as point of stress concentration and trigger or spread of failures, it is also necessary to optimize thermomechanical processes and surface treatments.

3.5. *Surface properties*

The effectiveness of implant treatment is significantly influenced by surface characteristics including composition and topography. During the first stage of integration, implants require a surface that promotes the proper mineralization and osteogenic differentiation.[89] Surface characteristics of implants, such as surface roughness, surface energy, and specimen elements, have either a direct or indirect effect on the adherence of cells to bone apatite.[90,91] The alloy should maintain a hydrophilic surface while also keeping the topographic micro-roughness for dental implants.[92] Surface morphology and surface roughness of an implant at micrometer (0.11–100 μm) and nanoscale (1–100 nm) scales[93] are compatible as shown in Fig. 2. According to some evidence, the bone reaction is affected at the nano- and micrometer scales,[94] leading to optimal cell attachment, greater implant/bone contact, and stronger torque release resistance.[89,95] Easy-to-ingest, physiologically important nanometric proteins have been found to positively affect cellular responses.[96] It has been demonstrated that rougher topographies are related to differentiation of an osteoblast, whereas smoother surfaces are associated with cell proliferation. The autophagic process may allegedly physically start the interaction of cells with a rough surface which might lead to cell differentiation.[97] A surface that encourages cell maturation is necessary to improve and speed up osseointegration. Figure 2 illustrates how, at the nanoscale, the topography displays a decrease in water contact angle and a corresponding increase in surface energy and hydrophilicity.

A Ti–6Al–4V alloy with different surface roughness levels was created in prior study.[93] Better outcomes were obtained with more roughness. Cell attachment and proliferation are impacted by

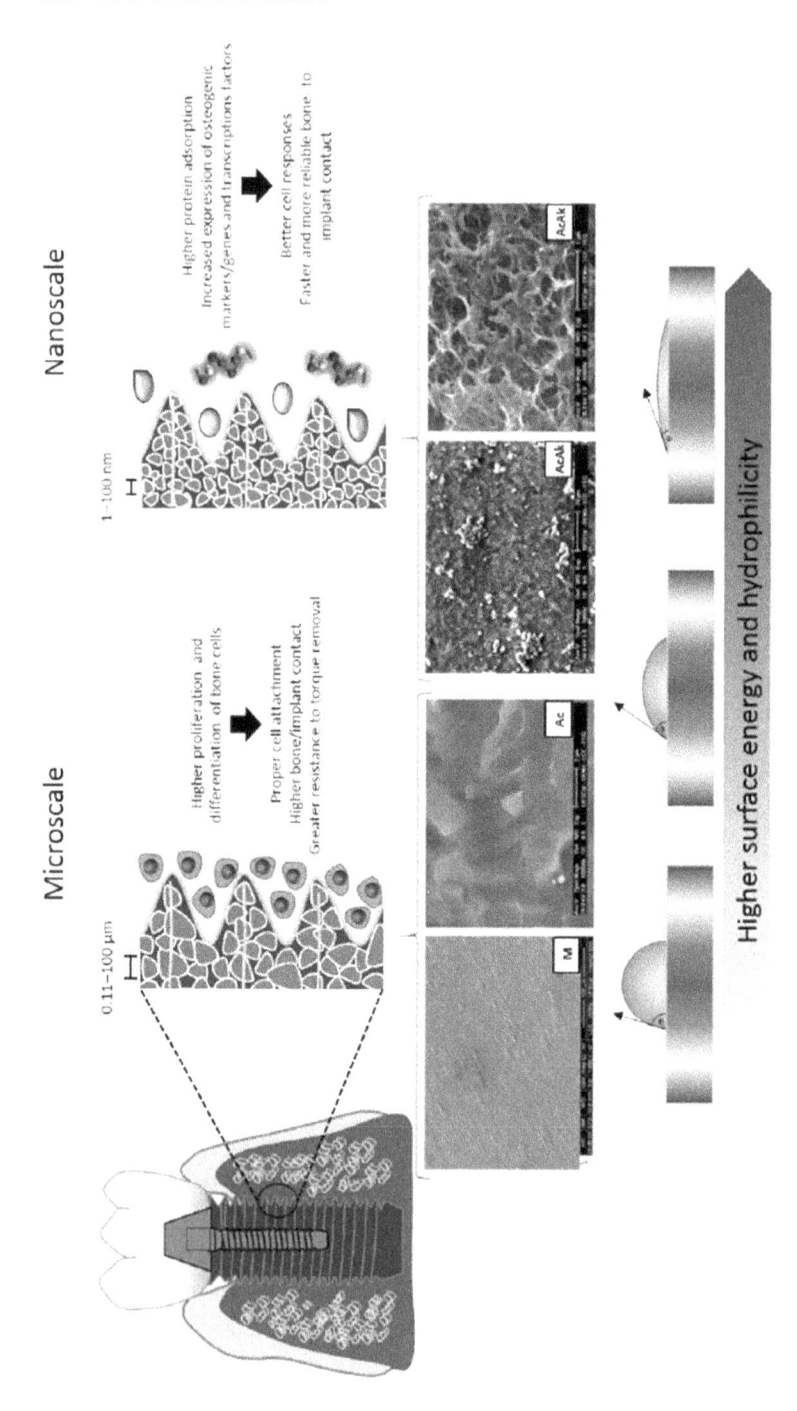

Fig. 2. Diagram of how implants affect topographic characteristics at the micro- and nanoscales. Photographs of microscale surfaces can be seen in specimens that were fabricated from Ti–6Al–4V-ELI and specimens that were subsequently acid-etched in HCl (Color online).[5]

variations in topographies. Fibronectin can attach to a rougher substrate greater by 10 times as it clings to a surface that is smooth, based on an investigation which additionally showed that rougher surfaces had a greater overall protein attachment percentage. The impact of minute differences in surface roughness (Ra 0.50 μm) on cellular response and protein adsorption must be made clear.[91] Enhancing wettability is also required to boost cell proliferation and adhesion on titanium alloy's implant surfaces.[98]

The alloy's surface chemical and physical properties can influence wettability. If water contact angle is less than 90°, the term "hydrophilic" or "hydrophobic" is used to describe the surface, and vice versa.[99] Ti–45Nb was discovered to have greater hydrophilicity because its contact angle was shorter (81.75°) than cpTi's (96.46°) .97. However, compared to cpTi (34 mN/m, 0.20 μm) and the Ti–50Nb alloy (32 mN/m, 0.46 μm), the Ti–50Zr alloy exhibited a higher surface energy and reduced surface roughness (37 mN/m, 0.17 μm).[91] Ti–Zr alloy exhibited a favorable *in vitro* biological effect because of the properties of the substrate composition. Although the alloy had a larger surface area than Ti–Ta–Nb–Zr, Ti had a rougher topology. Using various treatment approaches, the results of surface modifications have also been excellent. Mechanical friction treatment produced different grain sizes on the Ti–25Nb–3Mo–3Zr–2Sn alloy.[100] The nano-grained alloy had the maximum wettability and the least surface roughness when compared to the fine- and coarse-grained alloys (7.4 nm, 64.1°, and 7.1 nm, respectively). In comparison to grains that were 90 m and 180 nm in size, nanoscale grains (30 nm) shown enhanced cell responsiveness and absorbed additional protein.[100] By combining two acid-etched surfaces using an alkaline treatment, Ti–6Al–4V-ELI created surface micro-roughness (0.14–0.48 μm) and increased wettability (from 70° to 35°).[101] It was observed from the alkaline solution-treated specimens' efficient osteoconductive performance (AcAk) that their morphology has transformed from an acid etching (Ac) to a machined surface (M).

A nanotopography was created by electrochemical anodizing a Ti–6Al–7Nb alloy in a previous study.[98] The contact angle was dramatically reduced from 61.4° (machined surface) to 14.8° (treated surface). However, in comparison to the roughness and contact angle of a machined surface (123 nm; 59°), the nanoporous surface (having pores lesser than 15 nm) of the anodized Ti–25Nb–25Zr alloy was unaffected.[102] Notwithstanding, the changed shape and surface content were able to significantly increase the motility, adhesion, proliferation, and mineralization of mesenchymal cells. Microscale bonding strength strengthens the biomechanical properties of porous and rough surfaces, which enhances the connection of the implant with the cell environment.[89] This is due to the greater area of contact between the implant surface provided by these surfaces and the newly formed bone. Using a hydrothermal technique that uses a Ca-containing alkaline solution and a straightforward post-heat treatment, increasing cells' viability and differentiation on the surface of Ti–13Nb–13Zr looks to be an effective strategy without significantly altering surface shape.[103] This is because it produces anatase structures, improves the hydrophilicity of the surface, and maybe expands the surface area at the nanoscale.[104] It was discovered that the anatase phase was essential for increasing surface energy and reactivity.[99] A cpTi surface with submicron surface roughness and considerable hydrophilicity was previously shown to promote cells differentiation and maturation with osteogenic potential.[104,105] Few studies have compared the way various metal surfaces impact the properties of the metal's constituent parts and how tissues respond to them. In most studies, surface features and surface treatments are solely taken into account. As a result, additional study is necessary to completely determine how the chemical composition of different alloy's surfaces may impact wettability, surface texture, and biological aspects.

3.6. Corrosion behavior

It is commonly accepted that the development of a stable passive coating, predominantly composed of TiO_2, is the cause of titanium and similar alloys' improved corrosion resistance under various conditions.[105–107] It is feasible to swiftly recreate the passive coating on the Ti samples, regardless of any deterioration. Early biomedical applications frequently use α-titanium and (α + β) titanium alloys. Corrosion behavior in metallic materials is frequently significantly influenced by the application environment, alloy compositions, and microstructure.[108–110] Notwithstanding being true that human temperature, chemistry of the surroundings, and pH may rarely change the inside conditions of human body, such as those that lead to inflammation and allergies.

Alves et al.[111] conducted an examination of the corrosion performance of the materials in SBF at two different temperatures and found that for Ti–6Al–4V and cpTi, 25°C provided superior resistance to corrosion in comparison to 37°C. As a result, cpTi and Ti–6Al–4V exhibit different corrosion behaviors depending on the temperature. Similarly, Ti–6Al–4V corrosion performance is influenced by pH level.[112] Ti–6Al–4V in the neutral Ringer's solution exhibits strong corrosion resistance while having a limited passive range at pH = 8.[112]

Ti pitting corrosion typically happens in the oral surrounding since there is a greater amount of oxygen available and acidic food is more readily available.[19] In Ringer's solution, Ti–15Mo, cpTi, and Ti–6Al–4V alloys have greater resistance to corrosion, nevertheless only Ti–15Mo consistently gives a passive film.[113] A greater fluoride solution will always be present in a human body's surroundings, such as when cleaning teeth.[114,115] As a result, compared to β-Ti alloys, α-Ti and its (α+β) Ti alloy display reduced corrosion resistance under these circumstances. Fretting corrosion must also be taken into account for metal orthopaedic implants.[115] Fretting corrosion, which frequently appears at modular connections, is diminished by the development of a protective oxide layer.[115] Thus, it is necessary to employ Ti alloys with excellent corrosion resistance that are appropriate for the biomedical sector. As a result of their development, β-type Ti alloys have potential as biological application materials.

The chemical stability of the specimen yet has a considerable impact on the performance of the passive layer. When Ti–35Nb is made using mixed powder and SLM, heterogeneous microstructures including individual Nb grains frequently appear. In a work by Wang et al., the chemical homogeneity of the substrate was significantly increased by heating SLM-produced Ti–35Nb to 1000°C for 24 h in an Ar atmosphere. As a result, the Ti–35Nb produced by heat treatment has a higher corrosion potential, measuring −0.46 V (compared to SCE's −0.55 V). Additionally, one of the researchers[116] showed how the stability of the passive coating affects the corrosion resistance of β-Ti alloys. The passive current densities in Ti–10Mo alloy after heat treatment are noticeably substantially lowered, according to researchers. There have undoubtedly been several comparative studies on corrosion rates of three different kinds of Ti alloys with the goal of creating Ti alloys that are more suitable for biomedical usage. One of the researchers[113] looked at how well the alloys cpTi, Ti–6Al–4V, and Ti–15Mo corroded in Ringer's solution. In comparison to cpTi (145–1522 mV vs SCE) and Ti–6A–l4V (155–1460 mV vs SCE), they found that the passivation range of Ti–15Mo alloy is higher. One of the researchers studied the corrosion of Ti–Zr–Nb–Mo alloys with various Mo concentrations in the as-cast state. The findings revealed that constitutional undercooling caused by Mo results in a decrease in Ti–Zr–Nb–Mo alloy grain size with increasing Mo content. With an addition of 15 wt.% of Mo, the Ti–Zr–Nb–Mo alloy has the minimum passivation current density, measuring 2.31 ± 0.03 A cm^{-2}.

Individual additions of Fe, Sn, and Ag to Ti–25Zr–10Nb–10Ta in the Ringer's solution, Zareidoost et al.[117] observed that the material showed outstanding resistance to corrosion. Ag's standard electrode potential (0.799 V) is greater than Ti's (0.98 V), which results in a more stable passive film being created on Ti–25Zr–10Nb–10Ta. Ti25Zr10Nb10Ta1.5Ag shows greater corrosion resistance in Ringer's solution. Lin et al.'s[118] usage of different solution treatments and age treatment procedures resulted in changes to the microstructure of Ti–40Ta–22Hf–11.7Zr.

The findings revealed that the solution treated Ti–40Ta–22Hf–11.7Zr has a β+ω structure when it is casted, but after being heated to 900°C for 1 h, it changes to β phase. After being aged at 300°C for 15, 1.5, 12, and 24 h, respectively, the β-structure slowly transforms into "$\beta+\alpha$", "$\beta+\alpha+\alpha$", "$\beta+\alpha+\omega$" phases. With such a wide range of microstructures, Ti–40Ta–22Hf–11.7Zr alloys display a variety of electrochemical behaviors in Hank's solution. The sample with a single structure and solution treatment has the lowest current density, which is 0.49 ± 0.03 A cm^{-2}.

While EBM-manufactured Ti–6Al–4V alloys provide greater defence against corrosion in phosphate-buffered saline, SLM-manufactured Ti–6Al–4V alloys are more susceptible to pitting corrosion in 3.5% by weight NaCl solution compared to their wrought equivalents.[103,119] In light of this, the question "Do Ti alloys produced using different preparation techniques exhibit distinctive corrosion behaviors?" is raised. One of the researchers[120] examined the corrosion performance of Ti–24Nb–4Zr–8Sn created using SLM and traditional monolithic techniques to solve this problem. Although the microstructure of these two alloys differs, their chemical composition and monolithic phase are the same. Figure 3 illustrates almost similar potentiodynamic polarization curves and Nyquist plots of these alloys.[120,121] As a result, different production techniques that result in varied Ti–6Al–4V alloy corrosion behaviors are linked to the development of diverse phase compositions in the microstructure.

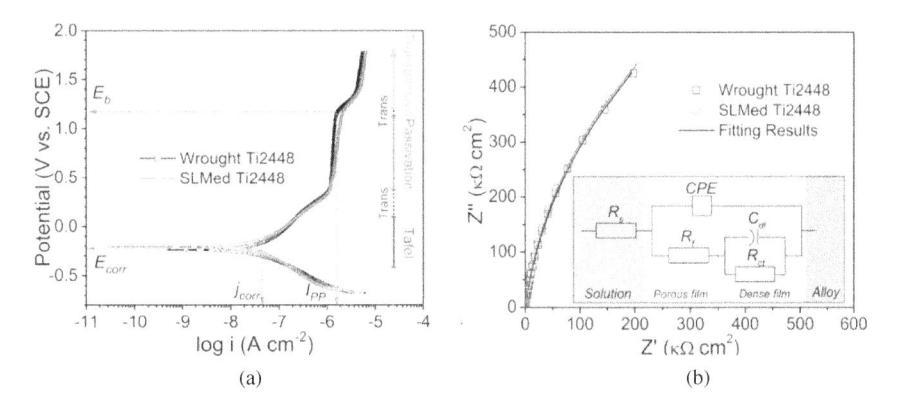

Fig. 3. Electrochemical studies of SLM-fabricated Ti–24Nb–4Zr–8Sn and wrought. Ti–24Nb–4Zr–8Sn was designated as Ti-2448 and is shown as (a) potentiodynamic polarization curves and (b) Nyquist plots (Color online).[121]

3.7. Tribocorrosion

The integrity of the passive oxide coating is severely impacted by mechanically aided corrosion events, such as tribocorrosion, which speeds up material deterioration. Despite the fact that Ti and its alloys are thought to be beneficial for implant applications, wear and corrosion nevertheless happen, especially in harsh environments like the human body.[122,123] The bodily fluids surrounding implants once they are implanted into a person provide a particularly hostile environment for corrosion because of the importance of pH, dissolved oxygen, and chloride levels. Further influencing the corrosion behavior of the implant is the oxide layer's ability to regenerate; the faster the film is mended; the fewer metal ions are discharged into the body. The release of metallic ions during tribocorrosion is crucial because it can negatively impact both the mechanical integrity and the biocompatibility.[124] The presence of wear debris, which when accumulated in periprosthetic tissues can cause osteolysis and subsequent aseptic loosening, is another sign of low wear resistance.[125]

According to investigations, wear particles made of Ti–6Al–4V alloys may produce inflammatory mediators that will impact the tissues around the prosthesis and result in osteolysis. The function of the β-stabilizing agents on the oxide coating is the major factor controlling how corrosion behaves in β-Ti alloys. For instance, Nb has been found to have a positive impact on the passivation behavior, which has been linked to its ability to reduce anion vacancies in the oxide layer, which is mostly made up of Ti- and Nb-oxides.[127] The wear behavior of Ti–6Al–4V has been extensively studied in both dry and wet environments,[128] while only sparse wear data is known for β-type Ti–Nb-based alloys.[129] Furthermore, for such alloy systems, the synergistic interaction between corrosion and wear is frequently disregarded. Experimental research is required since it might be difficult to anticipate the processes of corrosion and wear caused by particular combinations of materials, loadings, and conditions due to the complexity of the included phenomena.

3.8. Wear behavior

At the intersection of counterfeit joints, wear is prevalent. The bottom layer of bone passes through a cycle of change between the embed and the bone meddle resulting in osteolysis.[130] Titanium is a fantastic material for biomedical applications because of its exceptional corrosion resistance, general biocompatibility, and low Young's modulus, which is equivalent to conventional hardened steels and cobalt-based composites.[131,200] Because of frictional surface contact with opposed surfaces, hard particles released from dietary sources, food bolus, toothbrushes, and paste can all cause restorative Ti surfaces to be extracted during the chewing process.[132] The TiO_2 coating on titanium can be eroded by chewing friction, which might lead to loss of materials and ultimately the failure of dental implants and prosthesis. Metal particles are produced as a result of abrasion and corrosion, and ions originating from titanium have been found in nearby tissues and connected to inflammatory issues. Several tribocorrosion processes are depicted in Fig. 4 as they may occur when ductile metallic Ti and a strong inert counter-body (alumina) come into contact.[126]

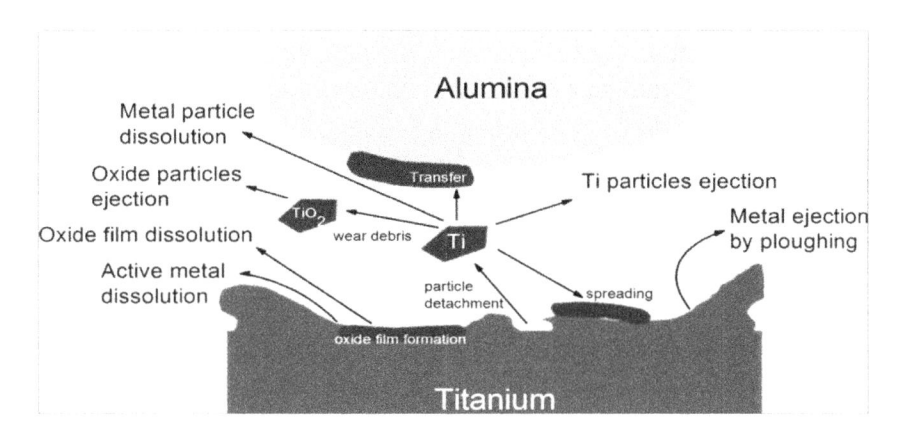

Fig. 4. Ti implants' tribocorrosion processes (Color online).[126]

3.9. Biocompatibility

Along with having excellent mechanical properties and good corrosion performance, implant materials also need to be highly biocompatible.[121,133] Without extra surgery, long-lasting implants composed of appropriate materials can be implanted into humans.[134] The materials would interact with biological fluid, proteins, and cells in a significant number of ways after being implanted. Traditional "$\alpha+\beta$" titanium alloys include constantly dangerous elements. This led to the recent development of titanium alloys, and an examination of their biocompatibility. The cytotoxicity of Ni–49.2Ti and Ti–26Nb was compared by McMahon et al.[135] and Ti–26Nb was shown to be less hazardous. In contrast to the Ni–Ti alloy, the Ti–19Zr–10Nb–1Fe alloy exhibits better hemocompatibility but equal cytocompatibility, according to Xue et al.[136] The lack of potentially harmful alloying components accounts for the higher biocompatibility of β-titanium. Today there is an urgent necessity for more research on β-type Ti alloys. Because titanium alloys are biologically inert, fibrous tissue capsules are quite much more probable to develop on the implant's surface.[105] This tendency will ultimately be seen in all types of Ti alloys. Because of their biological inertness, β-Ti alloys are secure but not bioactive. Therefore, osseointegration ought to be further enhanced even if β-Ti alloys do not include any potentially dangerous alloying elements. In order to increase the bioactivity of Ti alloys, there has generally been great consideration in surface modification. One of the researchers[137] utilized hydrothermal, electrochemical, or a combination of methods for alkali treatments to Ti–29Nb–13Ta–4.6Zr across various timeframes. The findings demonstrated that the surface of Ti–29Nb–13Ta–4.6Zr evolves into a mesh-like structure and has a substantial capability to promote the production of apatite independent of the techniques or variables used.

One of the researchers[138] used the sol–gel technique for coating Ti–29Nb–13Ta–4.6Zr with calcium phosphate/TiO2 composite coatings. It was observed that as titanium dioxide and calcium phosphate both are extremely sensitive to bone tissues, the material's biological activities might be greatly increased by the coating.

A lot of attention has been shown in organic coatings as well as inorganic coatings. Over the past few decades, studies using cpTi and Ti–6Al–4V have effectively immobilized extracellular matrix (ECM) proteins on the surface of titanium implants.[139] For instance, collagen-coated cp-Ti induces a more favorable response in human mesenchymal cells.[139] As an example, collagen-coated cpTi causes human mesenchymal cells to respond more positively. Different coatings also provide results that are comparable.[140] Sadly, there isn't a lot of knowledge now accessible through research about the organic coatings on β-type Ti alloys. On the other hand, considering the impressive success of organic coating on many of Ti alloys, it is projected that β titanium alloys having bioactive coating may eventually evolve as a development for biomedical field.

3.10. *Bioactivity of Ti*

Inflammation or allergic responses are thought to be quite rare to result from implants.[141] The tissue response classifies implants as "bio tolerant," that exhibits remote osteogenesis (creation of bone with incidental relationship to the material), "bio-inert," that exhibits contact osteogenesis, and "bioactive," that exhibits attachment osteogenesis. Thrombosis, particularly associated with blood coagulation and platelet adherence to implant surfaces, and fibrous tissue encapsulation of implants placed in soft tissues are two characteristics that might impair biocompatibility despite the reality that implants are in contact with human tissues and fluids. The physiologic consequences that Ti implants have are illustrated in Fig. 5.

3.11. *Osseointegration*

For cell proliferation, attachment, and differentiation to take place properly throughout the osseointegration of implants, their surface characteristics and chemistry must be improved. According to a previous review of scientific research, cpTi and Ti–6Al–4V show comparable osseointegration and biomechanical fixing.[5,142] However, Ti–6Al–4V has been outperformed in animal testing by cpTi and a

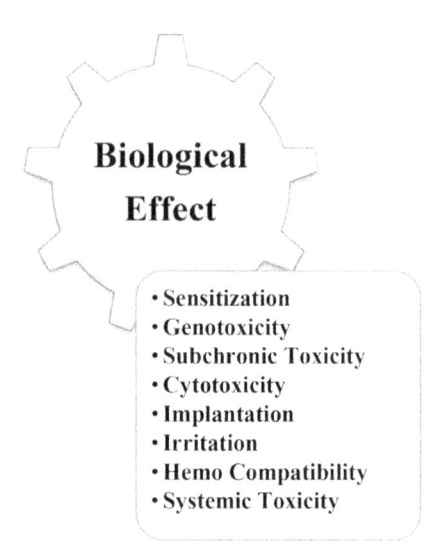

Fig. 5. How an implant affects the body (Color online).[130]

few alloys, such as Ti–15Mo–1Bi, Ti–15Zr, and Ti–24Nb–4Zr–7.9Sn (Table 3). It is clear that in the great majority of trials, alloys respond physiologically identically to cpTi when bonded to animal bones. According to one study, Ti–Zr implants worked better than cpTi implants when it came to generating new bone volume and lowering torque values.[9] Important alloys include Ti–Cu, Ti–Nb, and Ti–Nb–Ta–Zr since they showed bone tissue biocompatibility with no hindering of the growth of fresh bones.[67,143,144] The surrounding bone volume of the Ti–Nb–Ta–Zr alloy implant increased over time without altering the cpTi implants.[145,146] Rare human studies using titanium alloys are confined to the commercially available alloy TiZr.

Ti–Zr implants with very tiny dimensions have been used in studies[6,147,148] on humans to show that they display osseointegration comparable to cpTi. The rate of success rates and implant lifespan were equally excellent, based on this study. Despite the results, most investigations only offer at the most 1-year complement, that indicate continuing study is needed to prove the implant usefulness. Long-term research investigations looking into osseointegration in

Table 3. An overview of the properties of titanium alloy as well as data from animal models.

Alloy	Animal studies	Parameters	Follow up period	Results	Ref.
Ti–10.1Ta–1.7Nb–1.6Zr	Each set of 38 male Sprague-Dawley rats received 38 implants in the form of screws.	Assessments of removal torque, gene expression, and the bone-implant contact (BIC) and bone area (BA) ratios.	4 weeks	The RT between the Ti–Ta–Nb–Zr and Ti implants didn't vary considerably, the BA (41.2%) for the cpTi implants was higher as compared to the Ti–Ta–Nb–Zr implants ($p = 0.012$), the BIC proportion didn't change substantially, and the gene expression of an implant's adhered cell revealed that the Ti–Ta–Nb–Zr implant has approximately 3, 6, and 2 fold less proinflammatory, bone-forming countenance.	146
Ti–24Nb–4Zr–7.9Sn	Each group of white female rabbits have cylindrical shape implant.	Pull-out force; BV; Tissue Mineral Density (TMD);	12 weeks	After 12 weeks, the pull-out strength of the Ti–Nb–Zr–Sn groups was greater as compared to that of Ti-6A-14V group ($p = 0.05$), and the experimental group's BV and TMD were exceptionally greater at the same time as those of the control group ($p = 0.05$).	151
Ti–45Nb	Beagle tibia implant model; rabbits; 12 cylindrical implants each group.	BIC and BA, bone fusion, oral mucosal irritability, and the BV/TV ratio (the	12 weeks	The peri-implant bone volume and tissue volume were similar between the two groups; the impact of Ti–Nb alloy and cpTi on osseointegration was identical	67

(*Continued*)

Table 3. (Continued)

Alloy	Animal studies	Parameters	Follow up period	Results	Ref.
		proportion of bone tissue to total tissue volume).		and the bone area and bone contact of the Ti–Nb alloy enhanced promptly up to 12 weeks with no noticeable difference in Ti; neither group experienced oral pain or damage.	
Ti-15Mo-1Bi	Each group consists of 24 cylindrical implants and includes white female rabbits.	Formation of new bones	26 weeks	After 6 and 12 weeks post-implantation, there hasn't apparent distinctions among the groups; after 26 weeks, the bone areas of the Ti15Mo1Bi implant were higher as compared to the Ti-6Al-4V implant ($p = 0.001$), and the total bone areas of the Ti-15Mo-1Bi implant were roughly 249% that of the Ti6Al4V implants.	152
Ti-35Nb-2Ta-3Zr	Each group has 48 white male rabbits with implants.	Surface bone apposition ratio(BAR), pull-out force, and new BA	12 weeks	At no stage did the pull-out force, BAR, or BA appreciably vary across groups.	143
Ti-24Nb-4Zr-7.9Sn and nanotube-Ti-24Nb-4Zr-7.9Sn	Eight cylindrical implants are present in each group of white female rabbits.	BV/TV, BIC, BA, and ratios; Average trabecular thickness (Tb.Th); Average trabecular number	Up to 12 weeks	After 6 weeks, an implants' nanotube BIC and BA ratios compared with two groups were significantly higher ($p = 0.05$), and after 12 weeks, the bone area ratio compared to the other three	153

		(Tb.N); Trabecular mean separation (Tb.Sp)		categories was significantly higher ($p = 0.05$) for the nanotube-Ti–Nb–Zr–Sn implants. The highest values for the nanotube at 6 and 12 weeks were BV/TV, Tb.N, and Tb.Th. Ti–Nb–Zr–Sn
Ti–10Cu	96 implants are included in each package of white rabbits.	Mean optical density (MOD), Mineral addition rate (MAR), and BIC in the expression of TGF-b1 and BMP-2.	Up to 12 weeks	There were no variations in bone density, BIC, or MAR across groups at any point in time. However, after 1 and 4 weeks, the Ti–10Cu category's mean optical density value for BMP-2 was significantly higher as compared to cpTi group. TGF-b1 levels weren't different throughout groups at any point in time.	143
Ti–Zr, SLActive® Straumann Roxolid®	Each set of 13 female Bama minipigs (younger and older groups) received 13 screw-type implants.	Success and survival rates for implants; RT evaluation	8 weeks	In the geriatric population, the life expectancy rate of Ti–Zr implants was 85.7%, exactly the same as titanium implants; Ti–Zr implants worked better than titanium implants in terms of mean peak RT in the older group ($p = 0.250$), whereas titanium implants functioned better in terms of mean peak RT in the younger group ($p = 0.219$).	154

people are usually carried out on a tiny percentage of cases.[149,150] This is due to the difficulties in removing prejudices that the population being investigated is exposed to, such as biological variants, gender and age variants, and patient health conditions.

Before conducting such tests using Ti alloys, it is crucial to accurately establish the biological, chemical, mechanical, and electrochemical characteristics of such materials from prior research because *in vivo* studies are complicated and carry a danger of patient hurting. The development of *in vivo* research may also be impeded by costs and ethical issues. Supporting both human and animal research is essential to determine the efficacy and safety of biomaterials before they are utilized in medical treatments.

3.12. *Antimicrobial properties*

Strong human cell adhesion and osseointegration are necessary for dental implants to function well, thus it's critical that the materials used to make them resist bacterial colonization. Utilizing implant surfaces that prevent bacterial colonization improves the treatment's efficacy.[155] Ren *et al.*'s Ti–6Al–4V alloy was given antibacterial characteristics by using additional Cu.[157] In most cases, Cu particle discharged into the surrounding area stops the development of biofilms and eradicates microorganisms. However, with its high level of toxicity as a heavy metal, copper is nevertheless recognized as the alloy with the highest cytocompatibility and corrosion resistance. Similar results were obtained using a TiCu alloy that displayed strong antibacterial properties with no sacrifice of mechanical characteristics or corrosion performance of an alloy.[156] In addition to being incorporated into Ti to produce alloys, TiCuO coatings with promising *in vitro* antibacterial activity and biocompatibility have been made by sputtering Cu onto Ti–6Al–4V.[158,159] Ti–Ag alloys were also the subject of studies into their antibacterial properties.[159-161] The alloy's antibacterial qualities and corrosion resistance significantly improved by the rise in Ag content while preserving biocompatibility.[160-162] Both during surgery, when the tissue is vulnerable to pathogenic microorganisms, and immediately following osseointegration, when biofilms and the onset of peri-implantitis may develop, antibacterial qualities are essential.[144]

According to Chen et al.[160] a material must contain at least 3% by mass of Ag in order to exert a persistent and effective antibacterial effect on Streptococcus aureus. The specimen's capacity to release Ag ions into a solution may be related to its antibacterial properties.[163]

As a result, bacterial cell membranes may become more permeable, cells may become depolarized, and phosphate efflux may occur, allowing cell contents to seep out and preventing DNA replication.[161,164] Similar benefits were seen when adding Ag (nm size) to the outermost layer of specimen by surface modifications, which increased antibacterial properties while preserving the materials' biocompatibility.[163,165,166] There was a significant decrease in bacteria when compared to untreated surfaces. A previous study[167] found that 75 days after immersion, TiAg film still showed a small level of antibacterial properties, suggesting that they still retained some bactericidal potential. Another technique used to achieve antibacterial goals was the creation of a tantalum nitrate layer. During the procedure, the material's resistance to bacterial corrosion was also significantly improved. For antibacterial coatings to be successful, the substrate must degrade well, they must also be biocompatible, thinner, denser, and stiffer. Additionally, they must be inert, should not alter the material physical characteristics, and should not have any impact on the eventual cost of an implant.[158] Even if the results were encouraging, the ideal scenario could be for the materials to instantly exhibit bactericidal properties due to their individual components without impeding the development of bones and osseointegration, as was seen with Ti–Cu and Ti–Ag alloys. One option to perhaps get around toxicity difficulties is to create surfaces that can prevent bacterial adhesion and destroy germs immediately by touch without releasing any substance.[155,168]

4. Surface Modification of Ti Implants

When titanium implants are made using standard techniques, the surface layer is frequently oxidized, contaminated, stretched and plastically deformed, non-uniform, and exceedingly undefined.[169] Since such "natural" surfaces are obviously unsuited for biological purposes, surface modification is necessary. An essential additional explanation for surface modification is the requirement for certain surface properties that are distinct from those present on a large number of Ti medical implants.[4,170] For components that interact with blood, such as mechanical heart valve, blood level biocompatibility is crucial. Good wear and corrosion resistance is crucial for additional applications.

Surface modification is critical for overcoming these challenges. The ideal surface modification procedure enhances certain surface characteristics required for certain medical purposes while preserving the remarkable bulk characteristics of titanium and its alloys.[171,172] Fig. 6 displays a visual picture of many surfaces changing strategies which were suggested according to different therapeutic needs. The link between a bone and an implant can be strengthened by altering the implant's surface. The portions that follow go through several surface modification methods used to increase the biocompatibility and corrosion resistance of Ti alloys.

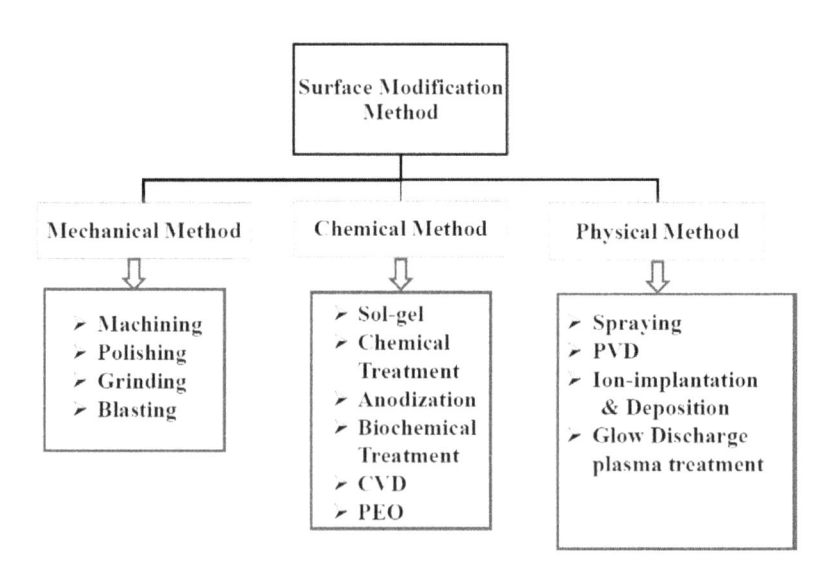

Fig. 6. Different techniques of surface modifications (Color online).[4]

These methods are classified into three categories: mechanical, chemical, and physical based on the way the changed layer behaves and grows on the material's surface.[170,173] The concept has received a lot of attention since it has been shown that surface treatment for dental implants has a greater survival rate than milling implants.[174,175] The major goals of the treatments include producing an osteoconductive surface that simulates bones, encouraging cell adhesion, improving corrosion resistance, and reducing ion leaking into the environment. Numerous methods, including mechanical friction,[100] isothermal oxidation,[176] and hydrothermal synthesis,[177] can change the surface of Ti alloys.

The method of sandblasting particles and then acid etching (SLA) an implant was one of the Ti surface modification methods with the highest level of effectiveness.[178] In addition to having strong contact with the bone in the early phases of osseointegration, SLA implant has better bone apposition and cell differentiation.[179,180] The TiZr alloy has gained access to SLActive® (Straumann), a sort of treatment that has been shown to be a great option for increasing the properties of the material. The effectiveness of the high-voltage anodic oxidation method known as MAO has been demonstrated in *in vitro* studies.[166,181] Using this method, a coating is created which is considerably porous, rougher, corrosion-resistant, and has greater surface area whereby an implant could make contact with the bone.[4,166,182] Additionally, the coating contains HAp, which accelerates up the process of mineralization and osteoblastic development processes and improves the durability and compatibility of the implant with the surrounding bone. The electrochemical anodizing technique improved cellular adhesion and proliferation on the surface of biomaterials in a manner that resembles the polished surfaces of Ti–6Al–7Nb and Ti–25Nb–25Zr alloys.[11,183]

The circumstances for cellular migration are improved when pores and nanotubes form on a material's surface.[183] For this type of therapy, Ti implants from Unite (Nobel Biocare) are now readily accessible. *In vivo* osteoconductive activity and survival rates have been shown to be superior than milled titanium.[184] Femtosecond

(FS) laser-produced nanostructured thin coatings have shown considerable cellular development and dispersion with a propensity to enhance the surface's wettability.[185,186] Findings were equivalent when using glow-discharge plasma, which enhanced Ti surface energy to enhance cellular adhesion.[89] Due to their enhanced electrochemical stability and protein adsorption results, which are both related to the specimen's chemical and physical surface alterations, plasma treatments are now employed extensively.[187,188] Niobium pentoxide and a second monolayer of graphene were combined in order to enhance the mechanical characteristics of Ti–Al–V alloy, particularly hardness and wear resistance to safeguard the implant's surface from corrosion.[189] By adding Nb-based coatings to stainless steel substrates, surface toughness and corrosion resistance were boosted, which also improved the material's biocompatibility.[58] Therefore, surface modifications to dental implants may be made to alter their biological, mechanical, physical, and chemical properties. By altering their surfaces, contemporary technology makes cytotoxic materials like stainless steel and Ti–Al–V alloy more desirable choices for therapeutic treatments. More research is needed on the readily available surface treatments for titanium implants so that Ti-based alloys may be treated without losing their characteristics.

5. New Advancements in Implant Materials Based on Titanium

It is essential for any implant to be able to continue in the body for a long period with no harmful adverse consequences, but this is true especially when implants are made from titanium alloys. In the process of making titanium alloys, it can be challenging to control the distribution, form, and size of the distinct phases of titanium. Despite having lower young's moduli as compared to various metallic biomaterials like Co–Cr alloys and stainless steel, but it is higher when compared with human bone, Ti alloys might be the source of the "stress shielding" issue due to this feature.[1,190] This continues to be the main problem and obstacle to the development of titanium alloy implants.

Given that it was observed that it promotes tissue growth and efficiently protected implants, a porous titanium alloy can be considered as a revolutionary discovery for the coming generations.[133,191] The development of titanium alloy implants faces a significant challenge when trying to stimulate the surface to form a bond to the insertion sites.[192,193]

Deposition, anodization, electrophoresis, ion implantation, sol–gel, MAO, acid, and alkaline treatment are a few examples of surface modification methods. Additional investigation is required into the capacity to develop an extra bio-responsive surface layer in order to arrive at a fresh and unanticipated conclusion. The advantages of enormous surface oxide nanotubes on Ti alloy make them an ideal target for more sophisticated techniques of surface modification.

In the important areas noted below, further research on titanium alloys for orthopaedics may be strengthened: (1) Computational materials science is needed to create innovative alloys for usage in the biomedical area. Using advanced techniques like phase field modeling, it is possible to describe how the microstructure changes at the time of manufacturing and how it impacts the mechanical properties. Phase field modeling, for instance, might be used to mimic the accumulation of precipitate during anti-aging therapy. The resulting evolving microstructure provides a starting point for forecasting mechanical properties. First-principles calculations can be used more frequently for creating new β Ti alloy compositions.

(2) It is critical to support the advancement of thermomechanical techniques that can improve the fatigue and wear resistance of alloys. High throughput techniques might make it easier to screen possible microstructures. These alloys can't be used in therapeutic settings since there isn't enough knowledge about their performance characteristics. Additive manufacturing should be looked at with the intention of developing both novel materials and individualized implants. (3) In order to evaluate long-term biocompatibility, large

animal models must be employed. These studies may primarily focus on leached metal ion concentrations and their impacts on cytotoxicity, corrosion rates, wear-related debris generation and the inflammatory reactions that result from it, osseointegration rates, and stress shielding effects. The efficacy of biomaterials should be evaluated utilizing commercially accessible biomedical implant simulators as well as to *in vivo* testing.

Due to its extraordinary advantages over other conventional manufacturing techniques, several AM approaches have garnered interest from researchers throughout the world in recent years. AM allows the production of actual parts from computer-aided design (CAD) files in a 3D printer by layering different materials.[194,195] A shorter manufacturing cycle, the ability to fabricate intricate components that are difficult to make through different conventional manufacturing methods, and a noticeable reduction in the use of energy, materials, and human resources are all benefits of this cutting-edge technology. AM methods may essentially be categorized based on the properties of the feedstock and the bonding method utilized to connect the material layers.[196] Rapid prototyping, complicated geometry, manufacturing of numerous integrated components, improved efficiency, and lower volume production are just a few of the reasons why AM is frequently utilized.[197] Due to a number of advantages, including increased automation, widespread access to CAM/CAD design software, and an expanding collection of printed materials, this approach, which is still gaining popularity, has drawn attention in the past 10 years.[198] This kind of fabrication necessitates in-depth familiarity with the microstructure and mechanical characteristics of fabricated components. It is important to remember that the qualities that are produced will be significantly influenced by the AM production method chosen. AM has capacity to create metals with specialized porosity structures that allow specialized cell morphologies permitting the cell proliferation and differentiation necessary for bone development, they are becoming popular in the biomedical area. Although various AM techniques adhere to the similar additive concept, regarding the materials that may be employed, every technique has benefits and drawbacks of its own, how they can be processed, and the scenarios in which they may be used.

The traditional PM metallurgy technique, also known as press-and-sinter, entails combining elemental or alloy powders with lubricants or additives to create a homogenous mixture. The part's machinability, wear resistance, or lubricity may all be enhanced by additives. The standard PM has been modified to make porous Ti structures,[34,35] despite the fact that the compacting pressure and sintering temperature have an effect on the microstructure parameters, particularly size, type, morphological and amount of porosity. One of the researchers[34] used the porous grade 4 cpTi samples for cortical bone replacement, which produced better stiffness results (20 to 25 GPa compared to 20 GPa of bone) at the lower values of both compacting pressure and sintering temperature, 38.5 MPa and 1000°C, yielding a material with about 40% porosity. Loose sintering (LS) is the process of creating porous structures with standard PM at minimal or negligible pressure. In comparison to specimens made with traditional PM, those created using this approach are more porous. The mechanical requirements of biomedical applications are not, however, guaranteed by the mean pore size of the porous Ti (17 m) and the mechanical strength (67 MPa).[35] SH approaches enable to manage porosity parameters like pore shape and percentage and ensure the strength of the component in order to address the limits of traditional PM.

Functional gradient materials (FGMs) are a class of modern materials that successfully mimic the hierarchy and gradient structure of biological structures while performing a range of activities. Since bone is an organic, natural FGM material and biomedical implants typically replace bone tissues, it makes sense to apply the FGM idea in these situations. FGMs have many benefits, as well as the capability to produce different gradations, such as composition, porosity, and size, this may be utilized to customize the anticipated biological and mechanical responses.[199] They can also reduce some drawbacks, like stress-shielding issues, enhance osseointegration, and improve electrochemical behavior and resistance to wear. Despite the fact that these are positive traits, there are still relatively few specific guidelines and specifications. The basic idea behind using FGMs in dental implants is that the properties of the implant may be carefully developed and changed to guarantee a full imitating of the peripheral bone tissue and to satisfy the biomechanical needs based on the particular location of the host bone. Therefore, using FGM dental implants is extremely advantageous and can improve implant stability and integration. The primary benefits of employing FGM components in dental applications include lowering the stress-shielding effect, enhancing biocompatibility, preventing thermal-mechanical failure, and meeting biomechanical criteria. The mechanical characteristics mismatch issues among implants and native biomaterials, which pose a serious problem since they might impede osseointegration and bone remodeling, can also be resolved with the help of these modern FGM implants. Since bone tissue is a common natural FGM structure, using FGM components to treat various bone-related conditions, such as orthopaedic implants, makes a lot of sense. Regarding implant applications, biomimetic FGM designs appear to be a promising approach.

6. Conclusion

Due to an imbalance between the elastic modulus of an implant material and the toxicity of some modern biomedical alloys, some biomaterials suffer stress shielding effects when implanted in humans. The answers to the two issues above will determine how metastable β-titanium alloys are designed for biological uses like orthopaedic and dental usage. Furthermore, β-Ti alloys have a lower modulus in comparison with modern metallic biomaterials Ti–6Al–4V (110 GPa), 316L stainless steel (200 GPa), and Co–Cr–Mo alloys (200–230 GPa).

The addition of metals like Mo, Nb, Ta, and Zr increases the strength and Young's modulus of titanium alloys. The titanium alloys Ti–12Mo–6Zr–2Fe and Ti–36Nb–2Ta–3Zr–0.3O were discovered to have the highest tensile strengths.

It has been established that the interstitial content of materials closely correlates with the strength and ductility of Ti alloys. Interstitial solutes form intermetallic phases that lessen the ductility of materials like titanium alloy while simultaneously reducing the cpTi hardness and improving the strength.

A lower elastic modulus was shown to enhance the uniformly distributing the stresses at the bone-implant contact while minimizing bone atrophy. Elastic modulus should be as close to the bone as is practical since "stress shielding" has been associated with osteoporosis and bone resorption close to implant placements. It was shown that if Young's modulus was too low, metallic implants would move erratically. This should not be done since it might make the implants loose and ultimately result in the prosthesis failing.

Because α-stabilizers act as substitutional solutes that lower the biomaterial atomic mobility or since the alloys are aged at lower temperatures, thereby encouraging the phase's precipitation, it has been noted that titanium alloys have higher hardness values and elastic modulus. This increases the material's hardness.

Ti–24Nb–4Zr—7.6Sn surpassed Ti–6Al–4V ELI in the strain-controlled fatigue testing in terms of fatigue resistance. Ti–15Zr discs and dental implants performed much better under fatigue compared to Ti-Grade IV. The ideal fatigue endurance of the alloy and cpTi was 560 MPa and 435 MPa, respectively.

Due to a smaller contact angle (81.75°) as compared to cpTi's (96.46°), Ti–45Nb is more hydrophilic than cpTi. Though, the Ti50Zr alloy had a higher surface energy and decreased surface roughness (37 mN/m, 0.17 m) compared to cpTi (34 mN/m, 0.20 m) and the Ti–50Nb alloy (32 mN/m, 0.46 m)67. The Ti–Zr alloy displayed a desirable *in vitro* biological profile (cell adhesion and proliferation) as a result of these properties of the substrate component. The alloy's surface area was greater than that of Ti–Ta–Nb–Zr, but Ti had a rougher topology.

Ti–26Nb is less cytotoxic than Ni–49.2Ti, according to research that compared the two materials' cytotoxicity. When compared to Ni–Ti alloy, the Ti–19Zr–10Nb–1Fe alloy exhibits better hemocompatibility but equal cytocompatibility. However, there exists an urgent requirement for additional *in vitro* studies on β titanium alloys. The lack of potentially dangerous alloying elements results in the enhanced biocompatibility of type Ti alloys. The biocompatibility of the alloys Ti–Cu, Ti–Nb, and Ti–Nb–Ta–Zr with bone tissue without preventing the formation of new bone was found to be highly intriguing.

Acknowledgments

The author(s) received no financial support for the research, authorship, and/or publication of this paper.

References

1. M. R. Hazwani, L. X. Lim, Z. Lockman and H. Zuhailawati, *Trans. Nonferr. Metals Soc. China* **32** (2022) 1.
2. L. C. Zhang and L. Y. Chen, *Adv. Eng. Mater.* **21** (2019) 1801215.
3. H. K. Uhthoff, P. Poitras and D. S. Backman, *J. Orthop. Sci.* **11** (2006) 118.
4. P. Pesode and S. Barve, *Mater. Today, Proc.* **46** (2021) 594.
5. J. M. Cordeiro and V. A. Barão, *Mater. Sci. Eng. C* **71** (2017) 1201.
6. M. Quirynen, B. Al-Nawas, H. J. Meijer, A. Razavi, T. E. Reichert, M. Schimmel, S. Storelli, E. Romeo and Roxolid Study Group, *Clin. Oral Implants Res.* **26** (2015) 831.
7. H. M. Grandin, S. Berner and M. Dard, *Materials* **5** (2012) 1348.
8. L. Mishnaevsky Jr., E. Levashov, R. Z. Valiev, J. Segurado, I. Sabirov, N. Enikeev, S. Prokoshkin, A. V. Solov'yov, A. Korotitskiy, E. Gutmanas and I. Gotman, *Mater. Sci. Eng. R, Rep.* **81** (2014) 1.
9. J. Gottlow, M. Dard, F. Kjellson, M. Obrecht and L. Sennerby, *Clin. Implant Dent. Relat. Res.* **14** (2012) 538.
10. H. J. Meijer, I. Naert, R. Persson, S. Storelli, M. D. Christiaan ten Bruggenkate and B. Vandekerckhove, *Clin. Implant Dent. Relat. Res.* **14** (2010) 896.

11. C. N. Elias, D. J. Fernandes, C. R. Resende and J. Roestel, *Dent. Mater.* **31** (2015) e1.
12. B. Mjöberg, E. Hellquist, H. Mallmin and U. Lindh, *Acta Orthop. Scand.* **68** (1997) 511.
13. D. Zaffe, C. Bertoldi and U. Consolo, *Biomaterials* **25** (2004) 3837.
14. A. C. Faria, R. C. Rodrigues, A. L. Rosa and R. F. Ribeiro, *J. Prosthet. Dent.* **112** (2014) 1448.
15. S. Datta, M. Mahfouf, Q. Zhang, P. P. Chattopadhyay and N. Sultana, *J. Mech. Behav. Biomed. Mater.* **53** (2016) 350.
16. M. Niinomi, M. Nakai and J. Hieda, *Acta Biomater.* **8** (2012) 3888.
17. S. Guo, J. Zhang, X. Cheng and X. Zhao, *J. Alloys Compd.* **644** (2015) 411.
18. S. Guo, Q. Meng, X. Zhao, Q. Wei and H. Xu, *Sci. Rep.* **5** (2015) 1.
19. M. Geetha, A. K. Singh, R. Asokamani and A. K. Gogia, *Prog. Mater. Sci.* **54** (2009) 397.
20. S. Bahl, S. Suwas and K. Chatterjee, *Int. Mater. Rev.* **66** (2021) 114.
21. P. Pesode and S. Barve, Magnesium alloy for biomedical applications, in *Advanced Materials for Biomechanical Applications* (CRC Press), (2022) pp. 133–158.
22. B. Thakur, S. Barve and P. Pesode, Magnesium alloy for biomedical applications, in *Advanced Materials for Biomechanical Applications* (CRC press), (2022) pp. 113–131.
23. L. Kunčická, R. Kocich and T. C. Lowe, *Prog. Mater. Sci.* **88** (2017) 232.
24. M. Long and H. J. Rack, *Biomaterials* **19** (1998) 1621.
25. Q. Chen and G. A. Thouas, *Mater. Sci. Eng. R, Rep.* **87** (2015) 1.
26. W. Nicholson, *J. Prosthesis* **2** (2020) 11.
27. M. Niinomi and M. Nakai, *Int. J. Biomater.* **2011** (2011) 836587.
28. R. B. Osman and M. V. Swain, *Materials* **8** (2015) 932.
29. A. Dalmau, V. G. Pina, F. Devesa, V. Amigó and A. I. Muñoz, *Mater. Sci. Eng. C* **48** (2015) 55.
30. X. J. Jiang, X. Y. Wang, Z. H. Feng, C. Q. Xia, C. L. Tan, S. X. Liang, X. Y. Zhang, M. Z. Ma and R. P. Liu, *Mater. Sci. Eng. A* **635** (2015) 36.
31. D. R. Correa, F. B. Vicente, T. A. Donato, V. E. Arana-Chavez, M. A. Buzalaf and C. R. Grandini, *Mater. Sci. Eng. C* **34** (2014) 354.
32. M. T. Mohammed, Z. A. Khan, M. Geetha and A. N. Siddiquee, *J. Alloys Compd.* **634** (2015) 272.

33. P. Manda, U. Chakkingal and A. K. Singh, *Mater. Character.* **96** (2014) 151.
34. I. Hacisalihoglu, A. Samancioglu, F. Yildiz, G. Purcek and A. Alsaran, *Wear* **332** (2015) 679.
35. E. Kobayashi, T. Yoneyama, H. Hamanaka, I. R. Gibson, S. M. Best, J. C. Shelton and W. Bonfield, *J. Mater. Sci. Mater. Med.* **9** (1998) 625.
36. K. J. Qiu, Y. Liu, F. Y. Zhou, B. L. Wang, L. Li, Y. F. Zheng and Y. H. Liu, *Acta Biomater.* **15** (2015) 254.
37. A. L. Ribeiro, R. C. Junior, F. F. Cardoso and L. G. Vaz, *J. Mater. Sci. Mater. Med.* **20** (2009) 1629.
38. E. C. Nelson, D. J. Fernandes, C. R. S. Resende and J. Roestel, *Dent. Mater.* **31** (2015) e.
39. A. T. Sidambe, *Materials* **7** (2014) 8168.
40. Y. S. Zhang, J. J. Hu, W. Zhang, S. Yu, Z. T. Yu, Y. Q. Zhao and L. C. Zhang, *Mater. Character.* **147** (2019) 127.
41. C. Andrew, L. C. Zhang, O. M. Ivasishin, D. G. Savvakin, M. V. Matviychuk and E. V. Pereloma, *Mater. Sci. Eng. A* **528** (2011) 1686.
42. L. Xiaofei, L. Dong, Z. Zhang, Y. Liu, Y. Hao, R. Yang and L.-C. Zhang, *Metals* **7** (2017) 131.
43. A. R. McAndrew, P. A. Colegrove, C. Bühr, B. C. D. Flipo and A. Vairis, *Prog. Mater. Sci.* **92** (2018) 225.
44. G.Xin, Y. Bai, J. Li, S. Li, W. Hou, Y. Hao, X. Zhang, R. Yang and R. D. K. Misra, *Corros. Sci.* **145** (2018) 80.
45. Z. Feng, J. Feng, W. Xiang, Q. Fu and W. Yuan, *Mater. Sci. Eng. A* **858** (2022) 144082.
46. D. Ruifeng, J. Li, H. Kou, J. Fan, Y. Zhao, H. Hou and L. Wu, *J. Mater. Sci. Technol.* **44** (2020) 24.
47. L. Tong, D. Kent, G. Sha, L. T. Stephenson, A. V. Ceguerra, S. P. Ringer, M. S. Dargusch and J. M. Cairney, *Acta Mater.* **106** (2016) 353.
48. G. Junheng, A. J. Knowles, D. Guan and W. M. Rainforth, *Scr. Mater.* **162** (2019) 77.
49. L. Tong, D. Kent, G. Sha, M. S. Dargusch and J. M. Cairney, *Scr. Mater.* **104** (2015) 75.
50. Y. Zheng, R. E. A. Williams, D. Wang, R. Shi, S. Nag, P. Kami, J. M. Sosa, R. Banerjee, Y. Wang and H. L. Fraser, *Acta Mater.* **103** (2016) 850.

51. W. Xiao, M. S. Dargusch, D. Kent, X. Zhao and C. Ma, *Materialia* **9** (2020) 100557.
52. P. Barriobero-Vila, G. Requena, S. Schwarz, F. Warchomicka and T. Buslaps, *Acta Mater.* **95** (2015) 90.
53. K.-K. Tseng, C.-C. Juan, S. Tso, H.-C. Chen, C.-W. Tsai and J.-W. Yeh, *Entropy* **21** (2018) 15.
54. H. Shixing, Q. Zhao, Y. Zhao, C. Lin, C. Wu, W. Jia, C. Mao and V. Ji, *J. Mater. Sci. Technol.* **79** (2021) 147.
55. K. R. Prakash and A. Devaraj, *Metals* **8** (2018) 506.
56. R. Davis, H. M. Flower and D. R. F. West, *J. Mater. Sci.* **14** (1979) 712.
57. E. S. K. Menon and R. Krishnan, *J. Mater. Sci.* **18** (1983) 365.
58. G. Ramírez, S. E. Rodil, H. Arzate, S. Muhl and J. J. Olaya, *Appl. Surf. Sci.* **257** (2011) 2555.
59. M. E. Hoque, N. N. Showva, M. Ahmed, A. B. Rashid, S. E. Sadique, T. El-Bialy and H. Xu, *Heliyon* (2022) e11300.
60. Y. Ikarashi, K. Toyoda, E. Kobayashi, H. Doi, T. Yoneyama, H. Hamanaka and T. Tsuchiya, *Mater. Trans.* **46** (2005) 2260.
61. S. Griza, D. H. G. de Souza Sá, W. W. Batista, J. C. G. de Blas and L. C. Pereira, *Mater. Des.* **56** (2014) 200.
62. Y. Shibata, Y. Tanimoto, N. Maruyama and M. Nagakura, *J. Prosdont. Res.* **59** (2015) 84.
63. D. Farlay, G. Falgayrac, C. Ponçon, S. Rizzo, B. Cortet, R. Chapurlat, G. Penel, I. Badoud and P. Ammann and G. Boivin, *Bone Rep.* **17** (2022) 101623.
64. Y. Li, C. Yang, H. Zhao, S. Qu, X. Li and Y. Li, *Materials* **7** (2014) 1709
65. J. Black and G. Hastings (eds.) *Handbook of Biomaterial Properties* (Springer Science & Business Media, 2013).
66. M. M. Dewidar, H. C. Yoon and J. K. Lim, *Metals Mater. Int.* **12** (2006) 193.
67. Y. Bai, Y. Deng, Y. Zheng, Y. Li, R. Zhang, Y. Lv, Q. Zhao and S. Wei, *Mater. Sci. Eng. C* **59** (2016) 565.
68. M. Abdel-Salam, S. El-Hadad and W. Khalifa, *Mater. Sci. Eng. C* **104** (2019) 109974.
69. P. Pralhad and S. Barve, *J. Eng. Appl. Sci.* **70** (2023) 1.
70. I. Kopova, J. Stráský, P. Harcuba, M. Landa, M. Janeček and L. Bačáková, *Mater. Sci. Eng. C* **60** (2016) 230.

71. M. U. Farooq, F. A. Khalid, H. Zaigham and I. H. Abidi, *Mater. Lett.* **121** (2014) 58.
72. L. M. Elias, S. G. Schneider, S. Schneider, H. M. Silva and F. Malvisi, *Mater. Sci. Eng. A* **432** (2006) 108.
73. S. E. Haghighi, H. B. Lu, G. Y. Jian, G. H. Cao, D. Habibi and L. C. Zhang, *Mater. Des.* **76** (2015) 47.
74. I. V. Okulov, S. Pauly, U. Kühn, P. Gargarella, T. Marr, J. Freudenberger, L. Schultz, J. Scharnweber, C. G. Oertel, W. Skrotzki and J. Eckert, *Mater. Sci. Eng. C* **33** (2013) 4795.
75. P. Pesode, S. Barve, S. V. Wankhede, D. R. Jadhav and S. K. Pawar, *Mater. Today Proc.* **72** (2022) 724.
76. S. X. Liang, X. J. Feng, L. X. Yin, X. Y. Liu, M. Z. Ma and R. P. Liu, *Mater. Sci. Eng. C* **61** (2016) 338.
77. M. K. Han, M. J. Hwang, M. S. Yang, H. S. Yang, H. J. Song and Y. J. Park, *Mater. Sci. Eng. A* **616** (2014) 268.
78. V. G. Pina, A. Dalmau, F. Devesa, V. Amigó and A. I. Muñoz, *J. Mech. Behav. Biomed. Mater.* **46** (2015) 59.
79. P. Pralhad, S. Barve, Y. Mane, S. Dayane, S. Kolekar and K. A. Mohammed, *Key Eng. Mater.* **944** (2023) 117.
80. M. Niinomi, *Mater. Sci. Eng. A* **243** (1998) 231.
81. F. Rubitschek, T. Niendorf, I. Karaman and H. J. Maier, *J. Mech. Behav. Biomed. Mater.* **5** (2012) 181.
82. R. A. Antunes and M. C. de Oliveira, *Acta Biomater.* **8** (2012) 937.
83. O. Karpenko, S. Oterkus and E. Oterkus, *Int. J. Fatigue* **162** (2022) 107023.
84. C. Gherde, P. Dhatrak, S. Nimbalkar and S. Joshi, *Mater. Today Proc.* **43** (2021) 1117.
85. S. J. Li, T. C. Cui, Y. L. Hao and R. Yang, *Acta Biomater.* **4** (2008) 305.
86. Y. Okazaki, S. Rao, Y. Ito and T. Tateishi, *Biomaterials* **19** (1998) 1197.
87. A. E. Medvedev, A. Molotnikov, R. Lapovok, R. Zeller, S. Berner, P. Habersetzer and F. Dalla Torre, *J. Mech. Behav. Biomed. Mater.* **62** (2016) 384.
88. C. W. Lin, C. P. Ju and J. H. Lin, *Biomaterials* **26** (2005) 2899.
89. I. S. Yeo, *Dent. Clin.* **66** (2022) 627.
90. X. Chen, A. Nouri, Y. Li, J. Lin, P. D. Hodgson and C. E. Wen, *Biotechnol. Bioeng.* **101** (2008) 378.

91. S. Sista, C. E. Wen, P. D. Hodgson and G. Pande, *J. Biomed. Mater. Res. A* **97** (2011) 27.
92. D. Mareci, G. Bolat, A. Cailean, J. J. Santana, J. Izquierdo and R. M. Souto, *Corros. Sci.* **87** (2014) 334.
93. D. D. Deligianni, N. Katsala, S. Ladas, D. Sotiropoulou, J. Amedee and Y. F. Missirlis, *Biomaterials* **22** (2001) 1241.
94. A. Wennerberg and T. Albrektsson, *Clin. Oral Implant Res.* **20** (2009)172.
95. J. M. Moreno, P. Osiceanu, C. Vasilescu, M. Anastasescu, S. I. Drob and M. Popa, *Surf. Coat. Technol.* **235** (2013) 792.
96. Y. S. Sun, J. F. Liu, C. P. Wu and H. H. Huang, *J. Alloys Compd.* **643** (2015) S124.
97. M. R. Kaluđerović, M. Mojić, J. P. Schreckenbach, D. Maksimović-Ivanić, H. L. Graf and S. Mijatović, *Cells Tissues Organs* **200** (2014) 265.
98. P. Vansana, K. Kakura, Y. Taniguchi, K. Egashira, E. Matsuzaki, T. Tsutsumi and H. Kido, *J. Dent. Sci.* **17** (2022) 1225.
99. J. M. Chaves, A. L. Escada, A. D. Rodrigues and A. A. Claro, *Appl. Surf. Sci.* **370** (2016) 76.
100. R. Huang, S. Lu and Y. Han, *Colloids Surf. B Biointerf.* **111** (2013) 232.
101. D. P. Oliveira, A. Palmieri, F. Carinci and C. Bolfarini, *Mater. Sci. Eng. C* **51** (2015) 248.
102. H. H. Huang, C. P. Wu, Y.S Sun, W. E. Yang, M. C. Lin and T. H. Lee, *Surf. Coat. Technol.* **259** (2014) 206.
103. J. W. Park, Y. Tustusmi, C. S. Lee, C. H. Park, Y. J. Kim, J. H. Jang, D. Khang, Y. M. Im, H. Doi, N. Nomura and T. Hanawa, *Appl. Surf. Sci.* **257** (2011) 7856.
104. M. O. Klein, A. Bijelic, T. Toyoshima, H. Götz, R. L. Von Koppenfels, B. Al-Nawas and H. Duschner, *Clin. Oral Implant Res.* **21** (2010) 642.
105. M. O. Klein, A. Bijelic, T. Ziebart, F. Koch, P. W. Kämmerer, M. Wieland, M. A. Konerding and B. Al-Nawas, *Clin. Implant Dent. Relat. Res.* **15** (2013) 166.
106. Y. Bai, X. Gai, S. Li, L. C. Zhang, Y. Liu, Y. Hao, X. Zhang, R. Yang and Y. Gao, *Corros. Sci.* **123** (2017) 289.
107. P. Qin, Y. Liu, T. B. Sercombe, Y. Li, C. Zhang, C. Cao, H. Sun and L. C. Zhang, *ACS Biomater. Sci. Eng.* **4** (2018) 2633.

108. P. Y. Guo, H. Sun, Y. Shao, J. T. Ding, J. C. Li, M. R. Huang, S. Y. Mao, Y. X. Wang, J. F. Zhang, R. C. Long and X. H. Hou, *Corros. Sci.* **172** (2020) 108738.
109. L. C. Zhang, Z. Jia, F. Lyu, S. X. Liang and J. Lu, *Prog. Mater. Sci.* **105** (2019) 100576.
110. B. Thakur, S. Barve and P. Pesode, *J. Mech. Behav. Biomed. Mater.* **138** (2023) 105641.
111. V. A. Alves, R. Q. Reis, I. C. Santos, D. G. Souza, T. D. Gonçalves, M. A. Pereira-da-Silva, A. Rossi and L. A. Da Silva, *Corros. Sci.* **51** (2009) 2473.
112. J. Loch, A. Łukaszczyk, V. Vignal and H. Krawiec, *Solid State Phenom.* **227** (2015) 435.
113. S. Kumar and T. S. Narayanan, *J. Alloys Compd.* **479** (2009) 699.
114. N. Dai, L. C. Zhang, J. Zhang, Q. Chen, M. Wu, *Corros. Sci.* **102** (2016) 484.
115. Simsek and D. Ozyurek, *Mater. Sci. Eng. C* **94** (2019) 357.
116. A. P. Alves, F. A.Santana, L. A. Rosa, S. A. Cursino and E. N. Codaro, *Mater. Sci. Eng. C* **24** (2004) 693.
117. B. Zareidoost and M. Yousefpour, *Mater. Sci. Eng. C* **110** (2020) 110725.
118. J. Lin, S. Ozan, K. Munir, K. Wang, X. Tong, Y. Li, G. Li and C. Wen, *RSC Adv.* **7** (2017) 12309.
119. N. Dai, L. C. Zhang, J. Zhang, X. Zhang, Q. Ni, Y. Chen, M. Wu and C. Yang, *Corros. Sci.* **111** (2016) 703.
120. P. Qin, Y. Chen, Y. J. Liu, J. Zhang, L. Y. Chen, Y. Li, X. Zhang, C. Cao, H. Sun and L. C. Zhang, *ACS Biomater. Sci. Eng.* **5** (2018) 1141.
121. L. Y. Chen, Y. W. Cui and L. C. Zhang, *Metals* **10** (2020) 1139.
122. A. L. Andrea, J. Vishnu, Y. Douest, K. Perrin, A.-M. Trunfio-Sfarghiu, N. Courtois, A. Gebert, B. Ter-Ovanessian and M. Calin, *Tribology Int.* **181** (2023) 108325.
123. K. K. Tae, M. Y. Eo, T. T. H. Nguyen and S. M. Kim, *Int. J. Implant Dent.* **5** (2019) 1.
124. A. C. Vieira, A. R. Ribeiro, L. A. Rocha and J.-P. Celis, *Wear* **261** (2006) 994.
125. V. Jithin and G. Manivasagam, *J. Bio-Tribo-Corros.* **7** (2021) 1.
126. J. C. M. Souza, M. Henriques, W. Teughels, P. Ponthiaux, J.-P. Celis and L. A. Rocha, *J. Bio-Tribo-Corros.* **1** (2015) 1.

127. V. Jithin, A. R. Ansheed, P. Hameed, K. Praveenkumar, S. Pilz, L. A. Alberta, S. Swaroop, M. Calin, A. Gebert and G. Manivasagam, *Appl. Surf. Sci.* **586** (2022) 152816.
128. F. Mohsen, K. Fallahnezhad, M. Taylor and R. Hashemi, *Tribol. Int.* **172** (2022) 107634.
129. Ç. Ihsan, A. Alves, C. Chirico, A. Pinto, S. Tsipas, E. Gordo and F. Toptan, *Metall. Mater. Trans. A* **51** (2020) 3256.
130. H. Md. Enamul, N.-N. Showva, M. Ahmed, A. B. Rashid, S. E. Sadique, T. El-Bialy and H. Xu, *Heliyon* **8** (2022) e11300.
131. A. Revathi, A. D. Borrás, A. I. Muñoz, C. Richard and G. Manivasagam, *Mater. Sci. Eng. C* **76** (2017) 1354.
132. L. Paul, E. Debels, K. V. Landuyt, M. Peumans and B. V. Meerbeek, *Dental Mater.* **22** (2006) 693.
133. P. Pesode and S. Barve, *Mater. Today Proc.* (2022). DOI: https://doi.org/10.1016/j.matpr.2022.11.248
134. J. T. Intravaia, T. Graham, H. S. Kim, H. S. Nanda, S. G. Kumbar and S. P. Nukavarapu, *Curr. Opin. Biomed. Eng.* **25** (2022) 100439.
135. R. E. McMahon, J. Ma, S. V. Verkhoturov, D. Munoz-Pinto, I. Karaman, F. Rubitschek, H. J. Maier and M. S. Hahn, *Acta Biomater.* **8** (2012) 2863.
136. P. Xue, Y. Li, K. Li, D. Zhang and C. Zhou, *Mater. Sci. Eng. C* **50** (2015) 179.
137. E. Takematsu, K. I. Katsumata, K. Okada, M. Niinomi and N. Matsushita, *Mater. Sci. Eng. C* **62** (2016) 662.
138. A. Dikici, M. Niinomi, M. Topuz, S. G. Koc and M. Nakai, *Sol-Gel Sci. Technol.* **87** (2018) 713.
139. M. Morra, C. Cassinelli, G. Cascardo, D. Bollati and R. R. Baena, *J. Biomed. Mater. Res. A* **96** (2011) 449.
140. A. Hoene, U. Walschus, M. Patrzyk, B. Finke, S. Lucke, B. Nebe, K. Schroeder, A. Ohl and M. Schlosser, *Acta Biomater.* **6** (2010) 676.
141. A. F. Williams, *Biomaterials* **29** (2008) 2941.
142. N. F. Nuswantoro, M. Manjas, N. Suharti, D. Juliadmi, H. Fajri, D. H. Tjong, J. Affi and M. Niinomi, *Ceram. Int.* **47** (2021) 16094.
143. Y. Guo, D. Chen, M. Cheng, W. Lu, L. Wang and X. Zhang, *Int. J. Mol. Med.* **31** (2013) 689.
144. B. Bai, E. Zhang, H. Dong and J. Liu, *J. Mater. Sci. Mater. Med.* **26** (2015) 1.

145. Z. Che, Y. Sun, W. Luo, L. Zhu, Y. Li, C. Zhu, T. Liu and L. Huang, *Mater. Des.* **223** (2022) 111118.
146. P. Stenlund, O. Omar, U. Brohede, S. Norgren, B. Norlindh, A. Johansson, J. Lausmaa, P. Thomsen and A. Palmquist, *Acta Biomater.* **20** (2022) 165.
147. F. Ghadami, M. A. Hamedani, G. Rouhi, S. Saber-Samandari, M. M. Dehghan, S. Farzad-Mohajeri and F. Mashhadi-Abbas, *J. Biomech.* **144** (2022) 111310.
148. P. Altinci, G. Can, O. Gunes, C. Ozturk and H. Eren, *Clin. Oral Implant. Res.* **18** (2016) 1193.
149. F. A. Shah, B. Nilson, R. Brånemark, P. Thomsen and A. Palmquist, *Nanomed. Nanotechnol. Biol. Med.* **10** (2014) 1729.
150. J. Zhang, B. Cai, P. Tan, M. Wang, B. Abotaleb, S. Zhu and N. Jiang, *J. Mater. Res. Technol.* **16** (2022) 1547.
151. L. Shi, L. Shi, L. Wang, Y. Duan, W. Lei, Z. Wang, J. Li, X. Fan, X. Li, S. Li and Z. Guo, *PLoS One* **8** (2013) e55015.
152. J. W. Lee, D. J. Lin, C. P. Ju, H. S. Yin, C. C. Chuang and J. H. Lin, *J. Biomed. Mater. Res. B, Appl. Biomater.* **91** (2009) 643.
153. X. Li, T. Chen, J. Hu, S. Li, Q. Zou, Y. Li, N. Jiang, H. Li and J. Li, *Colloids Surf. B, Biointerf.* **144** (2016) 265.
154. B. Wen, J. Chen, M. Dard and Z. Cai, *Clin. Implant.* **18** (2016) 120.
155. S. Ferraris and S. J. Spriano, *Mater. Sci. Eng. C* **61** (2016) 965.
156. A. Zhang, F. Li, H. Wang, J. Liu, C. Wang, M. Li, and K. Yang, *Mater. Sci. Eng. C* **33** (2013) 4280.
157. L. Ren, Z. Ma, M. Li, Y. Zhang, W. Liu, Z. Liao and K. Yang, *J. Mater. Sci. Technol.* **30** (2014) 699.
158. G. A. Norambuena, R. Patel, M. Karau, C. C. Wyles, P. J. Jannetto, K. E. Bennet, A. D. Hanssen and R. J. Sierra, *Clin. Orthop. Relat. Res.* **475** (2017) 722.
159. P. A. Pesode and S. B. Barve, *Mater. Today Proc.* **47** (2021) 5652.
160. M. Chen, E. Zhang and L. Zhang, *Mater. Sci. Eng. C* **62** (2016) 350.
161. M. K. Kang, S. K. Moon, J. S. Kwon, K. M. Kim and K.N Kim, *Mater. Res. Bull.* **47** (2012) 2952.
162. L. Yao, H. Wang, L. Li, Z. Cao, Y. Dong, L. Yao, W. Lou, S. Zheng, Y. Shi, X. Shen and C. Cai, *Mater. Des.* **224** (2022) 111425.
163. S. Ferraris, A. Venturello, M. Miola, A. Cochis, L. Rimondini and S. Spriano, *Appl. Surf. Sci.* **311** (2014) 279.
164. Q. Tang, X. Zhang, K. Shen, Z. Zhu, Y. Hou and M. Lai, *Colloid Interf. Sci. Commun.* **44** (2021) 100481.

165. R. Bright, D. Fernandes, J. Wood, D. Palms, A. Burzava, N. Ninan, T. Brown, D. Barker and K. Vasilev, *Mater. Today Bio.* **13** (2022) 100176.
166. I. D. Marques, V. A. Barão, N. C. da Cruz, J. C. Yuan, M. F. Mesquita, A. P. Ricomini-Filho, C. Sukotjo and M. T. Mathew, *Corros. Sci.* **100** (2015) 133.
167. L. Bai, R. Hang, A. Gao, X. Zhang, X. Huang, Y. Wang, B. Tang, L. Zhao and P. K. Chu, *Appl. Surf. Sci.* **355** (2015) 32.
168. L. M. Pandey, *Curr. Opin. Biomed. Eng.* (2022) 100423.
169. Y. Sasikumar, K. Indira and N Rajendran, *J. Bio- Tribo-Corros.* **5** (2019) 1.
170. A. Kurup, P. Dhatrak and N. Khasnis, *Mater. Today, Proc.* **39** (2021) 84.
171. R. Lin, Z. Wang, Z. Li and L. Gu, *Mater. Today Bio.* **15** (2022) 100330.
172. P. Pralhad, S. Barve, S. V. Wankhede and A. Chipade, *3c Empresa* **12** (2023) 392.
173. T. Perets, N. B. Ghedalia-Peled, R. Vago, J. Goldman, A. Shirizly and E. Aghion, *Mater. Sci. Eng. C* **129** (2021) 112418.
174. M. C. Goiato, D. M. Dos Santos, J. J. Santiago, A. Moreno and E. P. Pellizzer, *Int. J. Oral Maxillofac. Surg.* **43** (2014) 1108.
175. V. S. Muthaiah, S. Indrakumar, S. Suwas and K. Chatterjee, *Bioprinting* **25** (2022) e00180.
176. K. Aniołek and M. Kupka, *Mater. Chem. Phys.* **171** (2016) 374.
177. X. H. Liu, L. Wu, H. J. Ai, Y. Han and Y. Hu, *Mater. Sci. Eng. C* **48** (2015) 256.
178. X. Shi, L. Xu, K. B. Violin and S. Lu, *J. Mech. Behav. Biomed. Mater.* **53** (2016) 312.
179. D.Cochran, T. Oates, D. Morton, A. Jones, D. Buser and F. Peters, *J. Periodontol.* **78** (2007) 974.
180. X. Zhao, X. Ren, C. Wang, B. Huang, J. Ma, B. Ge, Z. Jia and Y. Li, *Surf. Coat. Technol.* **399** (2020) 126173.
181. G. Singh, S. Sharma, M. Mittal, G. Singh, J. Singh, L. Changhe, A. M. Khan, S. P. Dwivedi, R. T. Mushtaq and S. Singh, *J. Mater. Res. Technol.* **18** (2022) 1358.
182. H. A. AlMashhadani, A. A. Khadom and M. M. Khadhim, *Results Chem.* **4** (2022) 100555.
183. H. H. Huang, C. P. Wu, Y. S. Sun and T. H. Lee, *Thin Solid Films* **528** (2013) 157.

184. M. Degidi, D. Nardi and A. Piattelli, *Clin. Implant Dent. Related Res.* **14** (2012) 828.
185. Y. H. Jeong, H. C. Choe and W. A. Brantley, *Thin Solid Films* **519** (2011) 4668.
186. P. Bansal, G. Singh and H. S. Sidhu, *Mater. Chem. Phys.* **257** (2021) 123738.
187. T. Beline, I. D. Marques, A. O. Matos, E. S. Ogawa, A. P. Ricomini-Filho, E. C. Rangel, N. C. Da Cruz, C. Sukotjo, M. T. Mathew, R. Landers and R. L. Consani, *Biointerphases* **11** (2016) 011013.
188. R. C. Costa, J. G. Souza, J. M. Cordeiro, M. Bertolini, E. D. de Avila, R. Landers, E. C. Rangel, C. A. Fortulan, B. Retamal-Valdes, N. C. da Cruz and M. Feres, *J. Colloid Interf. Sci.* **579** (2020) 680.
189. M. Kalisz, M. Grobelny, M. Mazur, M. Zdrojek, D. Wojcieszak, M. Świniarski, J. Judek and D. Kaczmarek, *Thin Solid Films* **589** (2015) 356.
190. S. Prasad, M. Ehrensberger, M. P. Gibson, H. Kim and E. A. Monaco Jr., *J. Oral Biosci.* **57** (2015) 192.
191. N. Soro, E. G. Brodie, A. Abdal-hay, A. Q. Alali, D. Kent and M. S. Dargusch, *Mater. Des.* **218** (2022) 110688.
192. X. Y. Zhou, Z. H. Dou, T. A. Zhang, J. S. Yan and J. P. Yan, *Trans. Nonferrous Metals Soc. China* **32** (2022) 3469.
193. P. Mossino, *Ceram. Int.* **30** (2004) 311.
194. T. T. Sharon, S. O. Akinwamide, E. Olevsky and P. A. Olubambi, *Heliyon* **8** (2022) e09041.
195. A. Truong, P. Kwon and C. S. Shin, *Int. J. Mach. Tools Manuf.* **121** (2017) 50.
196. A. Dongdong and D. Gu, *Laser Additive Manufacturing of High-Performance Materials* (Springer, 2015), p. 15.
197. C. Ian, D. Bourell and I. Gibson, *Rapid Protyp. J.* **18** (2012) 255.
198. A. Bandyopadhyay, T. P. Gualtieri and S. Bose, *Add. Manuf.* **1** (2015) 9.
199. A. Rodriguez-Contreras, M. Punset, J. A. Calero, F. J. Gil, E. Ruperez and J. María Manero, *J. Mater. Sci. Technol.* **76** (2021) 129.
200. W. Sagar, P. Pesode, S. Gaikwad, S. Pawar and A. Chipade, *Mater. Sci. Forum* **1081** (2023) 41.

Electrochemical evaluation of Ti45Nb coated with 63s bioglass by electrophoretic deposition*

Yakup Uzun

Ataturk University, Faculty of Engineering,
Department of Mechanical Engineering,
Erzurum, Turkey
yakup.uzun@atauni.edu.tr

In this study, analyses were carried out to investigate the structural, mechanical and electrochemical behaviors of untreated and bioglass (BG)-coated Ti45Nb materials. The samples were coated with a mixture of 63s BG powder and phosphate ester (PE) at concentrations of 0.2, 0.4, 0.8, and 1 g using the electrophoretic deposition (EPD) method. Then, the structural, mechanical, and electrochemical properties of the untreated and coated samples were determined and characterized by using X-ray diffraction (XRD) and scanning electron microscope (SEM) devices. It was concluded that the 0.2 g coating provided resistance to corrosion for the Ti45Nb material more effectively than the others did.

Keywords: Ti45Nb; EPD; 63s bioglass powder; corrosion.

---

*To cite this article, please refer to its earlier version published in Surface Review and Letters, Vol. 30, No. 11 (2023) 2350083 (11 pages) DOI: 10.1142/S0218625X2350083X

1. Introduction

Orthopaedic bone implants are used to increase the function of the damaged part of hard tissue or replace some of the bone tissue. Such implants contain certain types of materials in particular,[1] namely 316L stainless steel, Co–Cr–Mo alloys, and Ti-based alloys, which are types of materials used as orthopaedic bone implants. Owing to their near-bone modulus of elasticity and high biocompatibility, β-Ti alloys have received a lot of attention recently. Ti45Nb, a β-Ti alloy, is used as an orthopaedic bone implant material with its modulus of elasticity close to that of bone (approximately 65 GPa). It is known that all these biomaterials may cause various problems due to possible ion release from their surfaces after implantation.[2,3] This requires the improvement of the surface properties of metals. Numerous studies have been conducted to improve implant surfaces with bioactive materials using surface modification techniques such as electrophoretic deposition (EPD).[4-12] Using the EPD technique and with inspiration from the composite structure of natural bone, biodegradable polymer/bioactive glass composite coatings for orthopaedic implants are obtained.[13] The EPD technique is widely used, especially in nanostructured bioactive coatings and biomedical products, to obtain a compact deposition of surface film with composite, polymeric, and ceramic coatings. The EPD method is a colloidal technique that electrophoreses charged particles in a suspension on an oppositely charged working electrode. EPD is carried out in two stages. The first stage involves charged particles that are suspended in liquid migrating to the oppositely charged electrode (i.e. electrophoresis). The build-up of particles on the counter electrode to construct the coating is the second stage (deposition). The technique is straightforward and inexpensive. It is also scalable and capable of producing dense, high-purity coatings on objects with complex shapes at ambient temperatures. The method can use aqueous or organic suspensions. Optimizing the colloidal suspension is critical for EPD. The transport properties of green deposits, including particle charge, viscosity, conductivity, structure, and cohesion, influence the outcome.

The main disadvantages of the process include a surface morphology with possible crack formation after firing at high temperatures and the potential to form coatings with limited adhesion to substrates.[13-15] Therefore, the selection of solvents and additives in the process is important. For electrophoresis, the electrostatic stabilization of colloidal particles in a solvent is essential. Water is an excellent dispersion medium for preparing high-zeta-potential stable suspensions. Phosphate esters (PEs), are electrostatic stabilizers which have been used successfully in such applications. By introducing protons to the powder surface, they help particles be charged positively within organic solvents.[16] Moreover, bioactive glasses (bioglass (BG)) are also intensively investigated in bone tissue engineering applications from the middle ear to dental defects to improve the osseointegration affinity of implant surfaces and impart antibacterial properties to these surfaces.[17-21] It has been proven by many researchers that nanostructured BG coatings mimic the mineralized components of natural bone, promote the repair of damaged bone and regeneration, and provide better integration of biocompatible and post-implantation metallic surfaces with the host bone tissue.[22,23] Many previous studies have also revealed that BG prevents the release of certain ionic products from biomaterials. Moreover, ionic dissolution products originating from BG increase new tissue formation, angiogenesis in antibacterial activity, enzyme activity, and the differentiation of mesenchymal stem cells. This way, the BG acts as a bioactive substrate on surfaces, providing the formation of the hydroxyapatite layer and increasing bone growth with the bone-implant contact interface.[24-30] BG also has antibacterial and inflammatory effects. When the particle sizes of some BG materials are reduced to nanoscale levels, they restrict demineralization reactions. Thus, they disintegrate more quickly than human bone remodeling and become brittle, restricting their potential of load-bearing.[14,25,26,30-33]

Chen et al.[34] prepared alginate/BG composite coatings from water-based suspensions using the AC-EPD approach. The researchers concluded that bioactive alginate/BG coatings were indeed potentially suitable for the bone tissue component of dental and orthopaedic implants. In the study performed by Floroian et al.,[33] titanium implants were thinly coated with BG and BG + poly (methyl methacrylate nanocomposite) films for the purpose of improving corrosion resistance and bone tissue bioactivity. Heise et al.[35] examined the effects of various compounds, including DME, CaP, HF, and NaOH, on an EPD CS-BG (chitosan BG) material. They discovered that the ternary-mix suspension produced positively charged composite particles, resulting in a cathodic EPD. A higher pH (6.5–7.5) results in a thicker coating; however, it also causes deep cracks. In their research, Höhlinger et al.[36] increased the bioactivity and hardness of a magnesium alloy (WE43) and stainless steel using two electrophoretic coatings of chitosan/BG/silica (316L). Even though the inclusion of silica reduced the coating material's adhesion to the substrate, it had no effect on coating adhesion in the Mg alloy. They also claimed that introducing silica particles increased the bioactivity and hardness of the coating whilst also keeping the chitosan concentration constant. Miola et al.[37] used EPD to create composite coatings of alginate and 45s BG particles doped with B, Zn, or Sr, and then, they mechanically and electrochemically characterized these coatings. The authors noted that the homogeneity of the coatings was good, they had enhanced adhesion ability, and they were protective toward the substrate. Uchikoshi et al. explored the stability and electrophoretic precipitation characteristics of alumina ethanol suspensions produced with PEs with three distinct alkoxy functional groups (ethyl, butyl, and butoxy ethyl).[16] Even though the zeta potentials of the three suspensions were quite comparable, the suspension prepared with butoxy ethyl acid phosphate had the highest stability. As it is understood based on this literature review, new studies are needed to understand the electrochemical properties of substrates with different structures by using the EPD method in BG coatings. In this study, a mixture of 63s BG powder and PE at concentrations of 0.2, 0.4, 0.8, and 1 g was coated on Ti45Nb material using the EPD method. Then, the structural, mechanical, and electrochemical properties of the untreated and coated samples were determined.

The samples were characterized by X-ray diffraction (XRD) and scanning electron microscope (SEM) analyses.

2. Experimental Details

2.1. *Materials*

This study employed a Ti45Nb alloy material prepared at the dimensions of 10 × 10 × 4 mm. Table 1 presents the chemical composition of the Ti45Nb alloy. The samples were sanded using SiC paper up to 2000-grit, they were washed using an ultrasonic cleaner with ethanol for 5 min and dried in a stream of hot air.

The study also included a commercially available 63s BG powder produced by Matexcel (SUITE 210, 17 Ramsey Road, Shirley, NY 11967/USA). Its powder features were as follows: cat no.: CER-0014, particle size: 0.2–20 μm, and purity: > 99%. The medium used to suspend the samples was reagent-grade ethanol (> 99.5%). PE supplied from Erca Group Kimya/Turkey was used as the electrostatic stabilizer (ERCAFOS 28 (Code: 370500, Chemical Name; Phosphoric acid, 2-ethylhexyl ester, Chemical Formula; $C_8H_{18}O^*(H_3PO_4)_x$/ $C_8H_{18}O^*(H_3O_4P)_x$ (Hill), AT number; 235-741-0,CAS number; 12645-31-7)).

Table 1. Chemical composition of Ti45Nb.

Alloying elements	%
Titanium	45
Niobium	44.94
Fe	<0.03
Cr	<0.01
Mn	<0.01
Mg	<0.01
Si	<0.1
K	<0.01
Na	<0.01
O	<0.095
N	<0.007

2.2. EPD

One percent dilute acetic acid solutions were prepared by magnetic stirring for 24 h at ambient temperature. 63s BG powder suspensions were produced by dispersing 0.2, 0.4, 0.8, and 1 g/L of powder using a sonicator. Then, PE at concentrations of 0.1–0.2% was added to the suspensions. The suspension was stirred magnetically prior to deposition to prevent particle coagulation and/or precipitation. Ti45Nb surfaces were deposited with EPD coatings. The process was carried out using a container and a 0.2-mm-thick stainless steel counter electrode. The distance between the electrodes was fixed at 10 mm. The potential was 20 V, and the reaction time was 20 min. Five different experimental conditions were created at concentrations of 0.2, 0.4, 0.8, and 1 g/L. Then, the samples were rinsed using distilled water and left at room temperature for 24 h to dry before characterization.

2.3. Characterization

The phase identification of the samples and 63s BG powder was accomplished by utilizing the XRD technique with an XRD-GNR-Explorer XRD unit and a Co Kα1 (λ = 1.7903 Å) source at 40 kV and 30 mA, using a scale of 2θ between 10° and 90°. Surface analyses were performed using an SEM FEI-Quanta 250.

2.4. Electrochemical corrosion

The electrochemical measurements in the study included the open circuit potential (OCP), potentiodynamic polarization tests (PPT), and electrochemical impedance spectroscopy (EIS). All tests were carried out at room temperature (~22°C) using freshly prepared samples and were performed in the standard practice for calculation of corrosion rates and related information from electrochemical measurements according to ASTM G102-89 Standard. A Gamry Series G750 potentiostat/galvanostat was used to perform the measurements in the simulated body fluid (SBF) solution (Table 2). Following the electrochemical corrosion tests, the data were analyzed

Table 2. Preparation of SBF for pH 7.25.[38]

No.	Reactant	Quantity
1	NaCl	7.966 g
2	$NaHCO_3$	0.350 g
3	KCl	0.224 g
4	$K_2HPO_4 \cdot 3H_2O$	0.228 g
5	$MgCl_2 \cdot 6H_2O$	0.305 g
6	1 kmol/m^3 HCl (35.4% HCl 87.28 ml, diluted)	40 cm^3
7	$CaCl_2$	0.278 g
8	Na_2SO_4	0.071 g
9	$(CH_2OH)_3CNH_2$	6.057
10	1 kmol/m^3 HCl	for pH adjustment, appropriate amount was used

using the Gamry Echem Analyst software. Electrochemical corrosion was facilitated in a three-electrode cell using the sample as the working electrode, Ag/AgCl as the reference electrode, and graphite as the counter electrode. The working electrode's process surface area was determined to be 1.76 cm^2. Initially, OCP tests were performed for 15,000 s at a scanning rate of 0.5 mV/s. Across the whole PPT, the samples were polarized in a scanning range of 1500 mV (final voltage vs V_{ref}) from −300 mV (initial voltage vs V_{ref}). For EIS, a sinusoidal excitation signal in a frequency range of 100 kHz to 0.01 Hz and an AC amplitude of 10 mV was chosen. The data-logging process of 10 electrochemical signals per frequency decade was used to trace the EIS spectra. Throughout the study, EIS readings for the electrochemical corrosion test were obtained after OCP stabilization and before PPT.

3. Results and Discussion

3.1. *Characterization of surface morphologies*

The XRD plots of all samples are given in Fig. 1. As seen in Fig. 1, the highest reflectance peak relative to all samples was β-Ti. Based on this result, it was seen that β-Ti peaks were obtained at

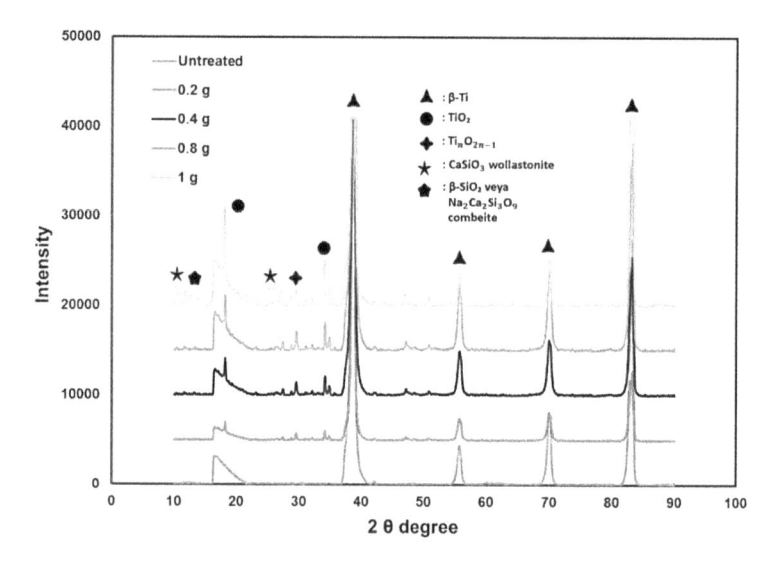

Fig. 1. XRD plots of untreated and coated samples in different suspensions (Color online).

$2\theta° = 38°$, $56°$, $69°$, and $83°$ of the Ti45Nb alloy in the samples. These peaks were compatible with the results reported in other studies in the literature.[39,40] After the coating process, homogeneous TiO_2, Ti_nO_{2n-1}, $CaSiO_3$ (wollastonite), β-SiO_2, or $Na_2Ca_2Si_3O_9$ (combeite) structures were obtained on the surfaces of the samples. The peaks were mostly amorphous because the coatings were relatively thin. As the powder density was different under other coating conditions, the elemental content remained the same under all conditions with the same type of powder content. This can be explained by the presence of overlapping peaks between the concentrations of 0.2 and 1 g.

Figure 2 shows the SEM images of the samples. Throughout all composite coatings, BG particles were seen to be dispersed in the polymeric matrix. As seen in Fig. 2(c) (0.8 g coating), clusters originating from agglomerated BG nanoparticles were formed. Additionally, it was seen in the SEM image that cracks had formed in this coating. At the 0.2 and 0.4 g concentrations, the lowest and highest agglomeration levels of BG nanoparticles were observed in the BG coatings. Again, in the SEM images, it was seen that the

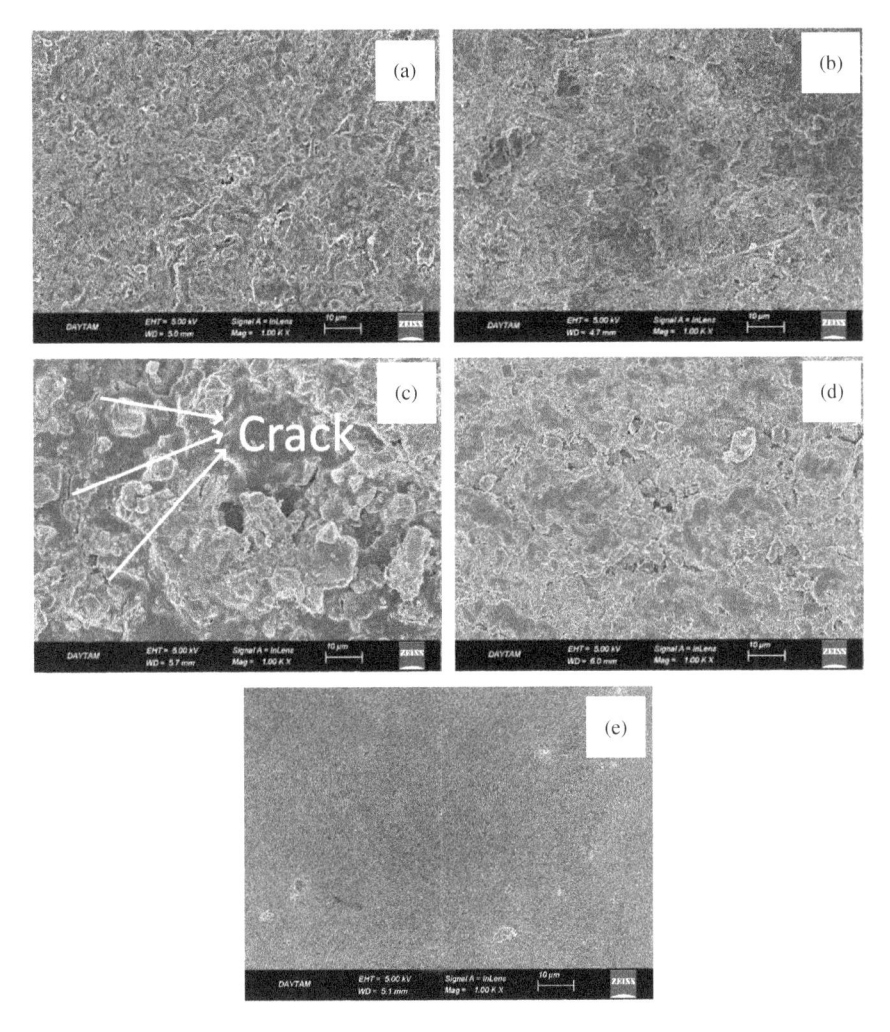

Fig. 2. SEM images of (a) 0.2 g, (b) 0.4 g, (c) 0.8 g, (d) 1 g, and (e) untreated samples (Color online).

BG coatings at the 0.2 and 0.4 g concentrations exhibited more homogeneous and smoother areas than the others.

The EDS results of the samples for the untreated, 0.2 and 1 g powdered suspensions are given in Fig. 3. The results confirmed the existence of Si, Ca, and P elements as the main components of the BG nanoparticles. Therefore, smooth-surfaced deposits were

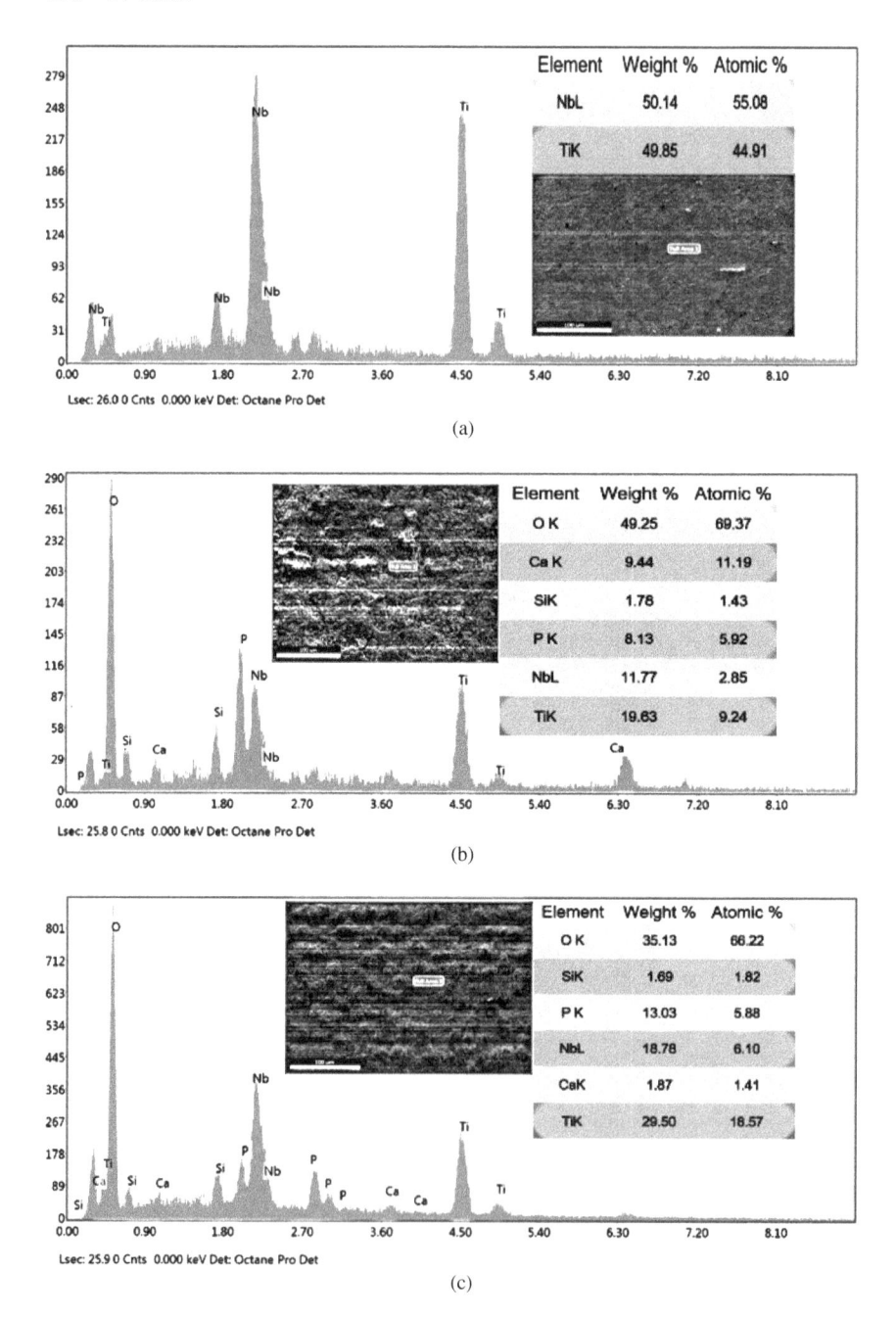

Fig. 3. EDS findings for samples in different conditions: (a) untreated, (b) 0.2 g, and (c) 1 g (Color online).

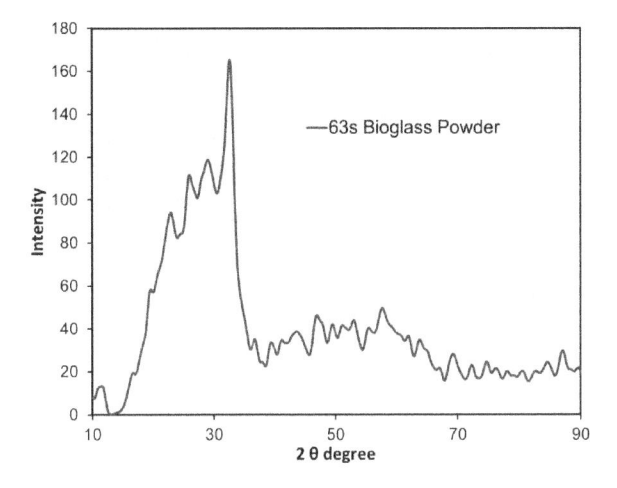

Fig. 4. The XRD plot of 63 s BG (Color online).

obtained from the suspensions prepared with PE. BG was suitably coated. Furthermore, a broad peak confirming the amorphous structure of the 63s BG powder can be seen in Fig. 4.

3.2. Corrosion behavior

Potentiodynamic polarization curves were obtained from all samples in the SBF solution. The cyclic polarization method is a method used to determine the susceptibility of a metal or alloy to pitting.[41] The cyclic polarization curves of the treated and untreated Ti45Nb alloy samples are given in Fig. 5. The cyclic polarization analyses were carried out after the samples reached equilibrium under open circuit conditions, starting from −250 mV below this value and up to 2000 mV by forward and backward scanning from the reached value.

The corrosion rates that were obtained were in the order of untreated > 0.8 g > 1 g > 0.4 g > 0.2 g (Table 3). The obtained corrosion rates were very low. Previous studies have revealed that for a metallic implant to not cause tissue damage and irritation, its corrosion rate must be lower than 2.5×10^{-4} mm per year.[41] All samples in this study met this requirement. In the plots shown in Fig. 5, it can be seen that when the scanning direction is turned

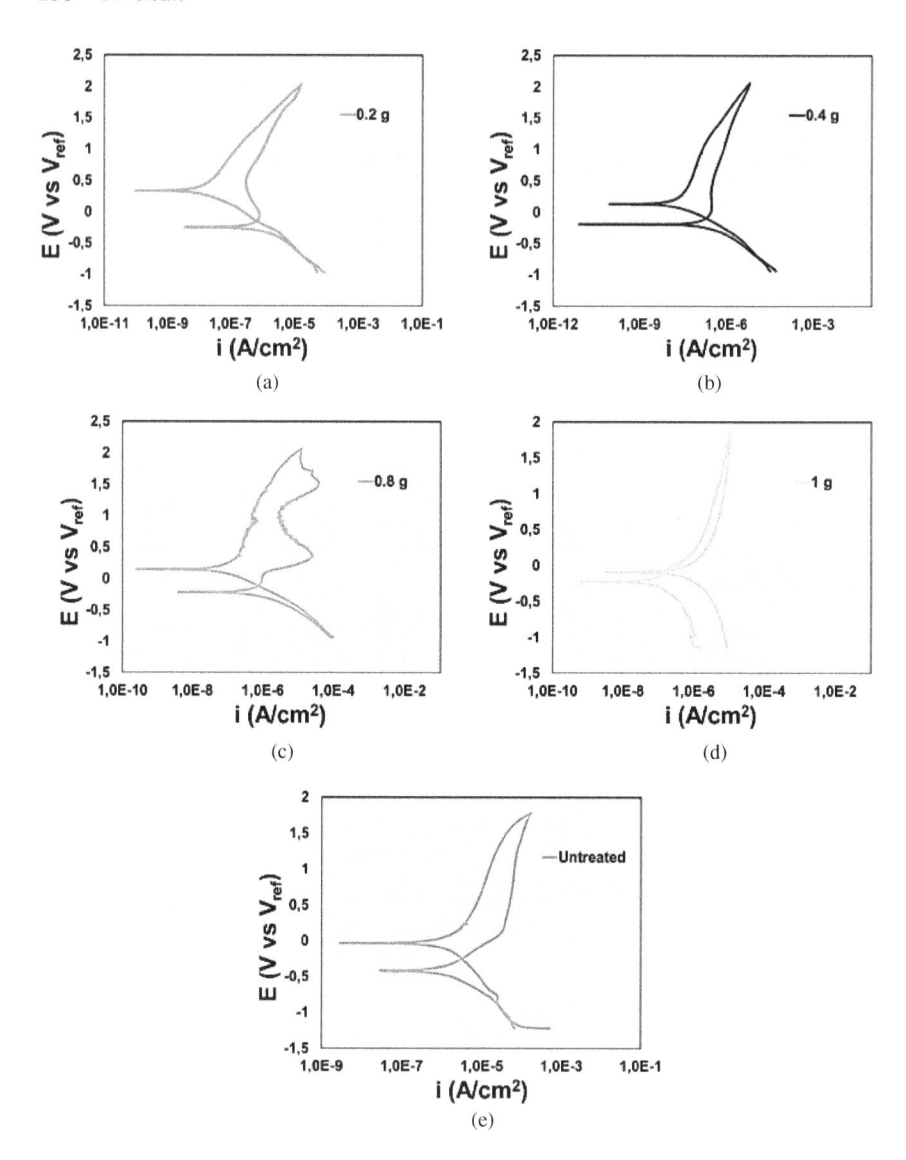

Fig. 5. Potentiodynamic polarization curves of the samples, at coating concentrations of (a) 0.2, (b) 0.4, (c) 0.8, (d) 1 g, and (e) untreated sample (Color online).

towards the left at the peak potential of 2 V, it started to move towards the low current density region. It has been reported in previous studies that this type of cyclic polarization curve is resistant to localized corrosion.[42]

Table 3. Test results for potentiodynamic polarization analyses.

Process	E_{corr} (mV)	I_{corr} (Acm^{-2})	Corrosion rate (mpy)	β_a mV/ decade	β_c mV/ decade	E_{pit} mV
Untreated	−520	396 × 10^{-9}	124.50 × 10^{-3}	375.7	61.9	213
0.2 g	−241	122 × 10^{-9}	38.50 × 10^{-3}	67.4	47.8	618
0.4 g	−186	123 × 10^{-9}	38.57 × 10^{-3}	303.8	151.3	412
0.8 g	−221	144 × 10^{-9}	45.40 × 10^{-3}	410.7	421.5	305
1 g	−218	142 × 10^{-9}	44.82 × 10^{-3}	82.3	57.4	110

The results on the electrochemical parameters are presented in Table 3. Among the E_{corr} values of the 0.2, 0.4, 0.8, 1 g, and untreated samples, the most positive E_{corr} value (approximately −186) was obtained in the 0.4 g sample. Among the corrosion rate and I_{corr} values, the highest values were obtained in the untreated sample (Corrosion Rate: 124.50 × 10^{-3} mmpy, I_{corr}: 396 × 10^{-9} Acm^{-2}), while the lowest values that were very close to each other were obtained in the 0.2 g and 0.4 g samples (Corrosion Rate: 38.50 × 10^{-3} mmpy, I_{corr}: 122/123 × 10^{-9} Acm^{-2}). Additionally, it was seen that the concentrations of 0.8 and 1 g had very close E_{corr}, I_{corr}, and corrosion rate values to each other, and these values were lower compared to those in the untreated sample. As a characteristic of corrosion at a constant rate,[43] while the current density in the untreated sample was approximately 10^{-6} A/cm^2, the treated samples had lower values, at 10^{-7} A/cm^2. The pitting potential (E_{pit}) was 0.618, 0.412, 0.305, 0.11, and 0.2 V for the 0.2, 0.4, 0.8, 1 g, and untreated samples, respectively. Cyclic potentiodynamic polarization tests are conducted to calculate pitting corrosion susceptibility. The higher the area of hysteresis in cyclic potentiodynamic polarization curves (if there is a significant increase in the current density values in the hysteresis cycle), the larger the susceptibility of the sample to pitting corrosion.[44] The 1 g coating provided an electrochemical potential of −0.1 V, which showed a less aggressive behavior than its E_{corr} value. In addition to this, it also showed the lowest passive current density and a small hysteresis cycle area. This small area was caused by its re-passivation behavior. This indicated higher pitting corrosion resistance.[45] On the other hand, although the untreated sample displayed a zone with a constant current density, it showed a larger area for its hysteresis cycle compared to other samples. This showed that the untreated sample had lower resistance to localized corrosion. The adhesion between BG nanoparticles in the BG coating strongly prevents the permeation of ions and water molecules. This way, the BG coating creates a strong barrier against the corrosion of the substrate material.[46] This situation also shows that the PE displays a good dispersion and a favorable surface. The behaviors of anodic and cathodic Tafel slopes (β_a and β_c) are dependent on the reaction mechanism taking place on the metal surface. As seen in Table 3, among the coated and untreated samples, the highest β_a and β_c values (0.410 and 0.421 V/decade) belonged to the 0.8 g coated sample. The lowest β_a and β_c values (0.0674 and 0.0478 V/decade), on the other hand, belonged to the 0.2 g coated sample. In this case, there was no clear trend in the Tafel curves. Anodic passivation can lead to passivation and dissolution that lead to a surface that can change the Tafel curve values on the electrode. Anodic polarization can sometimes lead to surface roughening, in addition to concentration effects due to passivation and dissolution. These effects can lead to deviations from the Tafel behavior.[43] These deviations were probably caused by the cracks formed on the 0.8 g coating. This is because cracks accelerate dissolution and produce concentration effects.

Figure 6 shows the EIS test results in terms of SBF ((a) impedance — frequency diagram, (b) phase angle–frequency diagram, (c) experimental Nyquist diagram, (d) equivalent circuit for modeling impedance parameters for the uncoated sample and (e) coated samples). In the Nyquist diagram (Fig. 6(c), it can be seen that the sample had a tendency to create capacitive semicircles centered on the real axis. Thus, it is clear that the corrosion process was controlled by anodic dissolution. The coated samples showed a greater impedance compared to the untreated sample. This, in turn, provided the coated samples with a superior diffusion barrier. The semicircle with the greatest diameter was found in the 0.2 g/l BG-coated sample. This result was in agreement with those in other studies in the relevant literature.[46]

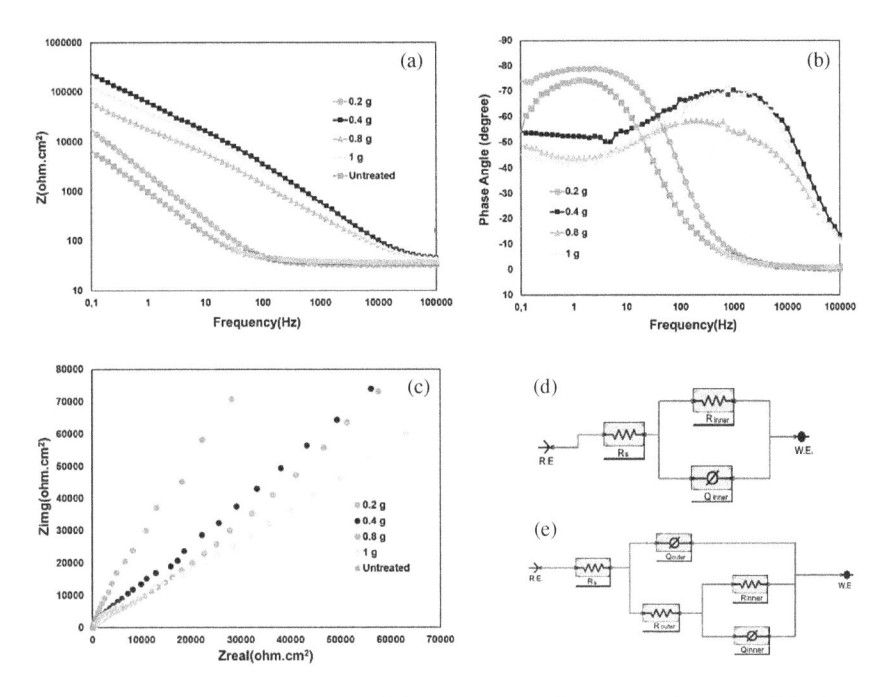

Fig. 6. EIS test results in SBF, (a) impedance—frequency diagram, (b) phase angle–frequency diagram, (c) experimental Nyquist diagram, (d) equivalent circuit for modeling impedance parameters for uncoated sample, and (e) coated samples (Color online).

The concentrations of four different metals (Si, P, Ca, Ti) in the test solution are given in Fig. 7. The dissolution amounts of these four elements were compared to the untreated sample and other coating conditions. Elements from the base material's chemical composition were dissolved in the test solution, and their concentrations decreased as the coating layer thickness increased. The metal concentration values obtained from the PPT were higher than those in the OCP test due to the effect of the corrosion mechanism. The Si, P, Ca, and Ti concentrations of the 0.2 g coated sample under the conditions of PPT and OCP were quite low compared to other ion emissions. The release was drastically reduced in the 0.2 g coated sample, and the protective film layer created a major barrier effect.

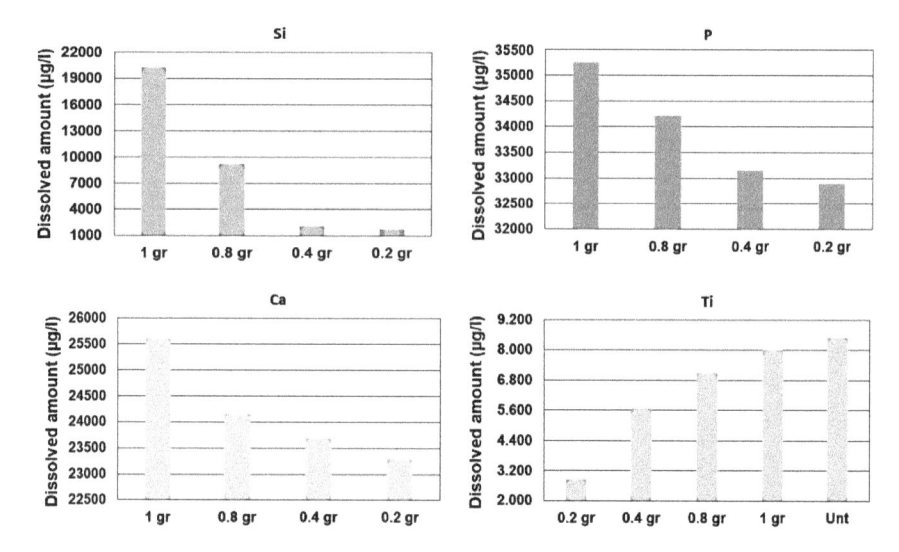

Fig. 7. ICP measurements of the untreated sample and samples coated at concentrations of 0.2, 0.4, 0.8, and 1 g (Color online).

4. Conclusion

This study investigated the structural, mechanical, and electrochemical characteristics of 63s BG-coated Ti45Nb samples at concentrations of 0 (untreated), 0.2, 0.4, 0.8, and 1 g in SBF solutions. The findings of the study can be summarized as follows:

- At the concentrations of 0.2, 0.4, 0.8, and 1 g, 63s BG Ti45Nb was successfully coated on the surface of the samples using PE, and cracks were formed only in the 0.8 g coated sample. It was determined that the BG coatings, especially at the 0.2 and 0.4 g concentrations, exhibited more homogeneous and smoother areas compared to the others.
- As the BG concentration increased, the peaks became less pronounced, confirming the amorphous structure of the BG particles. This verified the presence of the elements Si, Ca, and P as the primary components of the BG nanoparticles. It showed that the suspensions prepared with PE resulted in smooth-surfaced deposits, and the BG was properly coated onto the material.

- PE provided good dispersion and surface for coating.
- Among the coated samples, the 0.2 g concentration of the BG coating improved the corrosion resistance of the Ti45Nb material more effectively than the others did.

Acknowledgments

I would like to thank Atatürk University East Anatolia High Technology Application and Research Centre (DAYTAM) and Erzurum Technical University High Technology Application and Research Centre (YUTAM) for their assistance during the characterization of the samples.

References

1. R. Rojaee, M. Fathi, K. Raeissi and M. Taherian, *Ceram. Int.* **40** (2014) 7879.
2. F. Witte, N. Hort, C. Vogt, S. Cohen, K. U. Kainer, R. Willumeit and F. Feyerabend, *Curr. Opin. Solid State Mater. Sci.* **12** (2008) 63.
3. B. J. McEntire, B. S. Bal, M. N. Rahaman, J. Chevalier and G. Pezzotti, *J. Eur. Ceram. Soc.* **35** (2015) 4327.
4. A. Abdal-hay, N. A. M. Barakat and J. K. Lim, *Colloids Surf. A Physicochem. Eng. Asp.* **420** (2013) 37.
5. R. Guan, I. Johnson, T. Cui, T. Zhao, Z. Zhao, X. Li and H. Liu, *J. Biomed. Mater. Res. A* **100** (2012) 999.
6. X. Lin, L. Tan, Q. Zhang, K. Yang, Z. Hu, J. Qiu and Y. Cai, *Acta Biomater.* **9** (2013) 8631.
7. N. J. Shah, J. Hong, M. N. Hyder and P. T. Hammond, *Adv. Mater.* **24** (2012) 1445.
8. H. Fukuda and Y. Matsumoto, *Electrochim. Acta* **50** (2005) 5329.
9. S. B. Goodman, Z. Yao, M. Keeney and F. Yang, *Biomaterials* **34** (2013) 3174.
10. Q. Zhang, X. Lin, Z. Qi, L. Tan, K. Yang, Z. Hu and Y. Wang, *J. Mater. Sci. Technol.* **29** (2013) 539.
11. S. Kunjukunju, A. Roy, M. Ramanathan, B. Lee, J. E. Candiello and P. N. Kumta, *Acta Biomater.* **9** (2013) 8690.
12. Y. W. Song, D. Y. Shan and E. H. Han, *Mater. Lett.* **62** (2008) 3276.

13. S. Bakhshandeh and S. A. Yavari, *J. Mater. Chem. B* **6** (2018) 1128.
14. A. R. Boccaccini, C. Peters, J. A. Roether, D. Eifler, S. K. Misra and E. J. Minay, *J. Mater. Sci.* **41** (2006) 8152.
15. V. S. Saji, *J. Ind. Eng. Chem.* **103** (2021) 358.
16. T. Uchikoshi and Y. Sakka, *J. Am. Ceram. Soc.* **91** (2008) 1923.
17. J. Ballarre, I. Manjubala, W. H. Schreiner, J. C. Orellano, P. Fratzl and S. Ceré, *Acta Biomater.* **6** (2010) 1601.
18. A. Simchi, E. Tamjid, F. Pishbin and A. R. Boccaccini, *Nanomedicine Nanotechnol. Biol. Med.* **7** (2011) 22.
19. J. Ballarre, R. Seltzer, E. Mendoza, J. C. Orellano, Y.-W. Mai, C. García and S. M. Ceré, *Mater. Sci. Eng. C* **31** (2011) 545.
20. U. Brohede, S. Zhao, F. Lindberg, A. Mihranyan, J. Forsgren, M. Strømme and H. Engqvist, *Appl. Surf. Sci.* **255** (2009) 7723.
21. J. M. Gomez-Vega, E. Saiz, A. P. Tomsia, G. W. Marshall and S. J. Marshall, *Biomaterials* **21** (2000) 105.
22. S. R. Federman, V. C. Costa, D. C. L. Vasconcelos and W. L. Vasconcelos, *Mater. Res.* **10** (2007) 177.
23. N. Moritz, S. Rossi, E. Vedel, T. Tirri, H. Ylänen, H. Aro and T. Närhi, *J. Mater. Sci. Mater. Med.* **15** (2004) 795.
24. M. S. Bahniuk, H. Pirayesh, H. D. Singh, J. A. Nychka and L. D. Unsworth, *Biointerphases* **7** (2012) 41.
25. A. Balamurugan, G. Balossier, J. Michel and J. M. F. Ferreira, *Electrochim. Acta* **54** (2009) 1192.
26. A. A. Gorustovich, J. A. Roether and A. R. Boccaccini, *Tissue Eng. B Rev.* **16** (2010) 199.
27. M. C. Schausten, D. Meng, R. Telle and A. R. Boccaccini, *Ceram. Int.* **36** (2010) 307.
28. I. D. Xynos, A. J. Edgar, L. D. K. Buttery, L. L. Hench and J. M. Polak, *Biochem. Biophys. Res. Commun.* **276** (2000) 461.
29. T. Moskalewicz, S. Seuss and A. R. Boccaccini, *Appl. Surf. Sci.* **273** (2013) 62.
30. N. Drnovšek, S. Novak, U. Dragin, M. Čeh, M. Gorenšek and M. Gradišar, *Int. Orthop.* **36** (2012) 1739.
31. M. Mehdipour and A. Afshar, *Ceram. Int.* **38** (2012) 471.
32. K. Grandfield and I. Zhitomirsky, *Mater. Charact.* **59** (2008) 61.
33. L. Floroian, M. Florescu, F. Sima, G. Popescu-Pelin, C. Ristoscu and I. N. Mihailescu, *Mater. Sci. Eng. C* **32** (2012) 1152.

34. Q. Chen, L. Cordero-Arias, J. A. Roether, S. Cabanas-Polo, S. Virtanen and A. R. Boccaccini, *Surf. Coatings Technol.* **233** (2013) 49.
35. S. Heise, M. Höhlinger, Y. T. Hernández, J. J. P. Palacio, J. A. Rodriquez Ortiz, V. Wagener, S. Virtanen and A. R. Boccaccini, *Electrochim. Acta* **232** (2017) 456.
36. M. Höhlinger, S. Heise, V. Wagener, A. R. Boccaccini and S. Virtanen, *Appl. Surf. Sci.* **405** (2017) 441.
37. M. Miola, L. Cordero-Arias, G. Ferlenda, A. Cochis, S. Virtanen, L. Rimondini, E. Verné and A. R. Boccaccini, *Surf. Coatings Technol.* **418** (2021).
38. T. Kokubo and H. Takadama, *Biomaterials* **27** (2006) 2907.
39. G. V. Martins, C. R. M. Silva, C. A. Nunes, V. J. Trava-Airoldi, L. A. Borges and J. P. B. Machado, *Mater. Sci. Forum* **660–661** (2010) 405.
40. M. Taşdemir, F. Şenaslan and A. Çelik, *Gümüşhane Univ. J. Sci. Technol.* **10** (2020) 395.
41. R. Bhola, S. M. Bhola, B. Mishra and D. L. Olson, *Res. Lett. Phys. Chem.* **4** (2009) 574359.
42. D. C. Silverman, *In Proceedings of the Corrosion 98*, San Diego, 1998.
43. R. Vences-Hernández, F. Reyes-Calderón, J. C. Villalobos, H. J. Vergara-Hernández and J. A. Salazar-Torres, *J. Mater. Eng. Perform.* **29** (2020) 6520.
44. J. Caballero Sarmiento, E. Correa Muñoz and H. Estupiñan Duran, *Ingeniare* **25** (2017) 95.
45. S. Esmailzadeh, M. Aliofkhazraei and H. Sarlak, *Prot. Met. Phys. Chem. Surf.* **54** (2018) 976, S. Labbaf, *Prog. Org. Coatings* **147** (2020) 105803.
46. M. Alaei, M. Atapour and S. Labbaf, *Prog. Org. Coatings* **147** (2020) 105803.

Comparison of corrosion, tribocorrosion and antibacterial properties of silver coatings on Ti15Mo by magnetron sputtering*

E. Meletlioglu[†] and R. Sadeler

*Department of Mechanical Engineering,
Faculty of Engineering,
Ataturk University, Erzurum, Turkey*
[†]*emrahmeletli@atauni.edu.tr*

The aim of this *in-vitro* study was to evaluate the influence of Ag^+-ion-coated conditions on the corrosion, tribocorrosion and antibacterial properties of Ti15Mo alloy. The mean wear volume losses of all test specimens after tribocorrosion test procedures were determined using a noncontact 3D profilometer. The specimens' hardness, roughness values and microstructures were measured using the microhardness tester, surface profilometer, scanning electron microscopy (SEM) and X-ray diffraction (XRD) analysis. The mean wear volume loss of 30-min Ag^+-ion-coated Ti15Mo alloy was lower than the other specimens. In this study, correlations between the hardness, surface roughness and wear volume loss were found to be significant. The PVD coating process enhanced the anti-bacterial activity of Ti15Mo alloy owing mainly to the formation of silver film on the substrates.

Keywords: Ti15Mo; DC sputtering; corrosion; tribocorrosion; antibacterial property.

---

*To cite this article, please refer to its earlier version published in Surface Review and Letters, Vol. 30, No. 5 (2023) 2350027 (13 pages) DOI: 10.1142/S0218625X23500270
[†]Corresponding author.

1. Introduction

Implant materials with unique mechanical properties, corrosion strength and biocompatibility are needed in medical industry.[1] In the area of medical industry, titanium and its alloys are used for implant applications because of their low elasticity modulus, excelient strength-to-weight ratio, superior resistance to corrosion, easy fabrication, etc.[2,3] Due to its excellent features, Ti6AI4V alloy is widely utilized in biomedical industries such as in dental applications.[4] However, the major disadvantage of the Ti6AI4V alloy is its elastic modulus (~20 GPa), which is much higher than that of human bone (10–30 GPa). Also, the stress-shielding effect of Ti6AI4V titanium alloy due to the mismatch of the elastic modulus between live bone causes bone resorption.[5] Moreover, some studies noted that aiuminum and vanadium ions reieased from the Ti6Al4V alloy cause damage to the nervous system.[6,7] Hence, new improved titanium compounds with better corrosion and biocompatibility features compared to Ti6AI4V alloy are being examined. In view of this, Ti15Mo alloy has attracted interest of research community owing to its mechanical and biocompatibility properties.[8,9] Some studies reported that the Ti15Mo alloy provided good results in tribocorrosion and electromechanical tests.[10] However, it is still a problem that titanium alloy exhibits low wear strength and high friction coefficient when used as a metallic antagonist material in dental applications. When the implant is surgically placed in the mount, in the beginning the entire surface area may be in close contact with oral fluids.[11] The regional pH in the mouth can attain acidic values owing to the generation of fermentable carbohydrates (i.e. lactic, acetic or propionic acid),[12] and this influences the corrosion strength of Ti and its alloys.[13] It is inevitable for the implants placed in the mouth to be exposed to both two-body wear and corrosion mechanisms owing to the chewing movement. This phenomenon is called tribocorrosion and therefore, it can cause more damage than the individual effects of these mechanisms. In addition to this bad situation, bacterial infections related to implants cause inflammation on the surface of the implant or prosthesis, and finally implant loss occurs. Bacterial colonization may occur on the surface of implants.[14] To overcome these disadvantages, many investigations are continuing in an effort to improve new materials to be utilized for re-design of the available biomaterials (for implants and prostheses).[15,16] For those reasons, titanium implants need to be coated with antibacterial materials to prevent bacterial adhesion on the surface and also the tribocorrosion failure.

The surface of titanium implants is coated with various antibacterial materials (like Ag, Cu, Zn, etc.) in order to prevent bacterial adhesion and further reduce the implant infections.[17-19] Among these materials, silver stands out due to its extensive variety of antibacterial properties with fewer chances of developing the strength of the bacterial cell. The intriguing property of silver has been known since late antiquity. Due to the low toxicity of silver in the human body, it is assumed that it inhibits bacterial attachment during and after surgery.[20] Moreover, silver particles pass through the cell wall and cell membrane of bacteria and microorganisms and inhibit the DNA of these bacteria and microorganisms.[21,22] As a result, in the case of biomaterials, this type of silver surface functionalization is able to prevent bacterial colonization, especially through the contact-killing activity. Recently, silver deposition on a nanostructured film exhibited excellent antibacterial capability during the wound healing period.[23] The direct application of antibacterial coatings via surface modification is a simple and cost-effective approach. Some surface treatment and coating methods are frequently utilized to enhance the corrosion resistance, mechanical and wear features of biomaterials.[24] Particularly, DC magnetron sputtering has the advantages such as feasibility of producing uniform surface for titanium alloys, shorter deposition time intervals and the ability to be carried out at relatively lower temperatures compared with other surface modification methods. Optimization of the deposition process is critical to achieve the desired properties (biocompatibility, high corrosion strength) in silver films through sputtering. The effects of some deposition parameters such as target angle, argon pressure, distance and temperature on the various properties of titanium alloys have been reported in the literature.[25,26] However, various physical properties of these sputter-deposited silver films including microstructure, surface roughness and hardness, which are dependent on deposition parameters and sputtering techniques, should be investigated for suitable control over film quality. In protective applications such as corrosion resistance or biomedical coatings, the thickness of the films has a greater effect in terms of substrate bonding, structural and surface properties. Therefore, optimization of sputter-deposited silver film thickness/roughness for superior properties of coating and increasing the efficiency of these methods by determining the factors affecting corrosion, tribocorrosion and antibacterial properties are important problems in biomedical sciences.

The basic aim of this *in-vitro* research was to examine the effects of deposition time on the corrosion, tribocorrosion and antibacterial features of silver films on Ti15Mo using dual-axis computer-controlled chewing simulator device.

2. Materials and Methods

2.1. *Specimen preparation*

Ti15Mo alloy specimens with 13-mm thickness were cut from a cylindrical bar with a diameter of 20 mm by Electron Discharge Machine (EDM) and their compositions were obtained from their commercial grades and described elsewhere.[27] The SiC grit papers of different sizes starting from 200 to 1600 were utilized for mechanical polishing of the specimens. Afterward, acid etching with an acid solution by mixing 60% H_2SO_4, 10% HCl and distilled water at a ratio of 1:1:2, respectively, was performed in the fume hood in order to increase adhesion resistance between the coating and substrate materials. The etched specimens were cleaned with an ultrasonic bath using deionized water for 10 min. The specimens were dried in air and then placed into the PVD coating system (DC Magnetron Sputter DAYTAM; Center of East Anatolian High-Technology Research and Application). A silver target of 3.18 × 50 mm diameter with 99.99% purity was utilized for deposition.

High-purity (99.999%) argon was used as the sputtering gas. Process parameters were angle (target–substrate ~35°), distance (target–substrate = 11.8 cm), base pressure (6.8 × 10^{-3} mTorr), bias substrate (0 V), substrate temperature (50°C), substrate rotation (3 rpm), working gases (argon) and power (DC; 100 W). The actual deposition times for the films were 30, 45 and 60 min, respectively. Subsequently, the films were cooled to room temperature in vacuum. Three specimens were prepared under each condition.

2.2. *Microstructural analysis*

After the surface treatments, the specimen surface roughness was measured by utilizing a KLA Tencor Stylus Profile instrument. The phase structures of both untreated and silver-coated specimens were investigated by X-ray diffraction (XRD; GNR explorer) using Cu-Kα radiation (λ = 1.5418 Å) and JCPDS PDF-2 database. Scanning electron microscopy (SEM; FEI Quanta 250) examinations were conducted to observe the surface and cross-sectional morphology images of the coated and tested specimens. Hardness measurements were carried out by a microhardness tester (Wolpert Wilson Instruments) under an indentation load of 25 g and a loading time of 10 s. Seven different areas were chosen randomly and the result was an average value with standard deviation.

2.3. *Electrochemical tests*

The chemical composition of simulated artificial saliva (Butt's solution provided from the Department of Chemistry Education in Ataturk University) used in corrosion tests is shown in Table 1. In order to prepare artificial saliva, reagent-grade NaCl, KC1, $CaCl_2 \cdot 2H_2O$, $CO(NH_2)_2$, $NaH_2PO_4 \cdot 2H_2O$ and $Na_2S \cdot 9H_2O$ were dissolved in distilled water. Temperature and pH values were also controlled during the dissolution of the reagents.

All corrosion tests were performed utilizing the series potentiostat/galvanostat/ZRA system. A standard three-compartment cell consists of Ag/AgCl, graphite and the prepared specimens with a

Table 1. Chemical composition of the Butt's solution.[26]

Reagent	Composition
NaCl	0.4 g/L
KC1	0.4 g/L
$CaCl_2 \cdot 2H_2O$	0.9 g/L
$CO(NH_2)_2$	1 g/L
$NaH_2PO_4 \cdot 2H_2O$	0.7 g/mL
$Na_2S \cdot 9H_2O$	0.005 g/L

subjected surface area of 0.29 cm^2 used as the reference electrode, counter-electrode and working electrode in the solution, respectively. A new solution was utilized for each test. The specimens were left for 60 min until the open circuit potential (OCP) was reached. The beginning and ending potential values were quantified as open circuit potential (E_{oc}) in the Tafel experiments. The potentiodynamic polarization measurements were carried out by scanning from −2000 mV versus OCP +2000 mV versus Ref at a scan rate of 0.5 mV · s^{-1}.[28]

2.4. Tribocorrosion tests

Computer-controlled chewing simulator was designed and manufactured to evaluate the tribocorrosion behavior of Ti15Mo as illustrated in Fig. 1. Open circuit potential scannings were achieved by Gamry Echem Analyst G750 potensiostat/galvanostat electrochemical workstation with a three-electrode galvanic cell by specimen (working electrode), Ag/AgCl (reference electrode) and graphite (counter-electrode). Tribocorrosion tests were performed with 50-N mechanical force, 2-Hz wear frequency, 37°C ± 1°C temperature under atmospheric conditions, 6-mm-diameter steatite ball AI_2O_3, 0.7-mm lateral movement and for a time duration of 6000 s in pH artificial saliva (pH = 7), respectively. A modified Butt's artificial saliva (Table 1) was formulated in order to carry out the tribocorrosion tests.[29] Two-body wear tests were carried out at open circuit

Corrosion, tribocorrosion and antibacterial properties of silver coatings 313

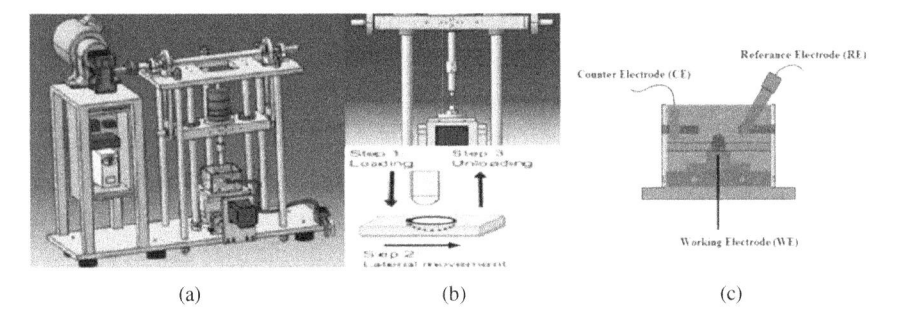

(a) (b) (c)

Fig. 1. Schematics of the two-body wear simulation test device (Color online).

potential, which is the difference in electrical potential spontaneously established between specimens and solution and at an applied passive potential of 0.2 V. The experimental series consisted of the following:

- allowing the system to stabilize at OCP during 2000 s,
- start the wear for 2000 s while the OCP is measured,
- wear test is stopped while the OCP is maintained for 1000 s.

At the end of the tribocorrosion tests, 2D and 3D profilometer images were taken and analyzed for volume loss on the wear surfaces (using Bruker-Contour GT 3D noncontact profilometer). The data obtained were analyzed using statistical software (SPSS Statics 20.0 for Windows 64-bit operating system; SPSS, Inc., Chicago, IL, USA). The means and standard deviations of Ra, HV and volume loss were calculated and analyzed using the one-way analysis of variance (ANOVA). The Games–Howell test was used for the post-hoc analysis. The significance level was set to $\alpha = 0.00001$.

2.5. Antibacterial activity

The plate counting process was utilized to quantify the antibacterial features of Ti15Mo alloy.[28] Two bacterial species, *Escherichia coli* (ATCC25922) and *Staphylococcus aureus* (ATCC25923), were selected for testing. Frozen powders of the bacteria were dissolved in a culture medium with a pH 1/4 7.2, cultivated at 37°C for 24 h. Afterward a bacterial suspension was obtained. Then, 100-μL volume of an inoculated bacterial solution with an approximate concentration of ~ 4×10^8 CFU/mL (colony forming units per milliliter) was used on each specimen. After a 24-h incubation time at 37°C, the number of colonies created by the growth of viable bacteria on the specimen was counted. The CFU/mL of culture solution was calculated with the formula: CFU/mL = (Number of colonies/ Capacity of culture in the plate) × Dilution factor. Dilution factor was defined as the ratio of final volume of bacterial suspension after dilution to the volume of bacterial suspension before dilution.

3. Results and Discussion

3.1. *Microstructures*

The variations in the XRD patterns of untreated and silver-coated Ti15Mo alloy treated for different times, are illustrated in Fig. 2. According to the XRD results, the untreated Ti15Mo alloy is illustrated in Fig. 2(a), where the diffractograms have peaks characteristic of a single crystalline structure (110, 200), which is typical of the β-phase (96-900-8555) of the alloys.[31,32] Figure 2(b) displays the XRD patterns of the silver films deposited at various deposition times of 30, 45 and 60 min. The XRD outcomes show that the silver phases (96-500-0219) were generated with a preferred orientation in the direction [111], with small contributions in the directions [200], [202] and parallel plane [311]. The [111] silver peak intensity was larger than those of the other peaks because the [111] direction in silver film has the lowest surface energy. Since silver has a cubic structure, it might be predicted that most of the nanoparticle-like structures are created via crystallites with cubic faces parallel to the substrate (single crystalline).[33] When the film is more than 1.6 μm thick (60-min coated), the structure of sputter-deposited silver films exhibited a preferred orientation in the [111] direction. The cubic phases (face-centered cubic) were parallel to the substrate surface. This preferred orientation in the [111] direction was

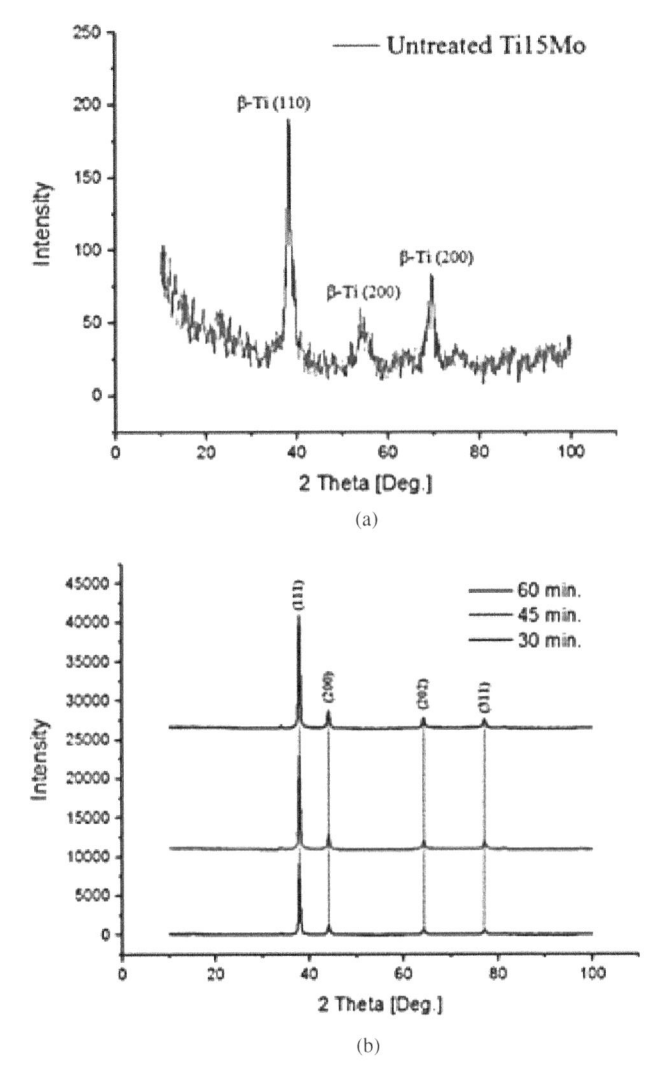

Fig. 2. XRD results: (a) untreated Ti15Mo and (b) silver-coated Ti15Mo (Color online).

also noticed for silver films deposited by sputtering.[34] XRD indicates that this is indeed the case. When the film thickness increases, there is a coalescence of these crystallites, but they do not interact in the same crystal planes. According to the Scherrer equation, this crashing coalescence contributes to large grains with crystallites growing in all directions, especially [111], as expected for a crystalline solid.[35] The high intensity and sharp peaks in XRD patterns confirm the highly oriented and polycrystalline nature of the silver films prepared in this study.

Figure 3 displays the cross-sectional SEM micrographs of all films. The increments in the coating time gave rise to merged nanocolumnar growth. This may be attributed to the surface-diffusion influence where a longer coating time (45–60 min) finally caused an increase in the temperature (20–50°C) that increased the adatom

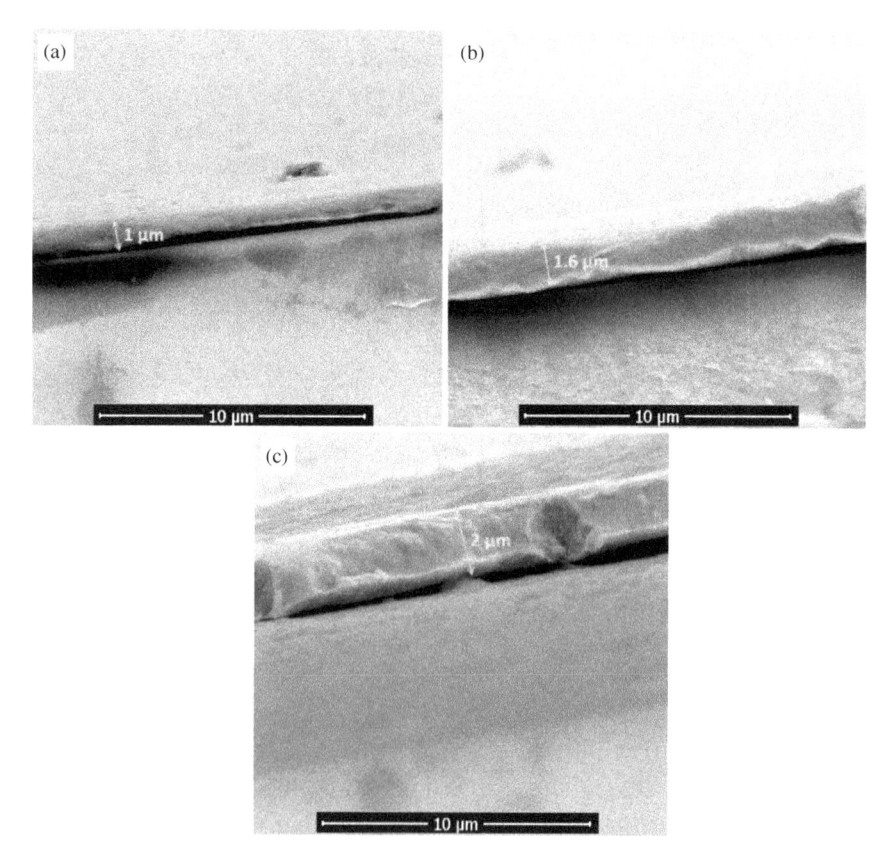

Fig. 3. Cross-sectional SEM micrographs observed for silver-coated alloy specimens with coating times of: (a) 30 min, (b) 45 min and (c) 60 min (Color online).

mobility. Also, the nanocolumnar growth of the silver films tends to bend with increasing film thickness.[36] According to the results, thin film thickness values were increased with the increase in coating times.

Surface hardness values of untreated specimens average HV = 230. A decrease of coating's hardness was observed with Ag^+-ion additions, owing to the incorporation of the soft silver on the structure. Surface roughness of untreated specimens was Ra = 0.152–0.241 μm. The average surface roughness values of the alloys decreased after the mechanical polishing process (Ra = 0.085–0.096 μm). The increase in surface roughness affects osseointegration positively and causes better adhesion of the coating to the implant. Surface roughness values increased after the acid-etching process (Ra = 2110–2126 μm). The acid-etching process leads to an increase in surface roughness, positively affecting osseointegration, allowing the coating to adhere better to the implant.[37] After the coating process, all specimens had low surface roughness. Moreover, the lowest surface roughness was observed for the 30-min coating time among all coatings because of finer grain sizes.

3.2. Corrosion resistance

Polarization curves recorded for untreated and silver-coated Ti15Mo alloy in artificial saliva at 37°C are presented in Fig. 4. Polarization analyses were carried out with scanning from a lower value compared to the OCP values. Corrosion parameters such as corrosion potential (E_{corr}), corrosion current density (I_{corr}) and corrosion rate (mpy) are shown in Table 2. As the polarization curves were evaluated, untreated Ti15Mo alloy showed a corrosion potential (E_{corr}) of –385 mV and all silver-coated specimens displayed a significant improvement in corrosion properties. For coated Ti15Mo alloy, the E_{corr} values for specimens with silver film deposited for 30, 45 and 60 min were measured as –197, –15 and 38 mV, respectively. Another important parameter for understanding the corrosion behavior is corrosion current density (I_{corr}). According to the results, the lowest I_{corr} values were obtained for silver film deposited for 30,

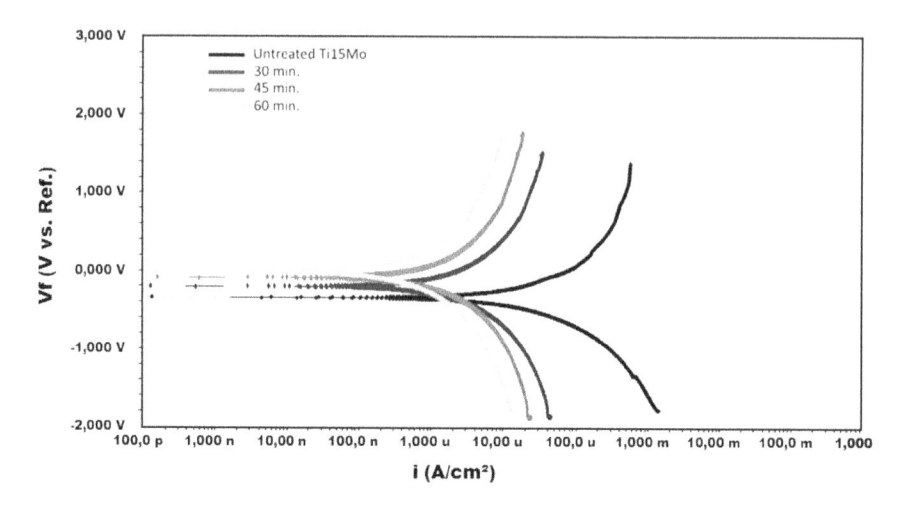

Fig. 4. Potentiodynamic polarization curves of untreated and silver-coated Ti15Mo alloy specimens for different processing times (Color online).

Table 2. Corrosion test results of Ti15Mo alloy specimens.

Process parameters	E_{corr}	I_{corr}	Corrosion rate (mpy)
Untreated	−385 mV	1.42×10^{-6} nA	0.280
30 min	−197 mV	749×10^{-9} nA	0.071
45 min	−15 mV	509×10^{-9} nA	0.054
60 min	38 mV	214×10^{-9} nA	0.039

45 and 60 min (749×10^{-9}, 509×10^{-9} and 214×10^{-9} A/cm^2), which were approximately lower than that for the untreated Ti15Mo alloy (1.42×10^{-6} A/cm^2).

The E_{corr} of the coated specimen shifted positively after the silver-coating process. This caused a dramatic reduction in thermodynamic tendency for corrosion. Second, after the coating process, the anodic section of the curves moved to lower I_{corr}. This implies that the anodic reactions were limited by the coating process and higher chemical stability and protection for the titanium alloy were provided.[38,39] SEM images of the Ti15Mo alloy after corrosion

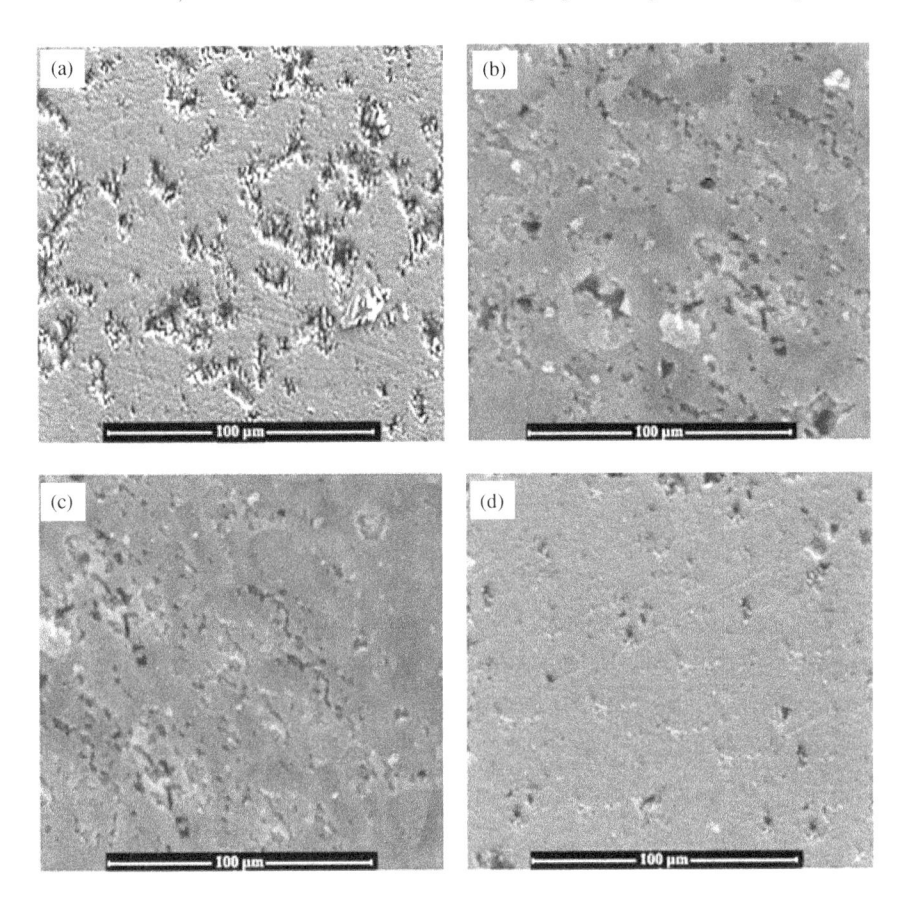

Fig. 5. Surface SEM images observed for alloy specimens after corrosion: (a) untreated specimen and specimens with depositon times of (b) 30 min, (c) 45 min and (d) 60 min.

tests are given in Fig. 5. In the untreated Ti15Mo alloy (Fig. 5(a)), the main corrosion mechanism was pitting, which is a type of localized corrosion. This situation can be explained by the metal ion release from the alloy to the environment. The presence of deep cracks in the coating layer of silver ion-coated Ti15Mo (Figs. 5(b) and 5(c)) is remarkable. The presence of such cracks has also been detected in other studies.[40] In addition, the presence of an oxide layer (white in color) on the silver-plated (30 min) surfaces is also noteworthy. In this case, saliva penetrating the cracks can come into direct contact with the substrate, accelerating the corrosion process. Silver ion-coated (60 min) Ti15Mo alloy (Fig. 5(d)) appears to be nearly undamaged on both surfaces. This situation is thought to be associated with protective and disincentive passive film (Ag oxide) on the alloy microstructure during the corrosion tests.[41]

3.3. Tribocorrosion behavior

The evolution of OCP with time of the Ti15Mo alloy is monitored during the tribocorrosion and is shown in Fig. 6. Before the starting of wear, the OCP value shows the presence of a passive film on the specimen surface. Immediately after the start of tribocorrosion, a sudden decrease in potential is observed. This fall in potential is owing to the mechanical segregation of the passive film (depassivation) and the exposure of the underlying raw, active Ti15Mo alloy to the solution. During the wear period, the potential shows transients of around 0.1 V in amplitude in the untreated specimen and up to 0.2 V in the coated ones. During the final portion of the wear,

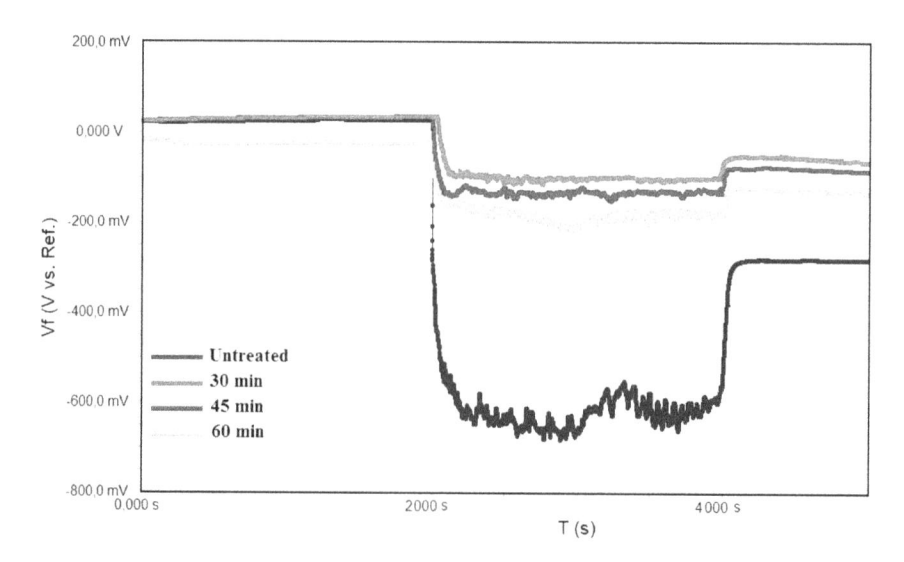

Fig. 6. Evolutions with time of the OCP during sliding of an Al_2O_3 ball against Ti15Mo for: (a) untreated specimen and specimens with deposition times of (b) 30 min, (c) 45 min and (d) 60 min (Color online).

the OCP shows an anodic shift due to repassivation of the active area. The recovery of the early OCP values is faster in the coated specimen, with the artificial saliva.

Table 3 summarizes the average hardness, surface roughness and mean volume loss values of the Ti15Mo alloy tested in this study. The highest mean volume value was measured in untreated Ti15Mo alloy, while the lowest mean value value was measured in 30-min Ag^+-ion-coated Ti15Mo alloy. In fact, it has been reported to exhibit better wear strength with added Ag^+ elements to titanium alloys.[42] However, increasing the Ag film thickness with its unique soft metallic property can reduce the hardness value, although it improves the toughness of the coatings.[43] Table 4 shows

Table 3. Hardness, surface roughness and mean volume loss values of the Ti15Mo alloy specimens tested in this study (standard deviation within parentheses).

Specimens	Untreated	30 min	45 min	60 min
Surface roughness (μm)	0.224	0.212	0.243	0.276
	(0.02)	(0.01)	(0.01)	(0)
Hardness (HV)	230	132	115	100.8
	(23.72)	(12.65)	(7.91)	(6.32)
Volume loss ($\times 10^{-1}$ mm^3)	1.01	0.2	0.1	0.41
	(0.09)	(0.05)	(0.03)	(0.08)

Table 4. Data of one-way ANOVA analyses for the mean wear volume loss of Ti15Mo alloy specimens.

Factor	Sum of squares	df	Mean square	F-ratio	Significance
Surface roughness (μm)	0.059	7	0.008	92.642	< 0.00001
Hardness (HV)	193,450.800	7	27,635.829	18.501	< 0.00001
Volume loss ($\times 10^{-1}$ mm^3)	5.604	7	0.801	136.261	< 0.00001

Notes: Regression analysis for the relation between surface roughness, hardness and the mean wear volume loss. df: Degrees of freedom and F-ratio: analysis of variance.

the one-way ANOVA results for the hardness, surface roughness and wear volume loss. In this study, correlations between the hardness, surface roughness and wear volume loss were found to be significant. The one-way ANOVA revealed significant differences between the groups ($p < 0.00001$). A study in the literature reported that Ti alloys presented lower wear volume loss that may be linked to its higher hardness values compared to the other alloys.[44] Two-dimensional noncontact profilometer volume loss analysis results of the Ti15Mo alloy after the tribocorrosion tests in the artificial saliva environment are given in Fig. 7. The highest wear depth was observed in untreated Ti15Mo alloy, while the lowest wear depth was observed in 30-min Ag^+-ion-coated Ti15Mo alloy. Figure 8 displays an instance of the three-dimensional noncontact profilometer test specimen and wear area analyses of the Ti15Mo alloy after the tribocorrosion tests in the artificial saliva environment. The surface quality of the coated Ti15Mo test specimens with no wear is very low, as shown in Figs. 8(b)–8(d). This result shows that there is no hydraulic degradation in coated Ti15Mo alloy due to artificial saliva through chewing test procedures. Based on the fact that implants remain in the human mouth for a long time, this behavior is a desired feature for titanium material.

3.4. Antibacterial properties

The plate counting process was used to evaluate the antibacterial effects of untreated and silver-coated titanium specimens. The optical photographs showing a range of viable bacteria found in each specimen are presented in Fig. 9. A colony is formed from a single living bacterium. Bacterial film formation was observed in untreated titanium specimen. This means that there are a large number of visible bacteria on the untreated titanium surface. On the other hand, the silver-coated titanium surfaces were prone to decreases in the growth of *E. coli* and *S. aureus* bacteria. According to the results, the colonies of bacteria gradually decreased with the increase in silver deposition time (Figs. 9(b)–9(d)). It can be said that the degradation of magnetron-sputtered silver coating (1 μm) layer may also take place allowing silver ion leaching into

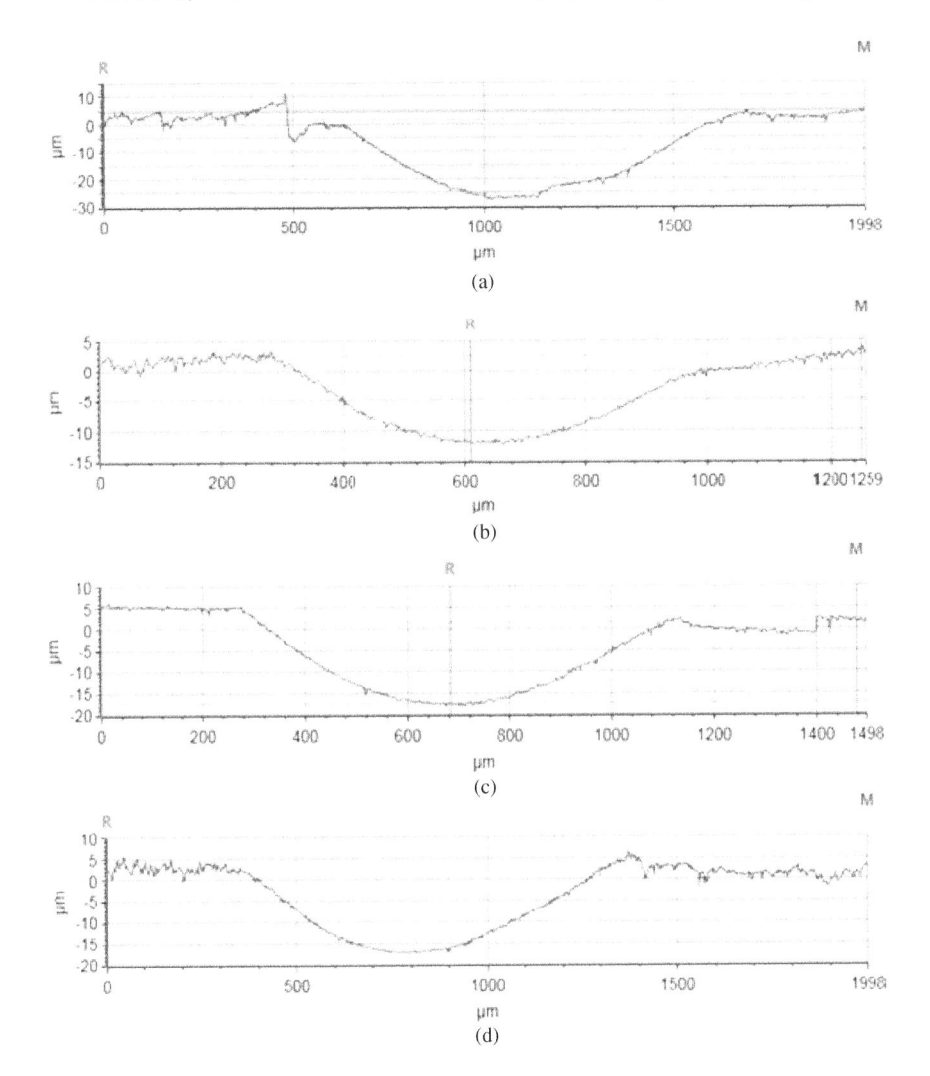

Fig. 7. Two-dimensional noncontact profilometer volume loss analysis results of the Ti15Mo alloy specimens after the tribocorrosion tests in the artificial saliva (pH = 7) environment: (a) untreated specimen and specimens with deposition times of (b) 30 min, (c) 45 min and (d) 60 min (Color online).

the bacterial solution during incubation as observed with thin silver layers. But, more uniform and conformal coating was obtained with increasing layer thickness and as a result, thicker films exhibited better antibacterial properties as bacteria could not reach the

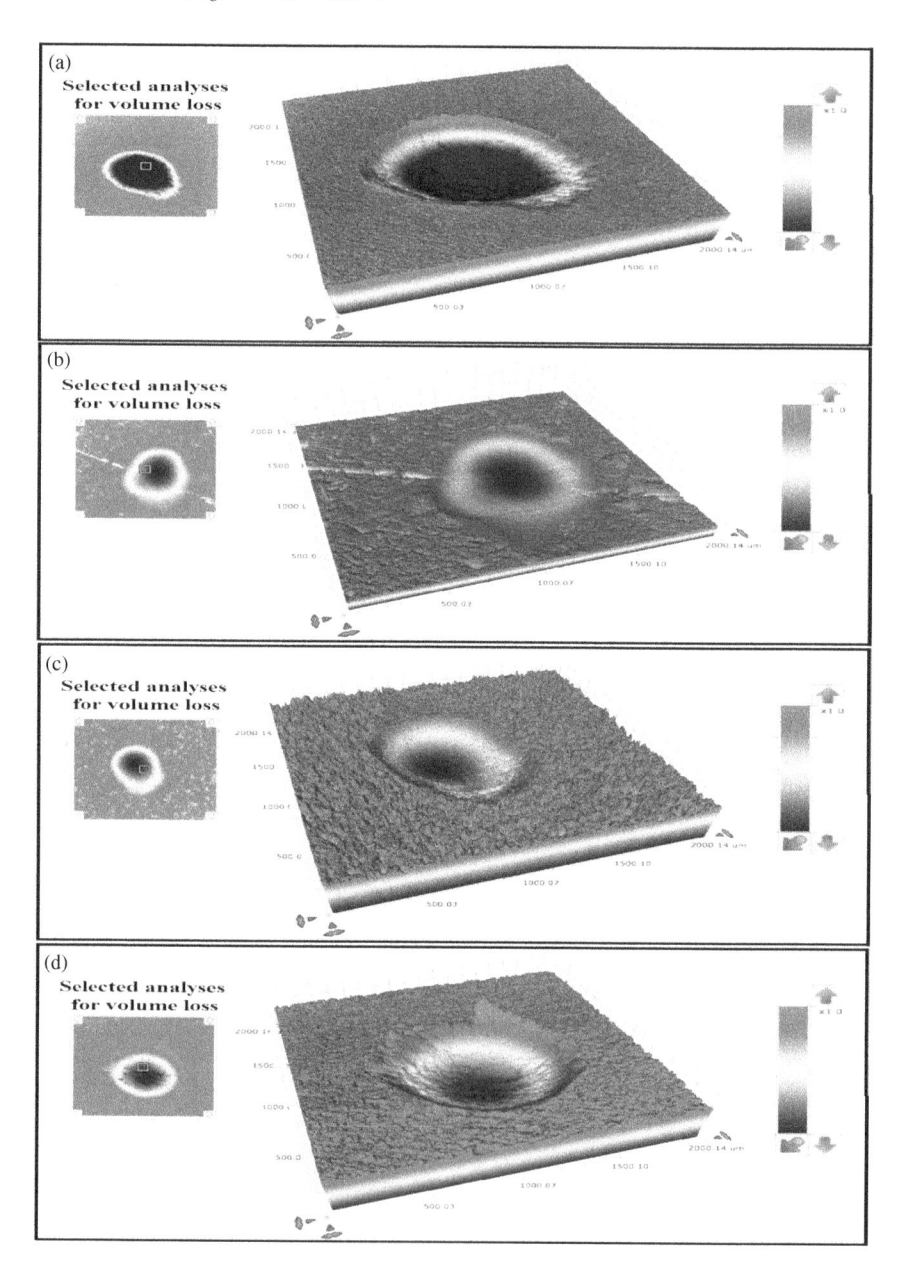

Fig. 8. Three-dimensional noncontact profilometer wear area analysis results of the Ti15Mo alloy specimens after the tribocorrosion tests in the artificial saliva (pH = 7) environment: (a) untreated specimen and specimens with deposition times of (b) 30 min, (c) 45 min and (d) 60 min (Color online).

Corrosion, tribocorrosion and antibacterial properties of silver coatings 325

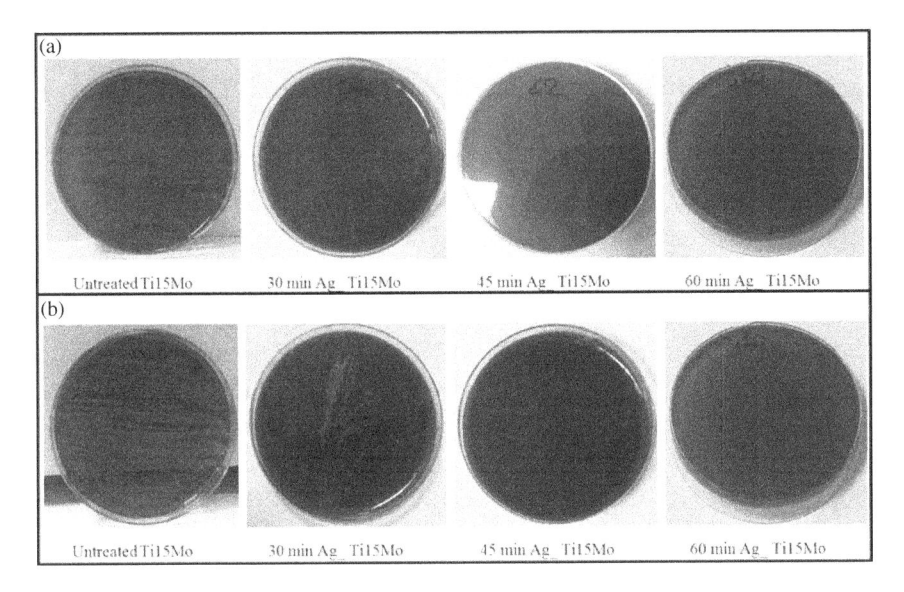

Fig. 9. Surface colonization results for (a) *E. coli* and (b) *S. aureus*, after 24 h (Color online).

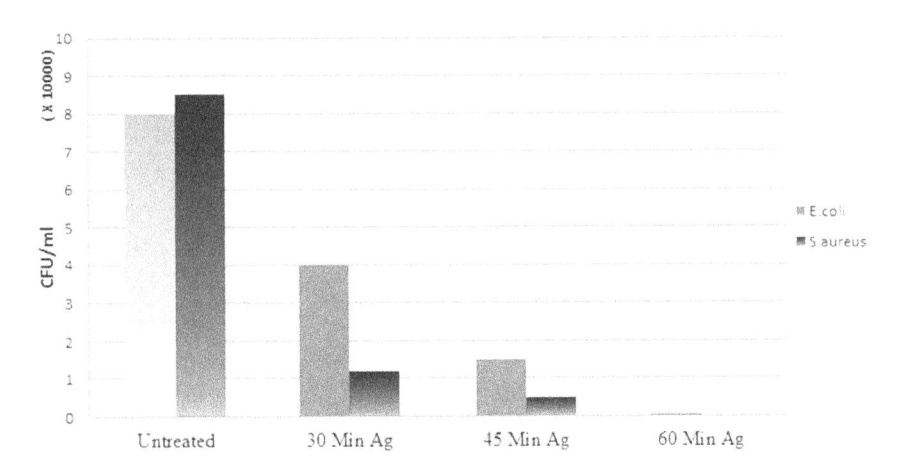

Fig. 10. The quantitative results (in CFU/mL) of antibacterial property of coated Ti15Mo surface against *S. aureus* and *E. coli* (Color online).

titanium surface. The CFU/mL value was calculated for each specimen and then illustrated in Fig. 10. On untreated titanium specimen surface, the numbers of bacterial colonies were substantially high compared to silver-coated titanium specimens. According to

Fig. 10, the ability of silver-coated Ti15Mo surface to suppress the growth of *E. coli* was less than that for *S. aureus*. This situation is thought to be associated with the spherical form of *S. aureus*, which provides a wide surface field for the Ag^+-ion to react.[45] There are two main reasons behind the good antibacterial property of the silver-coated Ti15Mo surface. The first reason is the role of the Ag^+-ion which comes into contact with the outer cell wall and causes oxidative stress with the help of reactive oxygen species and then destroys the cell membrane.[46] Once the membranes break down, the silver ion enters the cytoplasm and eliminates the intracellular structures. The other reason is the direct physical contact among bacterial cells and specimen nanoparticles present on the silver-coated Ti15Mo surface. This may be attributed to the direct contact surface causing the cell wall of the bacteria to become strained and damaged.

The analysis of these results allows concluding that the silver-coated specimens yield a reduction in the bacterial growth. The bactericide effect of the specimen with 60 min of coating was especially interesting, in which no bacteria colony was found at the dilution. As the antibacterial test conditions were the same for all the specimens, the differences in antimicrobial effect can be directly attributed to the Ag^+ release time into the environment. The ion release of Ag^+ is influenced by the parameters of sputter coating, but also affected by other parameters as film microstructure, particle size and distribution in the film thickness and target angle, which can limit the access of electrolyte and particle dissolution. Particularly, the segregation of silver on the surface forming aggregates over time is of extreme importance in order to estimate quantitatively the antibacterial effect, and additional experiments with statistical analysis are advisable. Nevertheless, a positive response of silver doping for the prevention of infection with *E. coli* and *S. aureus* was proved in all cases; however, considering that coating for above 45 min has demonstrated cytotoxic effects,[46] the silver doping should be restricted below this threshold.

4. Conclusion

Our study showed that while β-Ti phase peaks were seen on the uncoated alloy surface, Ag peaks were observed with diffraction on the coated alloy after the coating process. All alloy specimens had low surface roughness compared to acid-etched uncoated alloy specimen and the lowest surface roughness value was obtained for the alloy with silver deposition for 30 min. The thin film thickness values increased with increasing treatment time. This situation can be examined in terms of the effect of coating thickness on cytotoxicity in future studies. Within the limitations of this study, the hardness of the Ti15Mo alloy was significantly greater than the coated alloys.

Potentiodynamic polarization test results showed that the corrosion resistance increased after silver coating. As the process time increased, the corrosion resistance enhanced and the best corrosion properties were obtained for the alloy coated with silver film deposited for 60 min. The reason for this can be explained by the presence of pores and cracks in Ti15Mo alloy, reducing the protective properties of the Ag^+-film coating as a result of the difference between the large cathode surface (deposition film) and the small anode surface (the bottom of the pores).

All coated alloy specimens tested in this study had higher wear resistance in the artificial saliva environment compared to untreated alloy sample. The average wear volume loss for 30-min Ag^+-ion-coated Ti15Mo alloy sample was lower than other alloy specimens. This can be explained by the small surface roughness value of the alloy. Also, considering the fact that biomaterials stay in the human body for a long time, this positive behavior is a desired feature in titanium alloys. In this study, correlations between the hardness, surface roughness and wear volume loss were found to be significant.

The PVD coating process enhanced the antibacterial activity of Ti15Mo alloy owing mainly to the formation of the silver film on the substrate. The Ti15Mo alloy had strong long-term antibacterial activity (≥ 99%) even after being immersed in bacterial solution for up to 24 h. The ionic release of Ag^+ is influenced by the parameters of sputter coating, but also is affected by other parameters like film microstructure, particle size and distribution in the film thickness and target angle, which can limit the access of electrolyte and particle dissolution. Particularly, the segregation of silver on the surface forming aggregates over time is extremely important in order to quantitatively estimate the antibacterial effect, and additional experiments with statistical analysis are advisable.

Acknowledgment

This work was supported by Ataturk University's Coordination Unit of Scientific Research Projects (Grant No. PRJ7393).

References

1. H. Liu, J. Yang, X. Zhao, Y. Sheng, W. Li, C. L. Chang and X. Wang, *Corros. Sci.* **161** (2019) 108195, doi: 10.1016/j.corsci.2019.108195.
2. C. N. Elias, J. H. C. Lima, R. Valiev and M. A. Meyers, *JOM* **60**(3) (2008) 46, doi: 10.1007/sll837-008-0031-l.
3. V. A. R. Henriques, E. T. Galvani, S. L. G. Petroni, M. S. M. Paula and T. G. Lemos, *J. Mater. Sci.* **45**(21) (2010) 5844, doi: 10.1007/sl0853-010-4660-8.
4. X. Gong, Y. Cui, D. Wei, B. Liu, R. Liu, Y. Nie and Y. Li, *Corros. Sci.* **127** (2017) **101**, doi: 10.1016/j.corsci.2017.08.008.
5. V. S. Y. Injeti, K. C. Nune, E. Reyes, G. Yue, S. J. Li and R. D. K. Misra, *Mater. Technol.* **34**(5) (2019) 270, doi: 10.1080/10667857.2018.1550138.
6. A. Sargeant and T. Goswami, *Mater. Des.* **27**(4) (2006) **287**, doi: 10.1016/j.matdes.2004.10.028.
7. S. Nag, R. Banerjee and H. L. Fraser, *Mater. Sci. Eng. C* **25**(3) (2005) 357, doi: 10.1016/j.msec.2004.12.013.
8. J. R. S. Martins Júnior, R. A. Nogueira, R. O. D. Araújo, T. A. G. Donato, V. E. Arana-Chavez, A. P. R. A. Claro and C. R. Grandini, *Mater. Res.* **14** (2011) 107, doi: 10.1590/S1516-14392011005000013.

9. I. Hacisalihoglu, A. Samancioglu, F. Yildiz, G. Purcek and A. Alsaran, *Wear* **332** (2015) **679**, doi: 10.1016/j.wear.2014.12.017.
10. A. Brończyk, P. Kowalewski and M. Samoraj, *Wear* **434** (2019) 202966, doi: 10.1016/j.wear.2019.202966.
11. D. G. Olmedo, D. R. Tasat, P. Evelson, M. B. Guglielmotti and R. L. Cabrini, *J. Biomed. Mater. Res. A* **84**(4) (2008) **1087**, doi: 10.1002/jbm.a.31514.
12. J. D. Featherstone and B. E. Rodgers, *Caries Res.* **15** (1981) 377, doi: 10.1159/000260541.
13. M. Nakagawa, Y. Matono, S. Matsuya, K. Udoh and K. Ishikawa, *Biomaterials* **26** (2005) 2239, doi: 10.1016/j.biomaterials. 2004.07.022.
14. A. A. Yanovska, A. S. Stanislavov, L. B. Sukhodub, V. N. Kuznetsov, V. Y. Illiashenko, S. N. Danilchenko and L. F. Sukhodub, *Mater. Sci. Eng. C* **36** (2014) 215, doi: 10.1016/j.msec.2013.12.011.
15. D. Ao, X. Chu, Y. Yang, S. Lin and J. Gao, *Vacuum* **148** (2018) 230, doi: 10.1016/j.vacuum.2017.11.017.
16. C. N. Elias, D. J. Fernandes, C. R. Resende and J. Roestel, *Dent. Mater. J.* **31**(2) (2015) el, doi: 10.1016/j.dental. 2014.10.002.
17. J. A. Lichter, K. J. Van Vliet and M. F. Rubner, *Macromolecules* **42**(22) (2009) 8573, doi: 10.1021/ ma901356s.
18. G. Franci, A. Falanga, S. Galdiero, L. Palomba, M. Rai, G. Morelli and M. Galdiero, *Molecules* **20**(5) (2015) 8856, doi: 10.3390/molecules20058856.
19. Y. Zhang, X. Liu, Z. Li, S. Zhu, X. Yuan, Z. Cui and S. Wu, *ACS Appl. Mater. Interfaces* **10**(1) (2018) 1266.
20. B. W. Stuart, G. E. Stan, A. C. Popa, M. J. Carrington, I. Zgura, M. Necsulescu and D. M. Grant, *Bioact. Mater.* **8** (2022) 325, doi: 10.1016/j.bioactmat.2021.05.055.
21. A. Taglietti, C. R. Arciola, A. D'Agostino, G. Dacarro, L. Montanaro, D. Campoccia and L. Visai, *Biomaterials* **35**(6) (2014) 1779, doi: 10.1016/j.biomaterials.2013.11.047.
22. W. Yu, X. Li, J. He, Y. Chen, L. Qi, P. Yuan and X. Qin, *J. Colloid Interface Sci.* **584** (2021) **164**, doi: 10.1016/j.jcis.2020.09.092.
23. C. Mao, Y. Xiang, X. Liu, Z. Cui, X. Yang, K. W. K. Yeung and S. Wu, *ACS Nano* **11**(9) (2017) 9010, doi: 10.1021/acsnano.7b03513.
24. S. H. Ye, C. A. Johnson, Jr., J. R. Woolley, H. I. Oh, L. J. Gamble, K. Ishihara and W. R. Wagner, *Colloids Surf B, Biointerfaces* **74**(1) (2009) 96, doi: 10.1016/j.colsurfb.2009.06.032.

25. W. R. Fordha, S. Redmond, A. Westerland, E. G. Cortes, C. Walker, C. Gallagher and R. R. Krchnavek, *Surf. Coat. Technol.* **253** (2014) 52, doi: 10.1016/j.surfcoat.2014.05.013.
26. M. Del Re, R. Gouttebaron, J. P. Dauchot, P. Leclere, R. Lazzaroni, M. Wautelet and M. Hecq, *Surf. Coat. Technol.* **151** (2002) 86, doi: 10.1016/S0257-8972(01)01592-4.
27. I. Hacisalioglu, F. Yildiz, A. Alsaran and G. Purcek, *IOP Conf. Ser., Mater. Sci. Eng.* **174** (2017) 012055, doi: 10.1088/1757-899X/174/1/012055.
28. D. Mareci, R. Chelariu, D. M. Gordin, G. Ungureanu and T. Gloriant, *Acta Biomater.* **5**(9) (2009) 3625, doi: 10.1016/j.actbio.2009.05.037.
29. A. Butt, N. B. Lucchiari, D. Royhman, M. J. Runa, M. T. Mathew, C. Sukotjo and C. J. Takoudis, *J. Bio-Tribo-Corros.* **1**(1) (2015) 4, doi: 10.1007/s40735-014-0004-6.
30. Y. Zhu, H. Cao, S. Qiao, M. Wang, Y. Gu, H. Luo and H. Lai, *Int. J. Nanomedicine* **10** (2015) 6659, doi: 10.2147/IJN.S92110.
31. M. Sabeena, A. George, S. Murugesan, R. Divakar, E. Mohandas and M. Vijayalakshmi, *J. Alloys Compd.* **658** (2016) 301, doi: 10.1016/j.jallcom.2015.10.200.
32. M. Szklarska, G. Dercz, J. Rak, B. Łosiewicz and W. Simka, *Arch. Metall. Mater.* **60**(4) (2015) 2687, doi: 10.1515/amm-2015-0433.
33. J. E. D. Andrade, R. Machado, M. A. Macedo and F. G. C. Cunha, *PoKlieros* **23** (2013) 19, doi: 10.1590/S0104-14282013005000009.
34. N. Marechal, E. A. Quesnel and Y. Pauleau, *Thin Solid Films* **241** (1–2) (1994) **34**, doi: 10.1016/0040-6090(94)90391-3.
35. O. Çomakli, M. Yazici, T. Yetim, A. F. Yetim and A. Çelik, *J. Bionic Eng.* **14**(3) (2017) 532, doi: 10.1016/S1672-6529(16)60419-5.
36. W. Phae-ngam, C. Chananonnawathorn, T. Lertvanithphol, B. Samransuksamer, M. Horprathum and T. Chaiyakun, *Mater. Technol.* **55**(1) (2021) 65, doi: 10.17222/mit.2019.189.
37. B. R. Chrcanovic, A. Wennerberg and M. D. Martins, *Mater. Res.* **18** (2015) 963, doi: 10.1590/1516-1439.014115.
38. H. Liu, Q. Xu, X. Zhang, C. Wang and B. Tang, *Thin Solid Films* **521** (2012) 89, doi: 10.1016/j.tsf.2012.02.046.
39. M. Shokouhfar, C. Dehghanian, M. Montazeri and A. Baradaran, *Appl. Surf. Sci.* **258**(7) (2012) 2416, doi: 10.1016/j.apsusc.2011.10.064.

40. H. R. Bakhsheshi-Rad, E. Hamzah, A. F. Ismail, M. Aziz, M. Kasiri-Asgarani, E. Akbari and Z. Hadisi, *Ceram. Int.* **43**(17) (2017) 14842, doi: 10.1016/j.ceramint.2017.07.233.
41. T. Yetim, *J. Bionic Eng.* **13**(3) (2016) 397, doi: 10.1016/S1672-6529(16)60311-6.
42. M. Dong, Y. Zhu, L. Xu, X. Ren, F. Ma, F. Mao and L. Wang, *Appl. Surf. Sci.* **487** (2019) 647, doi: 10.1016/j.apsusc.2019.05.119.
43. J. F. Archard, *J. Appl. Phys.* **24** (1953) 981, doi: 10.1063/1.1721448.
44. W. H. Kao, Y. L. Su and Y. T. Hsieh, *J. Mater. Eng. Perform.* **26** (2017) 3686, doi: 10.1007/sll665-017-2815-3.
45. S. Chernousova and M. Epple, *Angew. Chem., Int. Ed.* **52**(6) (2013) 1636, doi: 10.1002/anie.201205923.
46. I. N. Mihailescu *et al*, *Int. J. Pharm.* **515**(1–2) (2016) 592, doi: 10.1016/j.ijpharm.2016.10.041.

Influence of mechanical and the corrosion characteristics on the surface of magnesium hybrid nanocomposites reinforced with HAp and rGO as biodegradable implants*

Venkata Satya Prasad Somayajula [†,‡], Shashi Bhushan Prasad [‡] and Subhash Singh [§,¶]

[†]*Department of Mechanical Engineering,*
VNR Vignana Jyothi Institute of Engineering and Technology,
Hyderabad, Telangana 500090, India
[‡]*Department of Production and Industrial Engineering,*
National Institute of Technology, Jamshedpur,
831014 Jharkhand, India
[§]*Department of Mechanical and Automation Engineering,*
Indira Gandhi Delhi Technical University for Women,
New Delhi, 110006 Delhi, India
[¶]*subh802004@gmail.com*

Magnesium composites stay relevant for the applications of biodegradable implant as they are harmless and possess characteristics such as density and elastic modulus analogous to the cortical bone in humans. But corrosion is one major issue associated with magnesium when the biomedical applications are contemplated. Moreover, load bearing abilities are also required in case of an orthopedic implant. In this study, to achieve the desired implant characteristics, hybrid nanocomposites (HNCs) of Mg–2.5Zn binary alloys such as metal matrix, hydroxyapatite (HAp), and reduced graphene oxide (rGO) as reinforcements were

---

*To cite this article, please refer to its earlier version published in Surface Review and Letters, Vol. 31, No. 3 (2024) 2450021 (20 pages) DOI: 10.1142/S0218625X24500215
§Corresponding author.

fabricated via the vacuum-assisted stir casting method. The overall weight percentage of the reinforcements was fixed at 3% and both the reinforcements varied in compositions by weight to prepare the samples S0 (Pure Magnesium), S1 (Mg–2.5Zn–0.5HAp–2.5rGO), S2 (Mg–2.5Zn–1.0HAp–2.0rGO), S3 (Mg–2.5Zn–1.5HAp–1.5rGO), S4 (Mg–2.5Zn–2.0HAp–1.0rGO), and S5 (Mg–2.5Zn–2.5HAp–0.5rGO), respectively. The influence of mechanical characteristics such as tensile strength, compressive strength, and microhardness as well as the corrosion over the surface of the nanocomposite in simulated body fluid (SBF) have been assessed for their suitability as biodegradable orthopedic implants. Results suggest that the fabricated nanocomposites exhibit superior characteristics in comparison to pure magnesium. Increasing the HAp from 0.5 wt.% to 2.5 wt.% enhanced the compressive strength and reduced the corrosion rate. On the other hand, increasing the rGO from 0.5 wt.% to 1.5 wt.% increased the tensile strength. The formation of apatite layer over the composites is observed in the SBF solution. Among all the fabricated hybrid nanocomposite samples, the sample S3 (Mg–2.5Zn–1.5HAp–1.5rGO) with equal wt.% of HAp and rGO exhibited 209.60 MPa of ultimate tensile strength, 300.1 MPa of ultimate compressive strength, and a corrosion rate of 0.91 mm/year thus making it the best suited and a prospective material for biodegradable implant application.

Keywords: Hybrid nanocomposites; corrosion; simulated body fluid; vacuum-assisted stir casting; orthopedic implant; surface analysis.

1. Introduction

The necessity of implants with high-quality metal-based biomaterials is on the rise and can further escalate based on the great demand they have as they are the key for recovery of fractured bones in humans.[1,2] Metal-based biomaterials like alloys of titanium or stainless steels are the most common among the orthopedic as well as craniofacial biomaterials that are in use because of their cyclic load withstanding abilities. But their dense nature and large elastic modulus result in stress shielding when employed for a cortical bone application.[3,4] There also exists a possibility of the liberation of metallic ions or particles that are toxic, at the time of corrosion or wearing and eventually leading to immune response at the implant site. Such complications involve secondary surgical procedures to remove the permanent implants made of metal which is a surgical risk and also a financial liability. Therefore, use of biodegradable implants has been looked into. The biodegradable implants gradually disintegrate and get disbursed or expelled within the body in humans and facilitate connection between tissue as well as implant eventually replacing it in order to eliminate the additional surgical procedure to get rid of the implants post-healing process.[5] The perfect implant that is biodegradable and used in orthopedics must possess osseointegration, low weight, strength greater than 200 MPa, elongation above 10%, and needs to degrade at a rate lower than 0.5 mm per year so as to have 90–180 days of life in SBF at 37°C in order to overcome diverse biomechanical forces.[1,6]

The metal-based materials of magnesium (Mg), zinc (Zn), and iron (Fe) with exceptional strength as well as ductility are extremely efficient as load bearing and biodegradable materials.[7,8] Magnesium is considered to be the common metallic biomaterial implant applied within human body for the engineering of bone tissue. Magnesium in its pure form degrades within the body fluids generated inside the human body. Hence, it is critical to keep a lid on the process of corrosion in magnesium. The biodegradable metal Mg has convincing biocompatibility and biodegradability as well as mechanical properties. Among the above materials, Mg and alloys of Mg are deemed to be worthier as they possess the desired characteristics such as a low 1.78 g/cm^3 density, 40–45 GPa elastic modulus, and 65–100 MPa yield compressive strength which are equivalent to the characteristics of the bone in humans which are 1.78 g/cm^3 density, 3–20 GPa elastic modulus, and 130–180 MPa yield compressive strength.[9,10] Also, Mg facilitates enhanced interactions with the cell.[11] Moreover, Fe and Zn with moduli of elasticities around 211.4 and 90 GPa, respectively, tend to create stress shielding effect over the bone.[1] But Mg that is pure, corrodes effortlessly in SBF solution of 7.4 pH and comprising of Cl-ions thereby losing its mechanical integrity making it an ideal implant.[12,13] In order to evade issues pertaining to corrosion and lower mechanical properties, multiple techniques such as modifying surfaces or coating, alloying with certain common elements such as rare earth elements, zinc, manganese, aluminum, calcium or forming composites (composting) have been devised to enhance the characteristics of material along with its resistance to corrosion.[14] These techniques would enhance the physical properties of magnesium and ensure that it is adequate for bone tissue repairs. Properties such as bioactivity, biocompatibility, mechanical as well as resistance to corrosion have been observed when bioactive nanocomposites such as nanohydroxyapatite (HAp) have been intermixed with magnesium.

Zinc when alloyed with MgS limits the evolution of hydrogen gas and also enhances the hardness, strength as well as fluidity of the alloy at room temperature because of the precipitation hardening effect.[15] Zn is necessary for the human body and as per studies, the daily intake recommended for adults is 8–11 mg/day. Zn when taken below the appropriate isn't as toxic as Al or Mn when dissolved *in vivo*, as it has an absorbability of 6.2 wt.% due to the biological activity in cells. Aluminum is detrimental to neurons and is linked to diseases such as dementia and Alzheimer's disease whereas manganese, for biomedical applications, excess in quantity isn't ideal since it results in a neurological disorder called "manganism" which is similar to the Parkinson's disease. Many research works have established the fact that adding little quantity of Zn (2 wt.%) to Mg alloy could enhance resistance against corrosion as well as tensile strength for biomedical applications.[16] Also, Zn has a 2.5 wt.% tolerance in magnesium. In the cast alloys of Mg, Ca with 1.34 wt.% solubility, refines grains and also lowers the oxidation thereby enhancing properties but when it comes to biomedical applications, only 1 wt.% of Ca needs to be alloyed with Mg because above that value, higher rates of corrosion have been observed.[1]

The toxic nature of many such elements in the alloys has forced the researchers to develop nanocomposites of magnesium metal comprising of special features reinforced with bioactive particulates to be used in biodegradable implants used for orthopedic applications.[17,18] It has been deduced previously that carbon-based nanomaterials exhibit diverse, unique features like outstanding elastic strength, aspect ratios as well as temperature-related properties.[19,20] The nanoplatelets of graphene are cost effective with sp^2 bonds of carbon and one atomic layer thickness, 2D-based honeycomb lattice structure has around 130 GPa internal strength, elastic modulus ranging till 1 TPa, 1 g per cm^3 density and a specific surface area that's about 2630 m^2 per gram,[21,22] biocompatible characteristics and are best suited as nanoreinforcements for metal-based nanocomposites. Biodegradation of nanoplatelets of graphene is via oxidized enzymes due to peroxidase, a naturally occurring oxidation process in biological environment resulting in disintegration.[6]

Even graphene with similar characteristics as that of its nanoplatelets, having high affinity towards electrons resists oxygen, helium as well as hydrogen molecules of gas.[23,24] Thus, corrosion can be avoided when combined with Mg-based alloys since it generates atomic hurdle.[25,26] The graphene showcased its nontoxic nature and has been termed safe at lower concentrations of about 50 µg per mL for applications in the biomedical field.[27,28] However, volume fraction shouldn't exceed 10% since there is the issue of high stiffness within the composite which is biomechanically undesirable for the natural bone.[6] Hence, graphene, its nanoparticles, graphene oxide or carbon nanotubes aren't safe to be an implant as solo materials and must be used only as reinforcements within a composite.

Speaking of the graphene family, the reduced graphene oxide (rGO) has the tendency of a better stability and is safer in the biological medium than the graphene oxide since the reduced form of graphene oxide lowers the dispersion of water because of lower oxygenated functional groups along its planes as well as the edges leading to lower cytotoxic potential and a much stable compound *in vivo*.[29,30] The same, i.e. suitability of rGO for cell culture and tissue engineering biomedical applications have been proven in research done earlier.[31]

The other types of reinforcements for magnesium which endorse osteogenesis, osteointegration and also positively impact the mechanical properties are phosphates of calcium, among which, fluorapatite (FAp) and HAp are the most common. The human cortical bone in nature is a complex structure with multiple levels that is associated with many processes that include redesigning and the major portion of it is made up of nanoHAp. The network within the body comprises of 2–5% of cells, 5% water, 25% calcified bone, and 70% of inorganic HAp.[32,33] The similarity of fabricated nano-HAp with cortical bone structure is quite intriguing in case of tissue engineering when bone development is considered. It gives the hardness needed in physicality for structural application in the recovery process of bone tissue which also endorses bonding as well as cell proliferation. When the aspect of corrosion is considered when HAp is added to magnesium, because of the occurrence of corrosion at surface level, HAp as Mg reinforcement controls the corrosion rate and also enhances the mechanical strength when the implant is either corroded or not. HAp also facilitates cell proliferation and adhesion when combined with Mg to form nanocomposites. In previous studies, nanocylindrical Hap-reinforced Mg–3Zn displayed enhanced resistance to corrosion, mechanical properties, cell activity, and viability because of reinforcement dispersion as well as strengthening of refined grains within the composite.[34] Similar type of results was also seen when 15–20 wt.% of HAp was added to Mg in pure form which showed enhanced mechanical properties.[35] However, adding HAp beyond 10 wt.% agglomerates the nanoparticles in metal matrix which leads to nonuniform disintegration. So, it is better if the content of HAp is lower than the specified limit. Moreover, uniform dispersion of the reinforcement particles in Mg nanocomposites enhances the desired properties and enhancing their suitability as biodegradable implants, knee, and hip joints replacement applications.[36]

This research work has been sketched with an objective to fabricate magnesium-based hybrid nanocomposites (HNCs) comprising of Mg–2.5Zn binary alloy as the metal matrix with 97 wt.% and varying wt.% of the reinforcements HAp and rGO constituting the 3 wt.% of the total nanocomposite, employing the vacuum-assisted stir casting technique. The fabricated HNCs have been analyzed for their mechanical properties and also their ability to resist corrosion in the simulated body fluid (SBF). Keeping in mind the suitability of the fabricated magnesium-based HNCs as biodegradable orthopedic implants, the objective has been defined.

2. Experimental Procedure

2.1. *Selecting the materials*

The ingots of magnesium and zinc in the purest form (99.9 wt.%) have been purchased from the dealer Parshwamani Metals, Mumbai in an attempt to combine them together to form a binary alloy needed for the matrix of the desired nanocomposite. The EDX spectroscopy was carried out on the samples of the metals obtained to check for the presence of impurities or any alloying compounds. The surface morphologies of the samples of the procured metals have also been observed under the scanning electron microscope (SEM). The SEM images of the morphologies along with the EDX spectroscopy of the procured metals are shown in Fig. 1. The surface of magnesium and zinc is slightly different in a way that Zn has a surface morphology that resembles a unidirectional structure, as seen in Figs. 1(a) and 1(c), respectively. The EDX spectroscopy of each of the specimens confirms the existence of the elements Mg and zinc also indicates that there are no impurities present within the procured metals of magnesium and zinc, respectively (Figs. 1(b) and 1(d)).

As for the reinforcements, rGO of nanoscale (99.9 wt.% pure) and HAp of nanoscale (99 wt.% pure) have been selected and procured at the centers of Ad-Nano Technologies Private Limited, Shimoga and Nano Research Lab, Jamshedpur, respectively. The HAp is a white powder with an average particle size lower than 50 nm whereas rGO is an extremely light, black colored fluffy powder with 1–3 layers and 0.8–2 nm thickness. The procured materials have also been observed under the SEM for their morphology and the EDX spectroscopy was done to confirm the elemental compositions of the procured powders, as shown in Fig. 2.

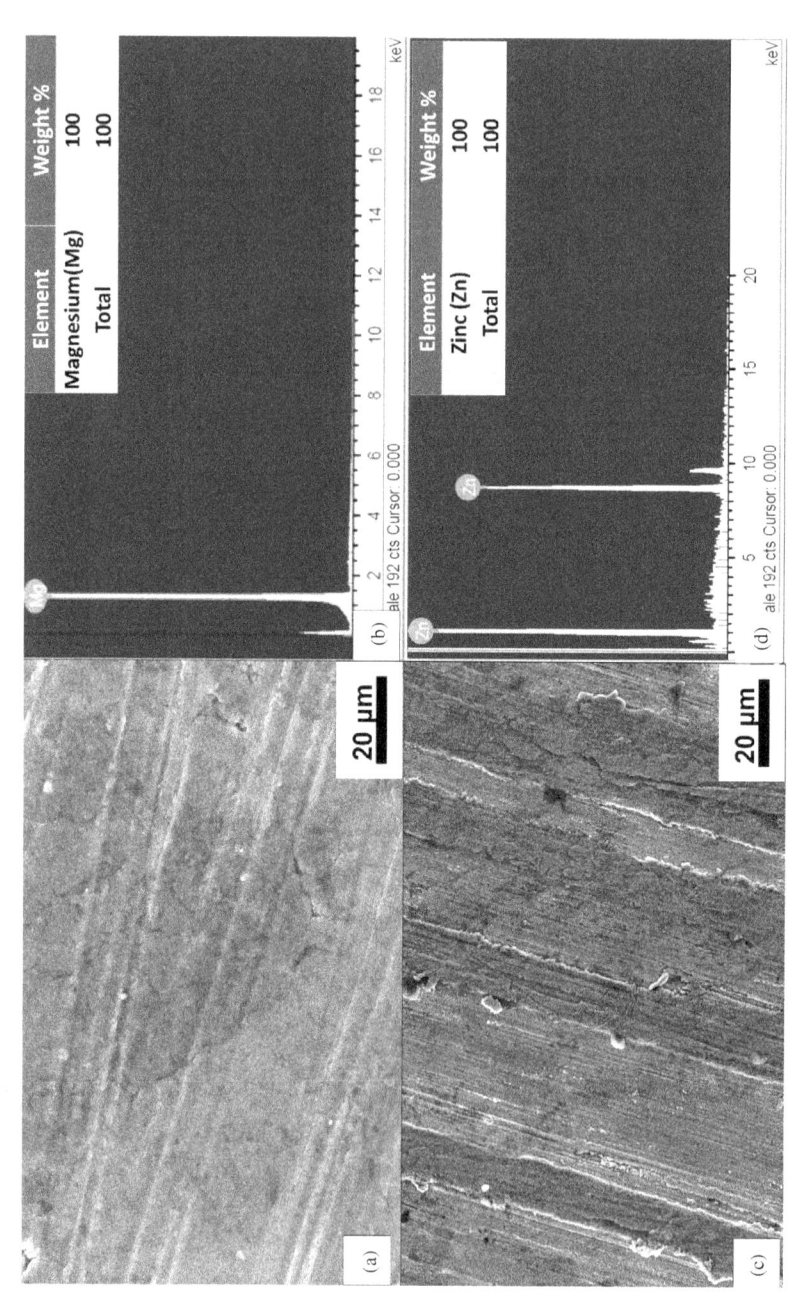

Fig. 1. (a) SEM image depicting the surface of pure magnesium. (b) EDX spectroscopy of pure Mg exhibiting the presence of Mg element. (c) SEM image depicting the surface of pure zinc. (d) EDX spectroscopy of pure Zn highlighting the presence of Zn element in the form of peaks (Color online).

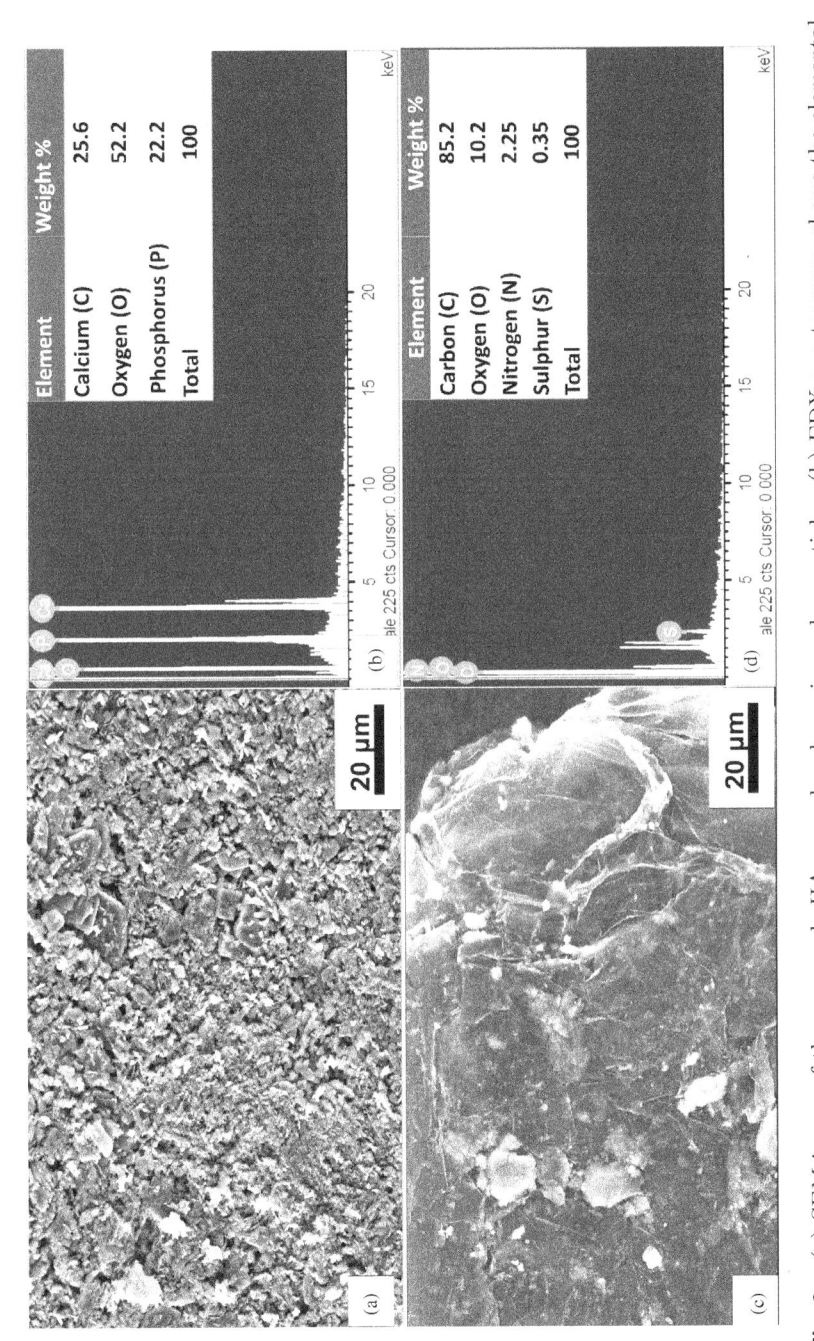

Fig. 2. (a) SEM image of the nanoscale HAp powder shows irregular particles, (b) EDX spectroscopy shows the elemental composition of nanoHAp powder, (c) SEM image of the nanoscale rGO shows wrinkled surface morphology, and (d) EDX spectroscopy of nanorGO powder representing the elemental composition (Color online).

The nanoHAp morphology is illustrated in Fig. 2(a) which is like irregular particulates whereas in Fig. 2(c), the rGO's wrinkled surface morphology is represented. The EDX spectroscopy confirms the presence of the elements of calcium, phosphorous, and oxygen in HAp (Fig. 2(b)) whereas in rGO, the EDX spectrograph confirms the existence of carbon, oxygen, nitrogen, and sulfur elements (Fig. 2(d)) which constitute the procured materials. The presence of sulfur is negligible since it is an extremely small percentage of 0.35 wt.% as compared to the overall weight percentage of the material.

2.2. Fabricating the hybrid nanocomposite

The fabrication of hybrid nanocomposite was carried out by the method of vacuum-assisted stir casting in which a layer of protective gas comprising of Argon and SF_6 in the ratio of 80:20 at 720°C working temperature. Pure Mg and Zn were melted together to produce magnesium-based binary alloy, Mg–2.5Zn. This being the matrix and the combination of nanoscaled reinforcements of rGO and HAp was subsequently added in varying compositions after preheating them at 200°C. Stirring was done for 15 min at 300 rpm, followed by ultrasonic vibration so as to avert reinforcement agglomeration and to facilitate uniform dispersion of the particulates of the reinforcement.[37] The melt of the hybrid nanocomposite was at a constant 660°C for the duration of the process. Prior to pouring of the composite melt into the die cavity, it was ensured that the die is pre heated to 200°C and also the entire air from the die cavity has been pumped out using the vacuum pump following which the molten melt from the furnace was dispensed into the preheated die cavity via the bottom pour valve. Six castings of the samples of dimensions 300 mm × 100 × 15 mm, S0 (Pure Magnesium), S1 (Mg–2.5Zn–0.5HAp–2.5rGO), S2 (Mg–2.5Zn–1.0HAp–2.0rGO), S3 (Mg–2.5Zn–1.5HAp–1.5rGO), S4 (Mg–2.5Zn–2.0HAp–1.0rGO), and S5 (Mg–2.5Zn–2.5HAp–0.5rGO), respectively, were made. A schematic diagram of the stir casting process is represented in Fig. 3.

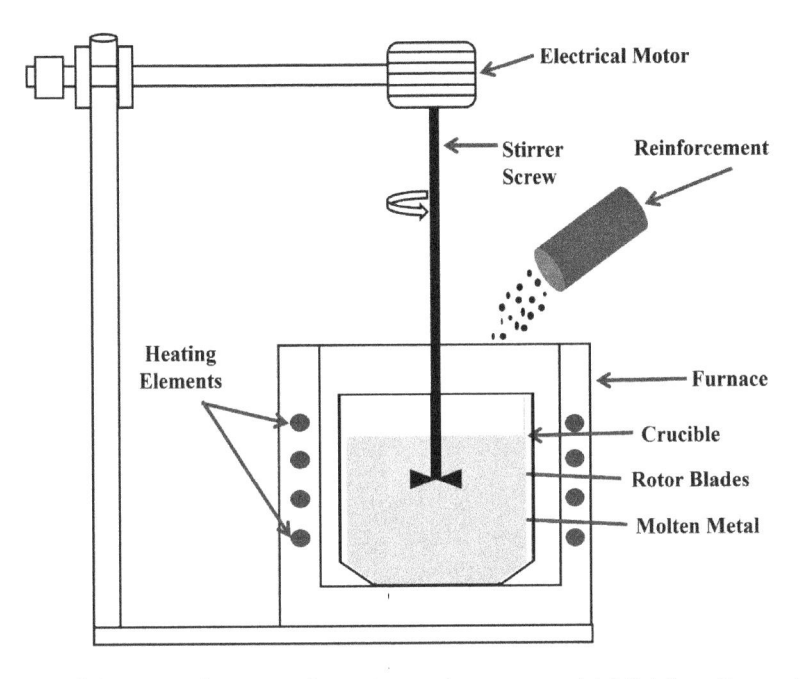

Fig. 3. Schematic diagram of a stir casting process highlighting its major components (Color online).

2.3. Mechanical testing

As shown in Fig. 4(a), the tensile test samples of the fabricated nanocomposite were cut into the dog bone shape, with dimensions as per the ASTM E8 standards. For high precision, the Wire-EDM was used to cut the samples. Subsequently, the tensile test was carried out on 20 kN capacity, FIE make, Universal Testing Machine (UTM) of UTES-60 HGFL model. Both the sample preparation as well as the tensile test were carried out at Adityapur Auto Cluster Lab, Jamshedpur, Jharkhand.

As shown in Fig. 4(b), the samples for compression test were cut into small cylinders, 14 mm long and 7 mm diameter as per ASTM E9 standards using the Wire-EDM for precision. The test for compressive strength followed and was done on UTM with 0.1 mm/min crosshead speed at ASR Metallurgy Private Limited, Jamshedpur.

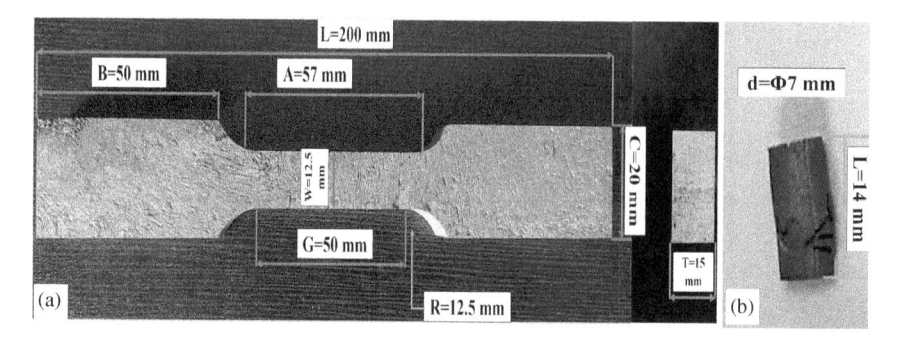

Fig. 4. (a) Image of the dog bone-shaped specimen along with dimensions prepared as per ASTM standards for tensile test and (b) Image of the specimen with dimensions prepared as per ASTM standards for compression test.

The Vickers Hardness Tester at Adityapur Auto Cluster was used for indentation tests to evaluate the microhardness of the prepared nanocomposite samples with a square based, 136° vertex angle diamond indenter at 100 gf force and 10 s dwell time. For every specimen, the microhardness values were observed at five different samples to obtain the average value. The sample preparation included polishing the samples with silicon carbide emery papers up to 2500 grit and subsequent surface finishing process with the help of 1 μm particle-sized diamond paste to obtain a mirror finish. All the procedures were in conjunction with the ASTM standards.

2.4. Microstructural characterization

Optical Microscope and SEM were used to observe and evaluate the microstructure of the nanocomposites. The optical microscopic observation was done only after the samples were prepared and etched as per the standards. For the sample preparation, the samples underwent diverse polishing steps which included initial rough grinding process over a belt grinder to eliminate basic scratches. This was succeeded by sample polishing via emery papers of SiC with different grit sizes ranging from 500 to 3000 to eliminate impurities as well as small scratches from the surface. In the next step, samples were polished over disk polishing machine on a velvet cloth using diamond paste of 1 μm as the polishing medium resulting in a shiny surface. Then, sample etching was done for 15 s by combining 2.5 g of 99.2% pure picric acid, 100 ml of 99.7% pure ethanol, 4 ml of 99.5% pure acetic acid, and 10 ml deionized water.[38,39]

For the X-ray diffraction (XRD) test, the samples were scanned using 20–80° incident angle of Cu-K$_\alpha$ radiation over 2_θ range to show various components of the phase. The operating parameters of the radiation were 45 kV voltage and 40 mA current. The EDX analysis was also done on the composites for verifying the elemental composition.

2.5. *Corrosion behavior*

The method of immersion testing in SBF was employed to evaluate the corrosion behavior of the fabricated HNCs. The SBF was fabricated as mentioned by Kokubo.[40] For 1 liter, the SBF has Na$^+$ 142 mM, K$^+$ 5 mM, Mg^{2+} 1.5 mM, Ca^{2+} 2.5 mM, Cl$^-$ 147.8 mM, HCO$_3^-$ 4.2 mM, 1 mM of HPO$_4^{2-}$, and 0.5 mM of SO$_4^{2-}$ ion concentrations, respectively, with a pH value of 7.4. The specimen preparation for immersion test included cutting the samples to the dimensions of 10 mm diameter and 5 mm thickness, grinding the samples with an emery paper of SiC based from 600 to 2000 grade, acetone bathing of the samples using ultrasonicator, and drying them post the bath in a hot air stream.

The corrosion rate was found out by the mass loss method, as per ASTM G31 standards. The samples were weighed prior to the immersion and then soaked in SBF for a duration of 168 h at 37°C temperature. The samples were cleansed with a solution of chromic acid and then weighed again. This procedure was repeated thrice to get optimum reproducible values. As mentioned in the ASTM standards, the formula used for finding out corrosion rate (C_r) is given in the following equation:

$$C_r = \frac{(K \times W)}{(A \times T \times D)}, \tag{1}$$

where C_r is the corrosion rate, K is the constant for 8.76×10^4 mm per year, W is the mass loss in grams, A is the area in sq. cm, T is the time of exposure in hours, and D is the density in g/cm^3.

3. Results and Discussions

3.1. *Microstructure analysis*

The optical microscopy images of all produced samples (S0–S5) are shown in Fig. 5. The grain size of sample S0 (pure magnesium) shows large, equiaxed structure with an average grain size of 212 μm (Fig. 5(a)), an indication of low strength. Only the α-Mg phase is

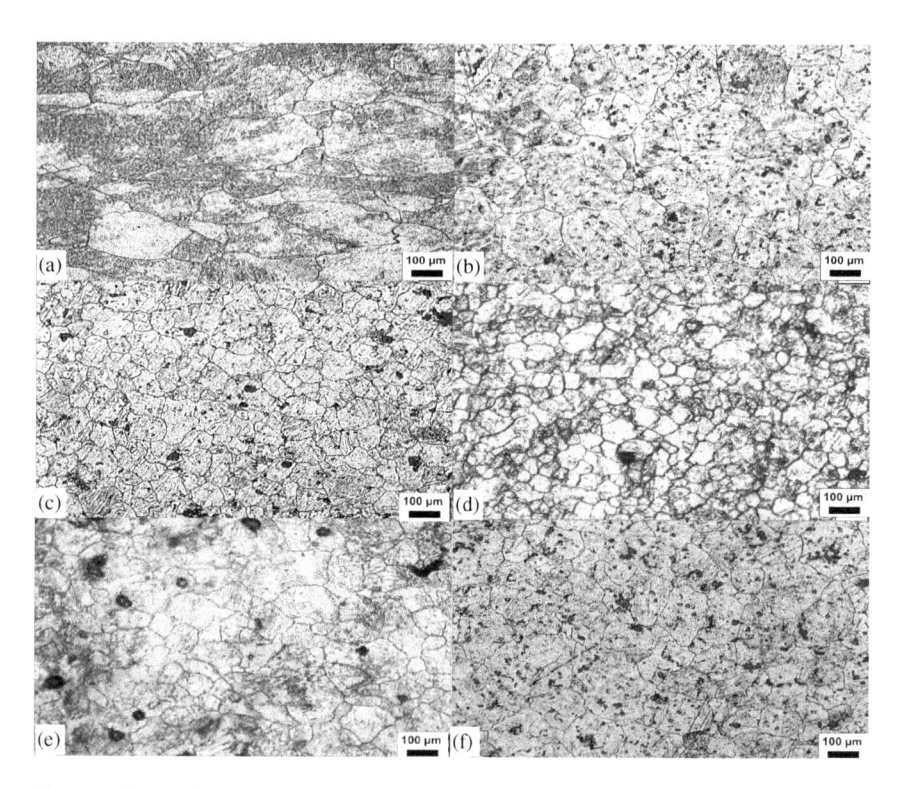

Fig. 5. Optical micrograph images of the surface of (a) pure magnesium sample S0 showing largest grain boundaries, (b) hybrid nanocomposite sample S1 (Mg–2.5Zn–0.5HAp–2.5rGO) showing large grain boundaries, (c) nanocomposite sample S2 (Mg–2.5Zn–1.0HAp–2.0rGO) showing smaller grain boundaries, (d) nanocomposite sample S3 (Mg–2.5Zn–1.5HAp–1.5rGO) showing the smallest grain boundaries, (e) nanocomposite sample S4 (Mg–2.5Zn–2.0HAp–1.0rGO) showing medium-sized grain boundaries, and (f) S5 (Mg–2.5Zn–2.5HAp–0.5rGO) showing medium-sized grain boundaries.

seen in magnesium that is pure.[15] The hybrid nanocomposite samples (S1–S5) on the other hand had revealed typical microstructure of an Mg-based alloy within the matrix in conjunction with β-phased grains near the boundaries (Figs. 5(b)–5(f)).

The nanocomposites in comparison to pure magnesium exhibit grain sizes that are smaller. Sample S1 with highest rGO content of 2.5 wt.% and least HAp content of 0.5 wt.% has an average grain size of 118.92 μm, S2 (Mg-2.5 Zn Matrix, 1.0 wt.% Hap, and 2.0 wt.% rGO) has 103.97 μm average sized grains, S3 sample with equal quantities of rGO and HAp (1.5 wt.% each) on an average has 96.6 μm grain size, S4 sample comprising of 2.0 wt.% HAp and 1.0 wt.% rGO has 121.69 μm of average grain size and S5 sample containing lest rGO content of 0.5 wt.% and highest HAp content of 2.5 wt.% has 135. 92 μm of average grain size, respectively. This is a clear indication of grain refinement in the HNCs (Samples S1–S5) when compared to pure magnesium (Sample S0). Sample S3 with equal quantities of rGO and HAp (1.5 wt.% each) displays maximum amount of refined grains and hence it has the least average grain size. Also, in samples S2 and S3 with 2 wt.% and 1.5 wt.% of rGO exhibited lower-sized grains in comparison to S1 sample with 2.5 wt.% rGO, the average grain size reduced with increasing concentration of rGO from 0.5 wt.% to 1.5 wt.% beyond which the average grain size increased again. Also, with increasing HAp content from 1.5 to 2.5 wt.%, the average grain size increased. This establishes the fact that rGO helps in grain refinement, whereas, HAp has a negative impact on the refinement of grains in a composite. As compared to the average grain size of pure Mg (212 μm), the average grain size of all the HNCs is lower. This has a positive influence over the mechanical properties of the nanocomposite in accordance to Hall–Patch effect. Thus, rGO has positive effect over the strength enhancement of a nanocomposite but to a certain limit. Among the composites, sample S5 with highest HAp content has the highest average grain size. The reason behind this may be agglomerated particles of the HAp reinforcement.

The microstructure of the HNCs was observed under a SEM as well and imaged, as seen in Fig. 6. The morphologies of the

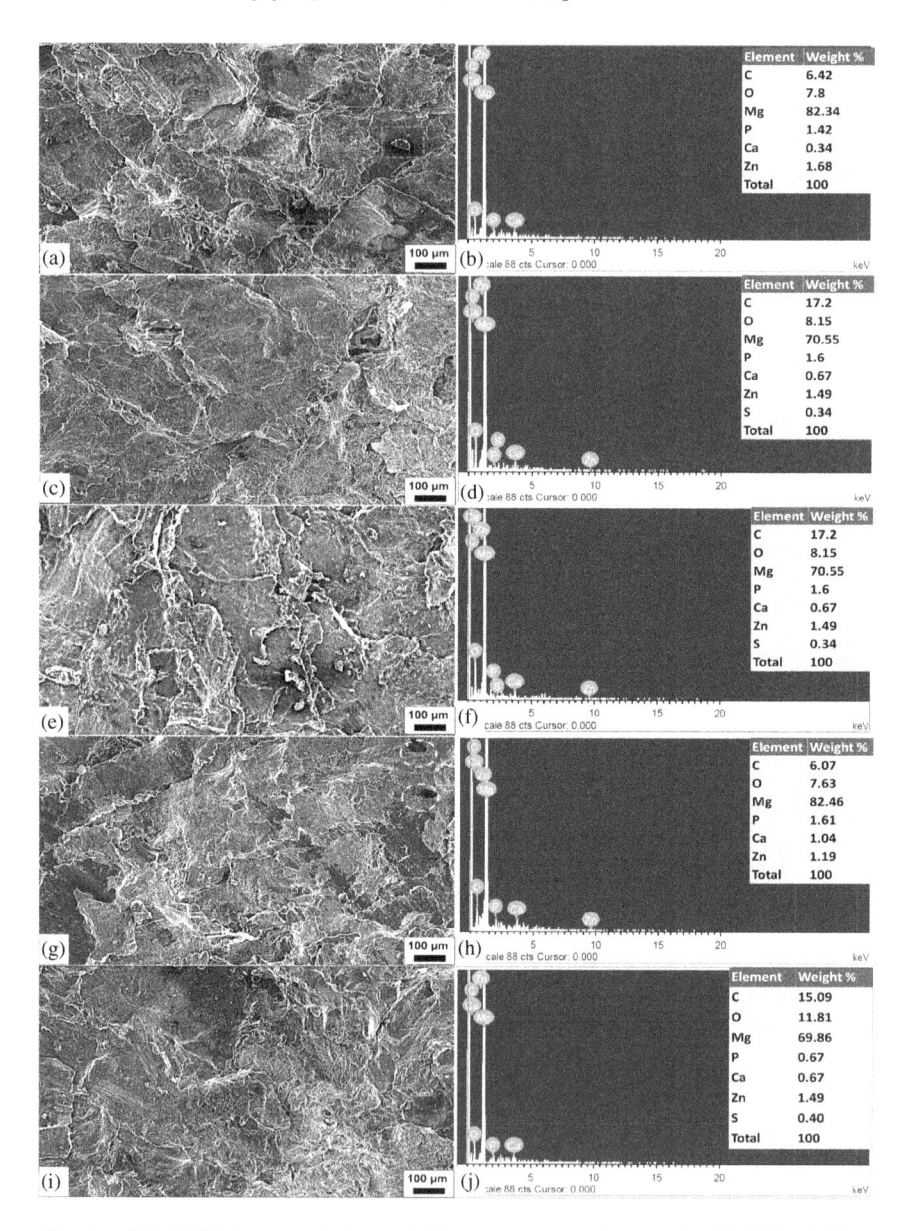

Fig. 6. (a) SEM image of the hybrid nanocomposite sample S1 (Mg–2.5Zn–0.5HAp–2.5rGO) showing agglomeration of the particles highlighted by a red rectangular box, (b) the corresponding EDS spectrum of the nanocomposite sample S1 (Mg–2.5Zn–0.5HAp–2.5rGO) showing the elemental composition of the nanocomposite, (c) SEM image of the hybrid nanocomposite sample S2

Influence of mechanical and the corrosion characteristics on the surface 349

fabricated nanocomposites are alike but varying with respect to the morphology of sample S0 (pure Mg) as seen in Fig. 1(a). Clearly, the reinforcements getting agglomerated is evident in the SEM images of S1 (Mg–2.5Zn–0.5HAp–2.5rGO) as well as S5 (Mg–2.5Zn–2.5HAp–0.5rGO) composite samples. The same has been marked via red rectangles in Fig. 6(a) which shows agglomerated rGO particles and Fig. 6(i) which exhibits agglomeration of Hap, respectively. The composites could be weakened if the agglomeration is high which shall negatively impact the mechanical strength. Extremely small cavities are seen in some of the composite samples which indicates their porous nature, especially samples S2 (Mg–2.5Zn–1.0HAp–2.0rGO) and S4 (Mg–2.5Zn–2.0HAp–1.0rGO) in SEM images showed cavities, circled in yellow (Figs. 6(c) and 6(g)). These cavities can hamper a composite's strength. In the nanocomposite sample S3 (Mg–2.5Zn–1.5HAp–1.5rGO), there are no cavities and also there is uniformly distributed reinforcement particle along the composite (Fig. 6(e)).

The EDS spectroscopy was also performed on the HNCs for elemental analysis. The EDS spectra of the composite samples S1–S5 along with the elemental weight percentages are presented in Figs. 6(b), 6(d), 6(f), 6(h), and 6(j), respectively. The presence of important elements such as magnesium, zinc, calcium, phosphorous, carbon, and oxygen within the nanocomposites is established

---

Fig. 6. (*Continued on facing page*) (Mg–2.5Zn–1.0HAp–2.0rGO) showing cavities circled in yellow, (d) EDS spectrum of the nanocomposite sample S2 (Mg–2.5Zn–1.0HAp–2.0rGO) showing the elemental composition of the nanocomposite, (e) SEM image of the hybrid nanocomposite sample S3 (Mg–2.5Zn–1.5HAp–1.5rGO), (f) EDS spectrum of the nanocomposite sample S3 (Mg–2.5Zn–1.5HAp–1.5rGO) showing the elemental composition of the nanocomposite, (g) SEM image of the hybrid nanocomposite sample S4 (Mg–2.5Zn–2.0HAp–1.0rGO) showing cavities circled in yellow, (h) EDS spectrum of the nanocomposite sample S4 (Mg–2.5Zn–2.0HAp–1.0rGO) showing the elemental composition of the nanocomposite, (i) SEM image of the hybrid of the nanocomposite sample S5 (Mg–2.5Zn–2.5HAp–0.5rGO) showing agglomeration of the particles, (j) EDS spectrum of the nanocomposite sample S5 (Mg–2.5Zn–2.5HAp–0.5rGO) showing the elemental composition of the nanocomposite (Color online).

via the EDS spectra. This supports the formation Mg–Zn-based HNCs with rGO and HAp reinforcements.

3.2. *XRD analysis*

Figure 7 depicts the XRD patterns of samples S0 (Pure Mg) and the HNCs sample S1 with 2.5 wt.% rGO and 0.5 wt.% HAp, S3 with 1.5 wt.% of rGO, HAp each and S5 with 0.5 wt.% rGO and 2.5 wt.% Hap, respectively.

The obtained XRD patterns of all nanocomposite samples confirm the presence of α Mg as the major phase. The intensity of peaks at 32°, 34°, and 37° of 2θ and the (1 0 0), (0 0 2), and (1 0 1) miller indices of crystallographic planes representing the prismatic, basal as well as the pyramidal planes, signify crystalline magnesium's

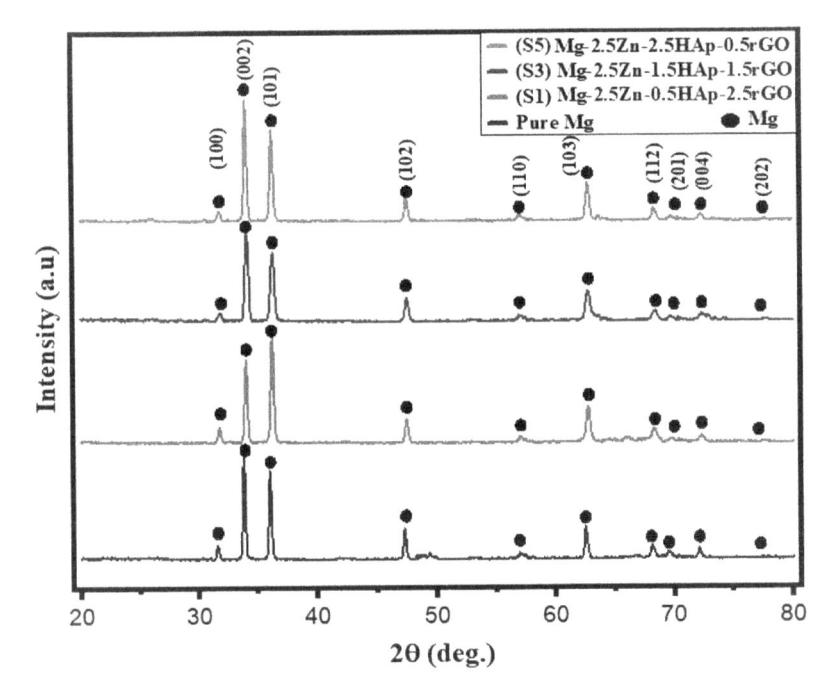

Fig. 7. The comparative XRD patterns of the samples S0 (Pure Mg), the hybrid nanocomposite samples of S1 (Mg–2.5Zn–0.5HAp–2.5rGO), S3 (Mg–2.5Zn–1.5HAp–1.5rGO), and S5 (Mg–2.5Zn–2.5HAp–0.5rGO) (Color online).

hexagonal close packing (HCP) structure. However, the XRD peak of HAp and the peak at 26.1° of 2θ representing the rGO's plane (0 0 2) hasn't been visible. The low content of the reinforcements in the hybrid nanocomposite might be the reason behind nondetection of HAp and rGO phases in the as-casted HNCs. Some of the recent research findings have testified analogous results pertaining to XRD in case of Mg-based composites reinforced with HAp and graphene.[39,41,42]

There aren't any intermetallic compounds detected as well which endorse the formation of high-quality nanocomposite. The high solubility of Zn in α-Mg up to 1.6 wt.% in ambient temperature conditions could be the reason behind Zn not being detected within the XRD peaks because, of the 2.5 wt.% Zn within the matrix, and 1.6 wt.% must have combined with magnesium. The solubility of Zn in Mg enhances the strength of the composite due to solid solution strengthening. The left-out content being negligible quantity couldn't be detected and the major gain from this disappearing phase of zinc and also the Mg–Zn phase in the nanocomposite is that, this could lower the galvanic corrosion and also successfully hamper the process of corrosion in Mg.[43,44]

3.3. *Mechanical characteristics*

3.3.1. *Tensile test analysis*

The tensile samples after undergoing the tensile test have been plotted for the variation in stress against strain, as shown in Fig. 8(a). The obtained values of yield tensile stress (YTS), ultimate tensile stress (UTS) as well as percentage elongation of the fabricated nanocomposite samples, respectively, are for S0, 43.4 MPa YTS, 95.56 MPa UTS and 6% elongation, for S1, 120.84 MPa YTS, 199.42 UTS and 8.98% elongation, for S2 123.62 MPa YTS, 202.04 UTS and 10.5% elongation, S3 has 130.48 MPa YTS, 209.6 MPa UTS and 10.4% elongation, S4 119.68 MPa YTS, 195.54 MPa UTS and elongation of 6.7% and in case of sample S5, YTS is 120.12 MPa, UTS is 188.75 MPa and 6.4% is the elongation.

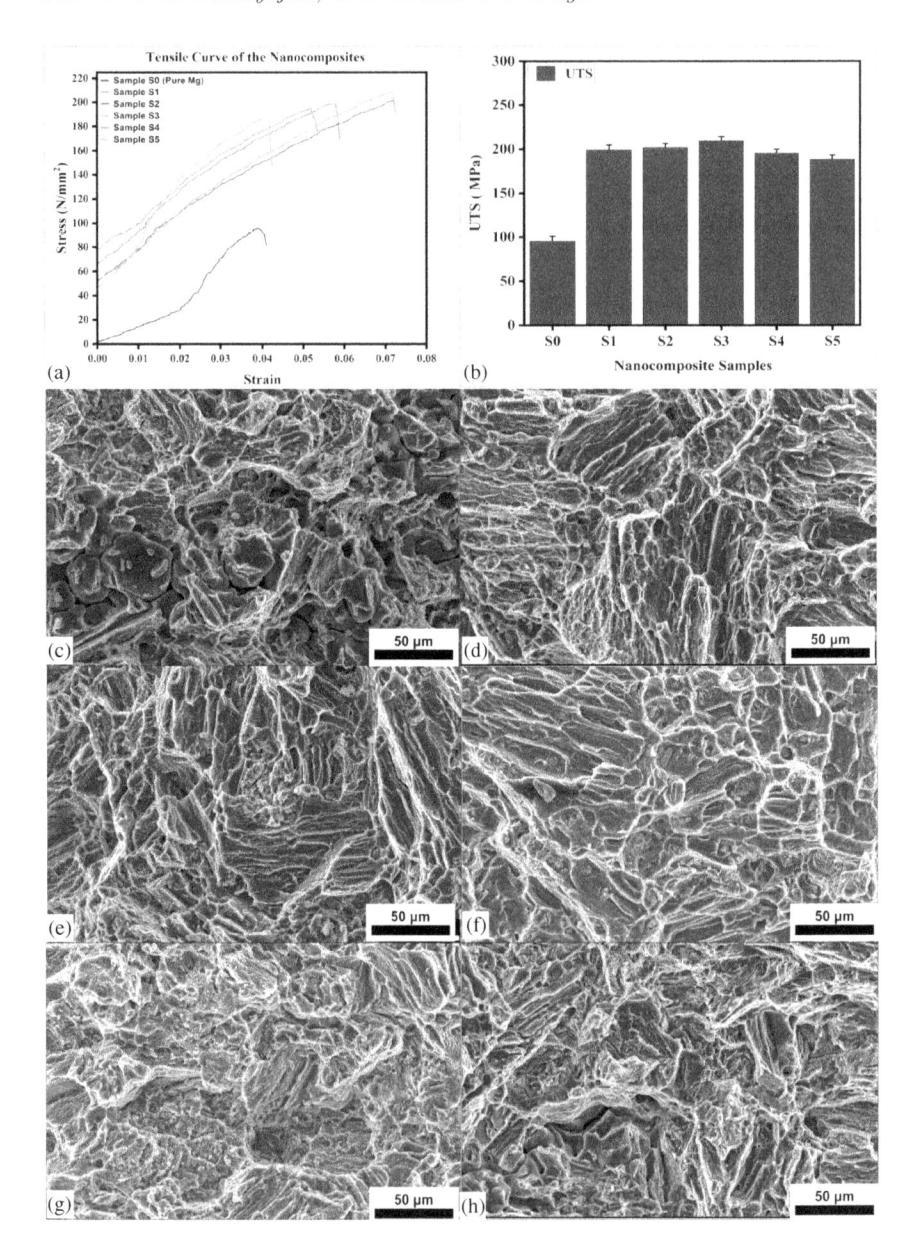

Fig. 8. (a) The comparison curve of the variation in the tensile stress w.r.t the corresponding strain of the fabricated samples S0 (Pure Mg) and the nanocomposite samples S1–S5, (b) Bar graphs indicating the ultimate tensile strengths of the fabricated samples, S0 (Pure Mg) and the hybrid nanocomposite

As per observations, the tensile strength of the fabricated HNCs is far superior to that of the pure Mg which proves that adding Zn, rGO, and HAp to Mg has elevated the strength within the composite. Refined grains are the reason behind such high values of YTS and UTS in the HNCs and this also abides the Hall–Patch relation in the following equation[45,46]:

$$\sigma_y = \sigma_o + K_y d^{-1/2}, \qquad (2)$$

where σ_y is YTS, K_y and σ_o are material constants, and d is the mean of the grain size. The slip systems quantity within a metal influences K_y. Metal with HCP system has more slip systems than face or body-centered cubic systems (FCC or BCC).[47] The hexagon close packed (HCP) structured Mg, greatly elevates the yield strength of a material. Therefore, the presence of more grain boundaries indicates that there are finer grains which elevate the yield strength of the fabricated composite.[48,49] The 1.6 wt.% of dissolved zinc in the composite is the reason for increased YTS and UTS of the nanocomposite. The alloying element zinc in the state of equilibrium condition possesses the highest solubility within the binary alloy and with that maximum solubility, it combines with the element magnesium leading to the formation of a solid solution strengthening which enhances the total strength of a given composite.[50,51]

Since rGO refines the grains and has the ability to strongly bond with the Mg matrix in the composite to increase its strength,[38] the fabricated nanocomposite sample S3 with rGO and HAp reinforcements in equal quantities (1.5 wt.% each) has the maximum

---

Fig. 8. (*Continued on facing page*) samples S1–S5. The SEM images of the tensile fracture morphology of the fabricated samples, (c) Sample S0 (Pure Mg) with dimples indicating ductile fracture, (d) Sample S1 (Mg–2.5Zn–0.5HAp–2.5rGO) indicating ductile fracture, (e) Sample S2 (Mg–2.5Zn–1.0HAp–2.0rGO) indicating ductile fracture, (f) Sample S3 (Mg–2.5Zn–1.5HAp–1.5rGO) indicating ductile fracture, (g) Sample S4 (Mg–2.5Zn–2.0HAp–1.0rGO) indicating brittle fracture, (h) Sample S5 (Mg–2.5Zn–2.5HAp–0.5rGO) indicating brittle fracture (Color online).

YTS and highest UTS (Fig. 8(b)). This is the same sample with 96.6 μm average grain size which happens to be the least among all the fabricated nanocomposite samples thereby confirming the positive influence of reduced grain size over the tensile strength. This is in conjunction with previous research where nanorGO platelets added to magnesium has refined grains and also improved mechanical properties of alloys of magnesium.[52-54] Sample S5 with highest HAp content (2.5 wt.%) and least rGO content (0.5 wt.%) has the least values of YTS and UTS. The higher quantity of HAp present in the nanocomposite has resulted in this because HAp comprises of a hard phase and with that hard phase and brittle nature, when it is combined with magnesium, the reinforcement HAp makes the overall composite to be brittle. It lowers the tensile strength as well as elongation in the composite.[55] For this precise reason, the sample S5 also has the least elongation of 6.4% which is only slightly higher than that of the S0 sample of pure magnesium with 6% elongation. But for some reason, the trend of lower strength with higher HAp content hasn't been much evident in case of the YTS. The evaluation of tensile strength of the fabricated samples revealed the highest YTS (130.48 MPa), UTS with 209.60 MPa, and elongation of 10.4% was in the nanocomposite sample S3 (Mg–2.5Zn–1.5HAp–1.5rGO). The sample also exhibited the least average grain size of 96.6 μm. This proves that the grain refinement within a composite is the primary parameter that influences the tensile strength of a specimen. The UTS increased with increasing concentration of rGO from 0.5 wt.% to 1.5 wt.%, beyond which it reduced. Also, the UTS increased with increasing quantities of HAp from 0.5 wt.% to 1.5 wt.% but there has been a restriction in the elongation as well as the yield strength. This is due to the brittle nature of HAp which hampers the tensile strength.[55]

The tensile fracture morphology of the samples S0, S1, S2, S3, S4, and S5 is represented in Figs. 8(c)–8(h), respectively. The presence of smaller dimples in the fracture morphology of S0 (pure Mg) is an indication of its tensile nature and a ductile fracture but with a lower tensile strength (Fig. 8(c)). But in case of the hybrid nanocomposite samples S1, S2, and S3 with higher amounts of rGO

(2.5 wt.%, 2 wt.%, and 1.5 wt.%, respectively) than HAp show the typical honeycomb structure with much deeper and larger dimples signifying extremely ductile fracture (Figs. 8(d)–8(f)). This is also a confirmation that rGO is the primary reinforcement responsible for high-tensile strength in the nanocomposites. Though there isn't much variation among the tensile strengths of samples S1 and S2, the highest tensile strength is seen in sample 3. Agglomeration of rGO particles or nonuniform reinforcement dispersion along with the influence of HAp might be the reason for this lower-tensile strength in S1 and S2. In Figs. 8(g) and 8(h), the tensile fracture morphology shows plastically deformed or torn ridges and surfaces with cleavages along with small dimples. These morphologies of the nanocomposite samples S4 and S5 with higher HAp content (2 wt.% and 2.5 wt.%, respectively) than rGO signify brittle fracture and lower tensile strength which confirms the HAp's brittle nature. The small and shallow dimples are a proof of the composite's tenacity and limited ductile nature.

3.3.2. Compression test analysis

The stress–strain graph of pure magnesium along with the nanocomposites tested for compression on the UTM is seen in Fig. 9(a) and the obtained UCS values of each sample are represented in the bar graph of Fig. 9(b). The obtained UCS values are 96.9, 288.2, 295.8, 300.1, 324.8, and 331.02 MPa for samples S0–S5, respectively. It can be inferred from the bar graph that among the composites, there is an increase in the compressive strength from samples S1 to S5 with an increase in the HAp content. The sample S5 with highest HAp content of 2.5 and 0.5 wt.% rGO records the highest UCS of 331.02 MPa and obviously the least UCS is of pure Mg (96.9 MPa). The least UCS (288.2 MPa) among the composites is observed in S1 with 0.5 wt.% HAp and 2.5 wt.% rGO. HAp being brittle is the reason for this which is in adherence to earlier experimentations done.[17,55] Also, if we clearly observe the curve in Fig. 9(a), among the fabricated nanocomposites, in samples S5–S2 failure strain slowly increases as rGO content increases from 0.5 wt.% to 2 wt.%

because rGO has a graphitic behavior at higher loads. Sample S1 again has a low-failure strain since rGO particulates agglomerate at high rGO quantity and this leads to nonuniform distribution, negatively impacting the mechanical properties of a nanocomposite.[38,53-55]

The compressive fractography images of the fabricated nanocomposites S0, S1, S2, S3, S4, and S5 are displayed in Figs. 9(c)–9(h), respectively. It can be seen from Fig. 9(c) that the fracture morphology of S0 (Pure Mg) reveals shear bands and cleavage fractures since magnesium is pure and has a brittle failure when subjected to compression. Among the nanocomposites (Figs. 9(d)–9(h)), the compression fracture morphology is similar since in every sample fracture is seen at 45° to the axis of loading. Cleavages are seen indicating the typical brittle fracture in the composites. The deformation in magnesium-based composites is determined by the shear bands that are associated with work hardening and deformation that is heterogeneous. Samples subjected to failure by shear bands have a higher rate of work hardening.[56] The cleavage cracks combined with grain boundaries that twist lead to "rivers" formation in a particular direction that assists spreading of the local cleavage cracks.[57] But in case of samples S3–S5 with lower quantities of rGO (1.5, 1, and 0.5 wt.%, respectively), the fracture morphology depicts shallow dimples in addition to cleavages and finer fractures, an example of mixed failure mode unlike the samples S0–S2. The reason is because higher quantities of rGO creates agglomeration due to solid van der Waals attractions among the atoms of carbon thereby negatively influencing the strength as well as mechanical properties in composites. Also, in samples S3–S5, appropriate quantities of HAp (1.5, 2, and 2.5 wt.%) are added, the brittle nature of HAp has made the mechanical properties of the nanocomposite much closer to the cortical bone with acceptable levels of solid solution strengthening in addition to high density.[32] So, in terms of mechanical characteristics, this composition of HAp can completely eliminate the stress shielding effect in an orthopedic implant and make it more robust. If higher wt.% of HAp is added to the composite, then this would lead to excessive agglomeration

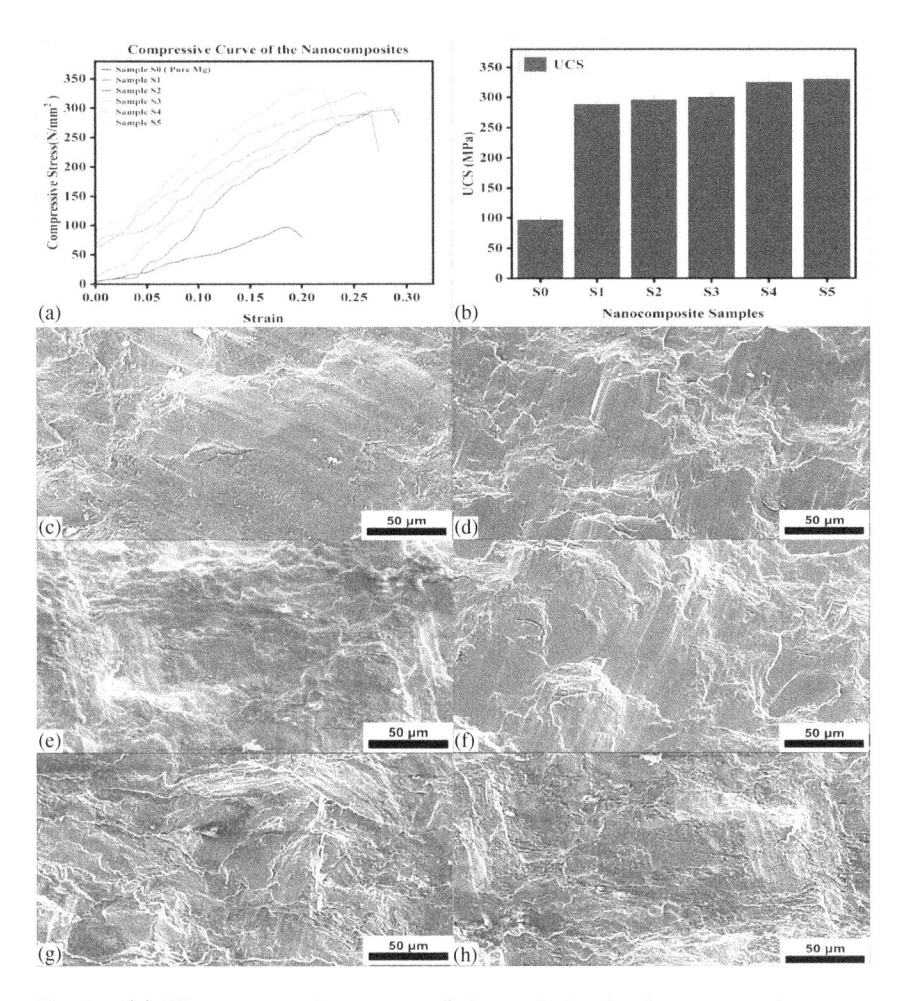

Fig. 9. (a) The comparative curves of the variation in the compressive stress w.r.t the strain in the fabricated samples S0 (Pure Mg) and the hybrid nanocomposite samples S1–S5. (b) Bar graphs representing the ultimate compressive strengths of the fabricated samples S0 (Pure Mg) and the hybrid nanocomposite samples S1–S5. The SEM images of the tensile fracture morphology of the fabricated samples. (c) Sample S0 (Pure Mg) indicating brittle fracture. (d) Sample S1 (Mg–2.5Zn–0.5HAp–2.5rGO) indicating brittle fracture. (e) Sample S2 (Mg–2.5Zn–1.0HAp–2.0rGO) indicating brittle fracture, (f) sample S3 (Mg–2.5Zn–1.5HAp–1.5rGO) indicating brittle fracture, (g) sample S4 (Mg–2.5Zn–2.0HAp–1.0rGO) indicating brittle fracture, and (h) sample S5 (Mg–2.5Zn–2.5HAp–0.5rGO) indicating brittle fracture (Color online).

of HAp particles which would make the composite extremely brittle and this will result in wider cracks and rougher surfaces of fracture which is undesirable in an orthopedic implant. To summarize, in case of the sample S0 (Pure Mg), the compressive strength exhibited very low 96.9 MPa in comparison to the HNCs which signifies the positive influence of the added elements within the composites. The highest UCS (331.02 MPa) was exhibited by sample S5 (Mg–2.5Zn–2.5HAp–0.5rGO) with 2.5 wt.% of HAp, the highest among all the other fabricated samples. The fracture morphologies of all the samples primarily exhibited brittle fracture. The samples S1 and S2 with higher rGO content (2 and 2.5 wt.%) exhibited mixed morphologies of small dimples and rough ridges but this wasn't the case for samples S4 and S5 which is a clear indication of elevated compressive strength. Moreover, the contribution of rGO in enhancing the compressive strength of the composite could not be significantly identified due to the presence of HAp and its direct influence on the compressive strength.

3.3.3. Microhardness analysis

The average microhardness values of the samples on Vickers hardness tester are represented in Fig. 10. All the obtained microhardness values 43.1 Hv, 56.5 Hv, 55.4 Hv, 62.1 Hv, 55.1 Hv, and 62.8 Hv for the samples S0, S1, S2, S3, S4, and S5, respectively, are in a close range to one another.

The highest microhardness is seen in sample S5 (62.8 Hv) which has 2.5 wt.% of HAp which is the highest quantity among all samples. The microhardness of a composite is enhanced when HAp's quantity in it is increased.[17,58] Sample S3 with equal quantities of rGO and HAp (1.5 wt.% each) has the second-best microhardness of 62.1 Hv among the fabricated nanocomposites. This is because rGO also like HAp, in higher quantities, enhances the microhardness of a composite by refining the grains and improving the strength.[38] Hence, in this case both the reinforcements, HAp and rGO have contributed to the microhardness of the nanocomposite sample S3. As expected, the next highest microhardness value of

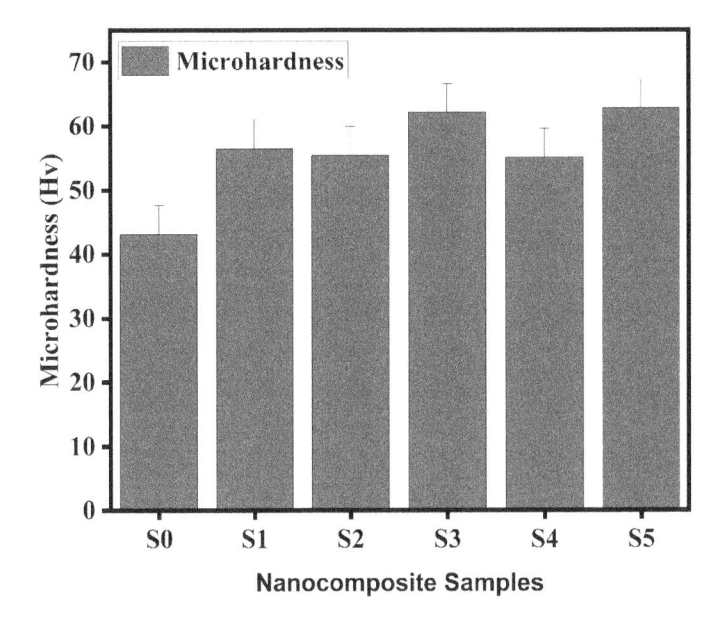

Fig. 10. Bar graph representing the average values of Vickers Microhardness of the fabricated samples S0 (Pure Magnesium), S1 (Mg–2.5Zn–0.5HAp–2.5rGO), S2 (Mg–2.5Zn–1.0HAp–2.0rGO), S3 (Mg–2.5Zn–1.5HAp–1.5rGO), S4 (Mg–2.5Zn–2.0HAp–1.0rGO), and S5 (Mg–2.5Zn–2.5HAp–0.5rGO) (Color online).

56.5 Hv is seen in sample S1 with 2.5 wt.% of rGO and the least microhardness value, 43.1 Hv, is seen in the sample S0 which is pure Mg.

3.4. Corrosion properties by immersion test

In an attempt to find the rate of corrosion of the fabricated samples in SBF, the samples were immersed inside a 60 ml of prepared SBF solution for a duration of 168 h. The intent of the experimentation was to find out the rate of corrosion by the weight loss method after dipping the samples inside SBF. This is an *in vitro* test which gives us an idea as to how long the fabricated samples survive in an environment the closely resembles the conditions within the human body. Before finding out the rate of corrosion in SBF, one has to understand the fact that, the rate of corrosion of a sample *in vitro* is much higher than the rate of corrosion *in vivo*.[59,60] Therefore, even if the rate of corrosion is slightly higher than the acceptable corrosion rates for an orthopedic implant, still it will be in the acceptable range for *in vivo*. Of course, the samples have to be tested again within the condition of *in vivo* for various other factors before being considered as an acceptable implant.

The FESEM images in Figs. 11(a)–11(f) exhibit the corroded surfaces of the fabricated nanocomposite samples after immersing them in the prepared SBF solution for a duration of 168 h. In Fig. 11(g), the average rate of corrosion calculated in mm per year

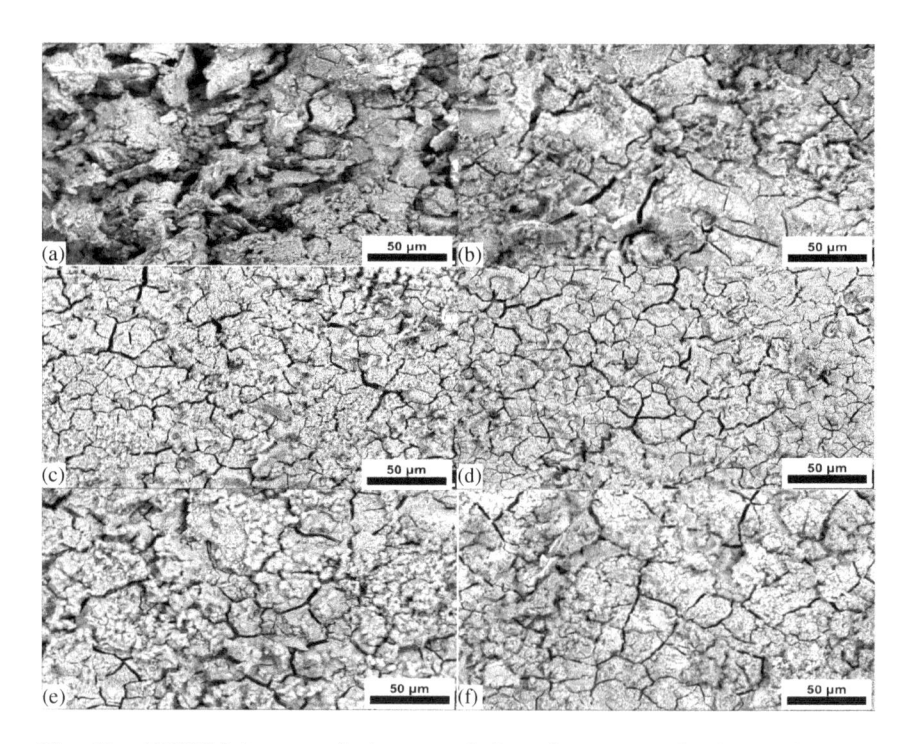

Fig. 11. FESEM images of the corroded surfaces post the immersion test carried out on the fabricated samples. (a) Sample S0 (Pure Mg), (b) Sample S1 (Mg–2.5Zn–0.5HAp–2.5rGO), (c) Sample S2 (Mg–2.5Zn–1.0HAp–2.0rGO), (d) Sample S3 (Mg–2.5Zn–1.5HAp–1.5rGO), (e) Sample S4 (Mg–2.5Zn–2.0HAp–1.0rGO) and (f) Sample S5 (Mg–2.5Zn–2.5HAp–0.5rGO), (g) Bar graph indicating the decreasing corrosion rate obtained after the immersion test carried out on the fabricated samples, S0 (Pure Mg) and the hybrid nanocomposite samples S1–S5 (Color online).

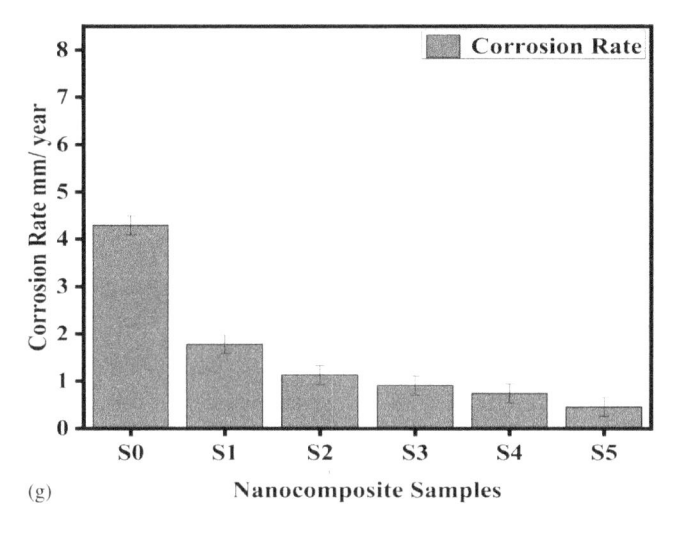

Fig. 11. (*Continued*)

for samples S0–S5 by the weight loss method, using Eq. (1) for three cycles, has been plotted as a bar graph against each sample. The obtained corrosion rates for the samples S0–S5 are 4.29 mm/year, 1.78 mm per year, 1.13 mm/year, 0.91 mm/year, 0.74 mm/year, and 0.46 mm per year, respectively. The highest rate of corrosion, 4.29 mm/year, is seen in the sample S0 of pure magnesium. Among the nanocomposites, it is clearly evident from Fig. 11(g) that the rate of corrosion is decreasing with increasing quantity of HAp. Therefore, it is the sample S5 with highest HAp content of 2.5 wt.% that exhibits the least corrosion rate of 0.46 mm per year.

After about seven days of immersion, the surface morphologies of the as cast samples reveal the typical corroded surfaces consisting of cracks and corrosion pits as shown in the FESEM images. In case of the sample S0, of pure magnesium the image in Fig. 11(a) shows extensive damage to the magnesium's surface exposing large corrosion pits and cracks. This indicates localized corrosion and also confirms the inability of pure Mg to withstand corrosion. The corrosion process is basically electrochemical kind in which oxidation of the metal occurs during the interaction phase with the aqueous medium of SBF. Usually, when the surface of magnesium is exposed to the aqueous medium, the mechanism of corrosion encompasses the microgalvanic coupling amid the cathode and anode regions which can be depicted via the following sequence of equations involving the various reactions that occur.[32,61,62]

The anodic reaction of magnesium is as follows:

$$Mg \rightarrow Mg^{2+} + 2e^-. \quad (3)$$

In the anodic reaction, at the anode, when atoms of magnesium oxidize to produce ionic species, then, the electrons are liberated which need to be received by alternative groups to conserve electroneutrality thereby resulting in the cathodic reaction given by

$$2H_2O + 2e^- \rightarrow H_2 + 2OH^-, \quad (4)$$

$$Mg^{2+} + 2OH^- \rightarrow Mg(OH)_2. \quad (5)$$

The overall process of corrosion in magnesium is the combination of the various partial reactions shown in the above equations. This corrosion process can be depicted by the following equation of reaction:

$$Mg + 2H_2O \rightarrow Mg(OH)_2 + H_2. \quad (6)$$

Therefore, hydrogen evolution is one of the major aspects that is associated with the process of corrosion in magnesium. This hydrogen evolution must be checked in order to improve the corrosion resistance in magnesium. The formed corrosion product, magnesium hydroxide (Mg (OH)$_2$), has the ability to act as a shielding sheet to prevent magnesium from corroding but its shielding effect is momentary because the magnesium hydroxide is a quasi-passive that combines with the Cl$^-$ ion existing within the SBF resulting in the formation of a soluble compound MgCl$_2$, thereby baring the surface of Mg to be corroded further more. The same process can be represented by the following reaction equation[63,64]:

$$Mg(OH)_2 + 2Cl^- \rightarrow MgCl_2 + 2OH^-. \quad (7)$$

One aspect that needs to be focused here is that, initially when all the samples were immersed into the SBF solution, there was

Influence of mechanical and the corrosion characteristics on the surface 363

excessive release of hydrogen bubbles. This is because the well-prepared surfaces of all the samples have been freshly exposed to the SBF solution and this has created a reaction between the exposed metal surface and the aqueous solution resulting in the liberation of hydrogen gas in the form of bubbles. This continued till the formation of protective layer over the surface.

Coming to the hybrid nanocomposite samples from S1 to S5, the rate of corrosion is much lower when compared to the pure magnesium sample S0. For the nanocomposite samples S1 and S2, consisting of 2.5 and 2 wt.% of rGO, the formations of cracks and microgaps are clearly evident. The cracks on the surface of the composite samples S1 and S2 are also larger compared to the remaining samples, this symbolizes a higher rate of corrosion. The existence of microgaps on the sample's surface could be due to the dissolution and falling off of the low quantity of nanoscaled HAp reinforcement (0.5 wt.%). When the overall reinforcement dispersion is considered in the composites, these higher amounts of rGO (2.5 and 2 wt.%) within the magnesium nanocomposite, agglomerate in certain areas becoming the cathodic regions which results in the occurrence of microgalvanic reactions.[65] The corrosion resistance of the composite is hampered due to these reactions. Also, one must understand the fact that the microgaps shall only be formed in the Mg–Zn alloy region where the reinforcements aren't present and these regions eventually shall be resulting in a localized corrosion. Hence, it can be seen from Figs. 11(b) and 11(c) that the cracks of S1 sample are much wider than in S2. The reason for lower cracks in S2 is because of the presence of 1 wt.% HAp apart from 2 wt.% of rGO. The HAp combines with SBF and forms an apatite layer that acts as a shield to the surface of Mg and prevents it from getting corroded.

With increasing quantities of HAp in a composite, up to 10 wt.%, the formation of apatite layer increases.[66] This is the scenario in the nanocomposite samples S3–S5 which consist of decent quantities of HAp (1.5, 2, and 2.5 wt.%) and lower quantities of rGO (1.5, 1, and 0.5 wt.%), respectively. The lower compositions of rGO disperse in a uniform manner within the matrix of magnesium and shield the composite's surface from coming in contact with the SBF solution and prevent corrosion from taking place.[38] This therefore increases the resistance of the metal composite against corrosion and lowers its corrosion rate. Hence, lower rates of corrosion are exhibited in these samples. When the nanocomposite is immersed into the SBF solution, HAp disintegrates into Ca^{2+}, HPO_4^{2-}, and PO_4^{3-} ions which electrostatically combine with the OH^- ion present in the SBF solution. The nanoscaled HAp also has the ability to offer the nucleation points to these freely and readily existing ions and it successfully forms a precipitation layer of calcium phosphate Ca–P.[67,68] This process of formation of the shielding precipitation layer can be represented by the following reactions:

$$10Ca^{2+} + 8OH^- + 6HPO_4^{2-} \rightarrow Ca_{10}(PO_4)_6(OH)_2 + 6H_2O, \quad (8)$$

$$10Ca^{2+} + 6HPO_4^{2-} + 2OH^- \rightarrow Ca_{10}(PO_4)_6(OH)_2, \quad (9)$$

$$3Ca^{2+} + 2PO_4^{3-} + 2OH^- \rightarrow Ca_3(PO_4)_2. \quad (10)$$

It is a proven fact that in comparison to magnesium hydroxide, the layer of Ca–P precipitated is extremely stable as it has resistance against the liberated hydrogen ions as well as the chloride destruction. Hence, the lower rates of corrosion in samples S3–S5 can be attributed to the formation of Ca–P layer due to the presence of HAp within the nanocomposite. The same can be seen in Figs. 11(c)–11(f). The images show small white particles over the surface of the sample, this is nothing but the apatite layer. This can be more prominently seen in Figs. 11(g) and 11(f) of the samples S4 and S5 which have higher quantities of HAp. An important aspect of HAp w.r.t. corrosion rate of a nanocomposite needs to be addressed i.e. if HAp is added in optimum quantities, then it disperses uniformly within the composite and significantly reduces the rate of corrosion by the formation of a shielding apatite layer.[69,70] If higher quantities of HAp beyond the optimum limit are added, then the particles agglomerate creating clusters eventually form clustering-related porosity all over the composite. The resulting pores as well as the agglomerated particles are extremely active regions for

Table 1. Mechanical properties of the as-cast samples.

Sample	Average grain size (μm)	UTS (MPa)	UCS (MPa)	Average microhardness (Hv)	Rate of corrosion (mm/yr.)
S0 (Pure Mg)	212	95.56	96.9	43.1	4.29
S1 (Mg–2.5Zn–0.5 HAp–2.5rGO)	118.92	199.42	288.2	56.5	1.78
S2 (Mg–2.5Zn–1.0 HAp–2.0rGO)	103.97	202.04	295.8	55.4	1.13
S3 (Mg–2.5Zn–1.5 HAp–1.5rGO)	96.60	209.60	300.1	62.1	0.91
S4 (Mg–2.5Zn–2.0 HAp–1.0rGO)	121.69	195.54	324.8	55.1	0.74
S5 (Mg–2.5Zn–2.5 HAp–0.5rGO)	135.92	188.75	331.02	62.8	0.46

corrosion as they become a secondary phase. This will enhance the rate of corrosion and will do more harm than good. Therefore, the above reasons endorse the decreasing rates of corrosion in the as-cast samples with increasing HAp content in addition to decreasing rGO content.

All the mechanical properties analyzed related to the as-cast samples are compiled and tabulated in Table 1.

4. Conclusions

The novelty of this work lies in the combination of the hybrid nanocomposite selected, in which the metal matrix is Mg–Zn binary alloy comprising of 2.5 wt.% of zinc and the reinforcements added are nanoHAp and rGO where not much work has been done. The overall composition of the reinforcements has been fixed to 3 wt.% and both the reinforcements have been varied. Six samples, one pure magnesium sample, S0, and five hybrid nanocomposite samples, S1(Mg–2.5Zn–0.5HAp–2.5rGO), S2(Mg–2.5Zn–1.0HAp–2.0rGO), S3(Mg–2.5Zn–1.5HAp–1.5rGO), S4(Mg–2.5Zn–2.0HAp–1.0rGO and S5(Mg–2.5Zn–2.5HAp–0.5rGO) were fabricated and the method employed was vacuum-assisted stir casting. The samples were then evaluated for their microstructural and mechanical properties along with their ability to resist corrosion in SBF. In the above analyses, the novel HNCs exhibited enhanced properties in comparison to pure magnesium (sample S0) thereby establishing their superiority over pure magnesium for orthopedic implant applications. Following are the conclusions from the above experimentation:

- The microscopic analysis revealed grain refinement on all of the hybrid nanocomposite samples in comparison to pure Mg. The highest refined grains with 96.6 μm average grain size were found in sample S3(Mg–2.5Zn–1.5HAp–1.5rGO).
- Among the reinforcements, rGO enhanced the tensile as well as yield strength, HAp hampered the yield strength of the composite but both reinforcements positively influenced the ultimate tensile strength. The highest YTS (130.48 MPa), UTS (209.60 MPa), and elongation of 10.4% were observed in the nanocomposite sample S3(Mg–2.5Zn–1.5HAp–1.5rGO). The tensile fracture morphologies displayed large dimples and honeycomb structures.
- The compressive strength of the composite is directly influenced by the quantity of HAp. The highest UCS (331.02 MPa) was exhibited by sample S5(Mg–2.5Zn–2.5HAp–0.5rGO). The fracture morphologies of all the samples primarily exhibited brittle fracture.
- The microhardness obtained from all the samples was extremely close to each other barring pure magnesium. The sample S5 exhibited the highest microhardness value of 62.8 Hv. The S3 sample exhibited the subsequent highest microhardness value of 62.1 Hv.
- In the hybrid nanocomposite, highest HAp and least rGO content brought the best resistance against corrosion. The corrosion rate of the sample S5(Mg–2.5Zn–2.5HAp–0.5rGO), immersed in a prepared SBF solution, was the least (0.46 mm per year). The FESEM images of the corrosion surfaces displayed corrosion cracks and apatite layer formation.

Therefore, considering the cumulative of all the material properties like the yield tensile strength (YTS), ultimate tensile strength (UTS), ultimate compressive strength (UCS), and rate of corrosion in SBF, the sample S3(Mg–2.5Zn–1.5HAp–1.5rGO) with 209.60 MPa UTS, 300.1 MPa UCS, and corrosion rate of 0.91 mm/year is the best suited and a prospective material for biodegradable implant application.

ORCID

V. S. P. Somayajula https://orcid.org/0000-0002-7800-4031
S. B. Prasad https://orcid.org/0000-0003-1743-4895
S. Singh https://orcid.org/0000-0002-4874-4805

References

1. R. Radha and D. Sreekanth, *J. Magnes. Alloy.* **5** (2017) 286.
2. T. Hanawa, *Sci. Technol. Adv. Mater.* **13** (2012) 064102; L. Tan, X. Yu, P. Wan and K. Yang, *J. Mater. Sci. Technol.* **29** (2013) 503.
3. B. Thakur, S. Barve and P. Pesode, *J. Mech. Behav. Biomed. Mater.* **138** (2023) 105641.
4. J. A. Jeffrey, S. S. Kumar, P. Hariharan, M. Kamesh and A. M. Raj, *Mater. Sci. Forum* **1048** (2022) 9.
5. C. Shuai, S. Li, S. Peng, P. Feng, Y. Lai and C. Gao, *Mater. Chem. Front.* **3** (2019) 544.
6. M. Shahin, K. Munir, C. Wen and Y. Li, *Acta Biomater.* **96** (2019) 1.
7. H. R. Bakhsheshi-Rad, E. Hamzah, A. F. Ismail, M. Daroonparvar, M. A. Yajid and M. Medraj, *J. Alloys Compd.* **658** (2016) 440.
8. A. H. Sanchez, B. J. Luthringer, F. Feyerabend and R. Willumeit, *Acta Biomat.* **13** (2015) 16.
9. A. Saberi, H. R. Bakhsheshi-Rad, A. F. Ismail, S. Sharif, M. Razzaghi, S. Ramakrishna and F. Berto, *Metals* **12** (2022) 207.
10. M. Shahin, C. Wen, K. Munir and Y. Li, *J. Magnes. Alloy* **10** (2022) 458.
11. M. Sankar, J. Vishnu, M. Gupta and G. Manivasagam, Magnesium-based alloys and nanocomposites for biomedical application, in

Applications of Nanocomposite Materials in Orthopedics, Woodhead Publishing Series in Biomaterials (Woodhead Publishing, 2019), pp. 83–109.
12. M. Haghshenas, *J. Magn. Alloys* **5** (2017) 189.
13. M. Ali, M. A. Hussein and N. Al-Aqeeli, *J. Alloys Compd.* **792** (2019) 1162.
14. P. Pesode and S. Barve, Magnesium alloy for biomedical applications, in *Advanced Materials for Biomechanical Applications* (CRC Press, 2022), pp. 133–158.
15. S. Cai, T. Lei, N. Li and F. Feng, *Mater. Sci. Eng. C* **32** (2012) 2570.
16. P. Pesode and S. Barve, *Mater. Today. Proc.* (2022).
17. S. Dutta, S. Gupta and M. Roy, *ACS Biomater. Sci. Eng.* **6** (2020) 4748.
18. S. C. Tjong, *Mater. Sci. Eng. R. Rep.* **74** (2013) 281.
19. S. Abazari, A. Shamsipur, H. R. Bakhsheshi-Rad, A. F. Ismail, S. Sharif, M. Razzaghi, S. Ramakrishna and F. Berto, *Materials* **13** (2020) 4421.
20. S. Abazari, A. Shamsipur, H. R. Bakhsheshi-Rad, S. Ramakrishna and F. Berto, *Metals* **10** (2020) 1002.
21. J. Geng, S. Chen and X. Chen, in *Adapting 2D Nanomaterials for Advanced Applications*, ACS Symposium Series, Vol. 1353 (American Chemical Society, 2020), pp. 253–293.
22. S. V. Satya Prasad, S. B. Prasad and S. Singh, in *Advanced Applications of 2D Nanostructures*, Materials Horizons: From Nature to Nanomaterials (Springer, Singapore, 2021), pp. 55–72.
23. L. Tsetseris and S. T. Pantelides, *Carbon N. Y.* **67** (2014) 58.
24. Y. Ann, A. F. Oliveira, T. Brumme, A. Kuc and T. Heine, *Adv. Mater.* **32** (2020) 2002442.
25. K. Munir, C. Wen and Y. Li, *J. Magnes. Alloys* **8** (2020) 269.
26. K. S. Munir, C. Wen and Y. Li, *Adv. Biosyst.* **3** (2019) 1800212.
27. Y. Talukdar, J. T. Rashkow, G. Lalwani, S. Kanakia and B. Sitharaman, *Biomaterials* **35** (2014) 4863.
28. W. C. Lee, K. P. Loh and C. T. Lim, *Biomaterials* **155** (2018) 236.
29. J. A. Quezada-Rentería, L. F. Cházaro-Ruiz and J. R. Rangel-Mendez, *Carbon* **122** (2017) 266.
30. K. Zhou, P. Yu, X. Shi, T. Ling, W. Zeng, A. Chen, W. Yang and Z. Zhou, *ACS Nano* **13** (2019) 9595.

31. M. Mehrali, A. R. Akhiani, S. Talebian, M. Mehrali, S. T. Latibari, A. Dolatshahi-Pirouz and H. S. Metselaar, *J. Eur. Ceram. Soc.* **36** (2016) 319.
32. H. Miao, D. Zhang, C. Chen, L. Zhang, J. Pei, Y. Su, H. Huang, Z. Wang, B. Kang, W. Ding and H. Zeng, *ACS Biomater. Sci. Eng.* **5** (2019) 1623.
33. V. Bazhenov, A. Lyskovich, A. Li, V. Bautin, A. Komissarov, A. Koltygin, A. Bazlov, A. Tokar, D. Ten and A. Mukhametshina, *Materials* **14** (2021) 7847.
34. G. Parande, V. Manakari, S. Prasadh, D. Chauhan, S. Rahate, R. Wong and M. Gupta, *J. Mech. Behav. Biomed. Mater.* **103** (2020) 103584.
35. H. S. AlSalem, A. A. Keshk, R. Y. Ghareeb, A. A. Ibrahim, N. R. Abdelsalam, M. M. Taher, A. Almahri and A. Abu-Rayyan, *Mater. Chem. Phys.* **283** (2022) 125988.
36. F. Khorashadizade, S. Abazari, M. Rajabi, H. R. Bakhsheshi-Rad, A. F. Ismail, S. Sharif, S. Ramakrishna and F. Berto, *J. Mater. Res. Technol.* **15** (2021) 6034.
37. S. S. Prasad, S. B. Prasad, K. Verma, R. K. Mishra, V. Kumar and S. Singh, *J. Magn. Alloys* **10** (2021) 1.
38. S. Abazari, A. Shamsipur and H. R. Bakhsheshi-Rad, *J. Magn. Alloys* **10** (2022) 3612.
39. D. Liu, G. Xu, S. S. Jamali, Y. Zhao, M. Chen and T. Jurak, *Bioelectrochemistry* **129** (2019) 106.
40. T. Kokubo and H. Takadama, *Biomaterials* **27** (2006) 2907.
41. M. Rashad, F. Pan, H. Hu, M. Asif, S. Hussain and J. She, *Mater. Sci. Eng. A* **630** (2015) 36.
42. P. Kumar, M. Kujur, A. Mallick, K. S. Tun and M. Gupta, *IOP Conf. Ser. Mater. Sci. Eng.* **346** (2018) 012001.
43. M. Cihova, E. Martinelli, P. Schmutz, A. Myrissa, R. Schäublin, A. M. Weinberg, P. J. Uggowitzer and J. F. Löffler, *Acta Biomater.* **100** (2019) 398.
44. Y. Li, J. Wang and R. Xu, *Vacuum* **178** (2020) 109396.
45. V. Roche, G. Y. Koga, T. B. Matias, C. S. Kiminami, C. Bolfarini, W. J. Botta, R. P. Nogueira and A. J. Junior, *J. Alloys Compd.* **774** (2019) 168.
46. E. O. Hall, *Yield Point Phenomena in Metals and Alloys* (Springer Science & Business Media, New York, 2012).

47. H. Yu, Y. Xin, M. Wang and Q. Liu, *J. Mater. Sci. Technol.* **34** (2018) 248.
48. H. Du, Z. Wei, X. Liu and E. Zhang, *Mater. Chem. Phys.* **125** (2011) 568.
49. S. Luan, L. Zhang, L. Chen, W. Li, J. Wang and P. Jin, *J. Mater. Res. Technol.* **23** (2023) 6216.
50. J. Geng, X. Gao, X. Y. Fang and J. F. Nie, *Scr. Mater.* **64** (2011) 506.
51. F. Rosalbino, S. De Negri, A. Saccone, E. Angelini and S. Delfino, *J. Mater. Sci. Mater. Med.* **21** (2010) 1091.
52. C. Shuai, B. Wang, S. Bin, S. Peng and C. Gao, *Mater. Des.* **191** (2020) 108612.
53. X. Du, W. Du, Z. Wang, K. Liu and S. Li, *Mater. Sci. Eng. A* **711** (2018) 633.
54. M. Rashad, F. Pan, J. Zhang and M. Asif, *J. Alloys Compd.* **646** (2015) 223.
55. M. Mehrjoo, J. Javadpour, M. A. Shokrgozar, M. Farokhi, S. Javadian and S. Bonakdar, *Mater. Exp.* **5** (2015) 41.
56. M. Rashad, F. Pan, D. Lin and M. Asif, *Mater. Des.* **89** (2016) 1242.
57. M. Rashad, F. Pan and M. Asif, *Mater. Sci. Eng. A* **649** (2016) 263.
58. J. Kubásek, D. Vojtěch, J. Maixner and D. Dvorský, *Sci. Eng. Compos. Mater.* **24** (2017) 297.
59. L. N. Zhang, Z. T. Hou, X. Ye, Z. B. Xu, X. L. Bai and P. Shang, *Front. Mater. Sci.* **7** (2013) 227.
60. P. Yin Yee Chin, Q. Cheok, A. Glowacz and W. Caesarendra, *Appl. Sci.* **10** (2020) 3141.
61. S. Z. Khalajabadi, M. R. Kadir, S. Izman and M. Marvibaigi, *J. Alloys Compd.* **655** (2016) 266.
62. Y. Yan, Y. Kang, D. Li, K. Yu, T. Xiao, Y. Deng, H. Dai, Y. Dai, H. Xiong and H. Fang, *Mater. Sci. Eng. C* **74** (2017) 582.
63. S. S. Prasad, S. Singh and S. B. Prasad, *AIP Conf. Proc.* **2341** (2021) 040008.
64. Z. Cui, W. Li, L. Cheng, D. Gong, W. Cheng and W. Wang, *Mater. Charact.* **151** (2019) 620.
65. Y. Yang, Y. Cheng, S. Peng, L. Xu, C. He, F. Qi, M. Zhao and C. Shuai, *Bioactive Mater.* **6** (2021) 1230.
66. A. Dubey, S. Jaiswal, S. Haldar, P. Roy and D. Lahiri, *J. Mater. Eng. Perform.* **28** (2019) 5702.

67. M. Lotfpour, C. Dehghanian, M. Emamy, A. Bahmani, M. Malekan, A. Saadati, M. Taghizadeh and M. Shokouhimehr, *J. Magnes. Alloys* **9** (2021) 2078.
68. H. R. Bakhsheshi-Rad, E. Hamzah, M. Daroonparvar, R. Ebrahimi-Kahrizsangi and M. Medraj, *Ceram. Int.* **40** (2014) 7971.
69. M. M. Iqbal, A. Kumar, R. Shabadi and S. Singh, *Surf. Rev. Lett.* **30** (2023) 2141008.
70. M. M. Iqbal, A. Kumar and S. Singh, *AIP Conf. Proc.* **2341** (2021) 040031.

www.ingramcontent.com/pod-product-compliance
Lightning Source LLC
Chambersburg PA
CBHW061255310325
24273CB00004B/64